P9-CKA-270

About Island Press

Island Press, a nonprofit organization, publishes, markets, and distributes the most advanced thinking on the conservation of our natural resources—books about soil, land, water, forests, wildlife, and hazardous and toxic wastes. These books are practical tools used by public officials, business and industry leaders, natural resource managers, and concerned citizens working to solve both local and global resource problems.

Founded in 1978, Island Press reorganized in 1984 to meet the increasing demand for substantive books on all resource-related issues. Island Press publishes and distributes under its own imprint and offers these services to other nonprofit organizations.

Support for Island Press is provided by The Geraldine R. Dodge Foundation, The Energy Foundation, The Ford Foundation, The George Gund Foundation, William and Flora Hewlett Foundation, The James Irvine Foundation, The John D. and Catherine T. MacArthur Foundation, The Andrew W. Mellon Foundation, The Joyce Mertz-Gilmore Foundation, The New-Land Foundation, The Pew Charitable Trusts, The Rockefeller Brothers Fund, The Tides Foundation, Turner Foundation, Inc., The Rockefeller Philanthropic Collaborative, Inc., and individual donors.

About Defenders of Wildlife

Defenders of Wildlife is a national, nonprofit membership organization headquartered in Washington, D.C. Founded in 1947, Defenders has nearly half a century of leadership in educating and advocating for the protection, restoration, and enhancement of all species of wild animals and plants in their natural communities.

Defenders' staff of wildlife professionals works to shape policy and programs at the state and federal levels through numerous partnership efforts with public agencies. Through conferences and publications—including the award-winning magazine *Defenders* and membership newsletter *Wildlife Advocate*—the organization works to educate people across the nation on a wide range of biodiversity-related issues.

Defenders gratefully acknowledges the support of more than 80,000 members, corporations, foundations, and a national network of devoted conservation activists.

SAVING NATURE'S LEGACY

SAVING NATURE'S LEGACY

Protecting and Restoring Biodiversity

Reed F. Noss

and

Allen Y. Cooperrider

Foreword by

Rodger Schlickeisen

Defenders of Wildlife

ISLAND PRESS

Washington, D.C.   Covelo, California

Library of Congress Cataloging-in-Publication Data

Noss, Reed F.
Saving nature's legacy : protecting and restoring biodiversity / Reed F. Noss and Allen Y. Cooperrider ; foreword by Rodger Schlickeisen.
p. cm.
Includes bibliographical references (p.) and index.
ISBN 1-55963-247-x (cloth).—ISBN 1-55963-248-8 (paper)
1. Ecosysten management—United States. 2. Biological diversity conservation—United States. 3. Ecosystem management. 4. Biological diversity conservation. 5. Biological diversity.
I. Cooperrider, Allen Y. II. Title.
QH76.N67 1994 93–48895
333.95′16′0973—dc20 CIP

Printed on recycled, acid-free paper

Manufactured in the United States of America
10 9 8 7 6 5 4 3

We dedicate this book to Dr. Ted LaRoe,
leader of the Cooperative Fish and Wildlife
Research Units of the National Biological
Survey, a visionary public servant,
and a friend.

CONTENTS

3. Conservation Strategies—Past, Present, and Future 67

4. Selecting Reserves 99

5. Designing Reserve Networks 129

FIGURES

TABLES

FOREWORD

Conservation policy is never static, but always evolving. Human nature being what it is, the process of building consensus is slow, typically occurring over the course of decades. In time, consensus gives rise to new policies, and new policies ultimately come to represent the status quo. But the status quo is never as secure as the expression implies; it is, in fact, constantly being challenged by a generation of new ideas and constituencies seeking to build consensus around those ideas. It is an ever-evolving cycle of change that is, perhaps, at its most exciting stage when consensus stands on the threshold of policy change. The implementation of policy, after all, gives satisfaction to the arduous process of consensus building.

Today we stand on the threshold of a revolution in wildlife conservation policy. A realignment of beliefs within the wildlife conservation community and an increasing sensitivity of the public to environmental issues lies at the root of this change. Even resource agencies, so long the captive audience of narrow commodity-oriented constituencies, have felt compelled in recent decades to accept broader constituencies. The concerns of an increasingly urban, better educated, and more diverse population have come to bear on resource agencies. What has emerged is a more holistic view of the relationships among plants, animals, and the environment. Where, as a nation, we have tended to focus on the pieces—our favorite places and preferred species—we now are moving increasingly toward a recognition that if we are to save the pieces, we must also protect the systems upon which they depend for their survival. Arriving at this understanding has been facilitated by our experience over the last twenty years with one of the nation's most compelling conservation laws—the Endangered Species Act.

A quarter of a century ago, the plight of specific species prompted Congress to legislate an Endangered Species Act. Not until 1973, however, when a groundswell of public sentiment arose in support of endangered species protection, did the act develop any real teeth. Fifteen years later, by 1988, nearly every state had enacted its own endangered species legislation and established its own program paralleling the federal program. Federal resource agencies all had developed active programs, and many conservation groups launched bold, new initiatives designed to stem the tide of extinction and make conservation the responsibility of us all.

But even as groups were leaping on the bandwagon of endangered species recovery, it was becoming increasingly apparent that the Endangered Species Act alone was not sufficient. While recovery efforts had won some notable battles (e.g., bald eagle and American alligator recovery), they were clearly losing the war. Essentially, species were in far greater danger than had been realized, and recovery efforts could not keep pace with the backlog of species in peril. Lack of adequate funding and commitment to enforcing the act were just a few of the problems that hampered recovery efforts during the eighties. Very disturbing, for example, was the realization that the act does not protect habitat sufficiently to sustain species once they have been successfully recovered and delisted. Although the law required that critical habitat be defined at the same time a species was listed, agencies in effect dealt with the requirement by ignoring it. In 1986, for example, 46 species were added to the list, but critical habitat designations were made for only four of those species.

Increasingly, scientists began to turn to the source of the problem, arguing for greater habitat protection in conjunction with species protection. Scientists and others reasoned that by improving habitat protection fewer species would become endangered. But others argued that enough land in the United States is already set aside. The reality, however, is that less than four percent of our lands are in protected status. Moreover, these lands inadequately maintain our nation's biological diversity largely because they were originally set aside not for their biological values, but for their esthetic, recreational, military, historical, or commodity values. We now know, too, that most of our national parks are not large enough to sustain viable populations of some larger mammals, like bears and wolves. And lastly, many activities practiced on our public lands—like timber harvesting, cattle grazing, and mineral extraction—are unsustainable at certain levels, and in some cases have hastened the demise of species and degraded the very habitats upon which they depend.

The challenge, then, has been to develop an affordable and expeditious method for determining which lands are in need of special protection, and to manage entire landscapes more sensitively to avoid mass extinction of species and destruction of ecosystems. Although many databases of varying capabilities existed that provided crucial information on the status of individual species, none had been developed on a broader, landscape scale until the Gap Analysis Project (GAP) was initiated in the late eighties. Conducted by the Cooperative Research Units of the U.S. Fish and Wildlife Service, GAP is an ecological inventory that integrates satellite imagery with data on individual species, land ownership patterns, and management practices in a computerized Geographic Information System. By looking at entire landscapes, GAP provides a sort of "score card" of a region's biodiversity, identifying "hot spots" of species richness and habitat types that are currently unprotected in a region's system of conservation lands.

Defenders of Wildlife immediately grasped the value of such an inventory and became GAP's leading proponent in Congress. Not only does GAP identify critical areas in terms of species and habitat, but it also provides a valuable framework for determining how lands should be managed. In essence, it lays the foundation for a coherent biodiversity conservation strategy to protect species and steer development to less sensitive areas. This does not imply that GAP proposes to "lock up" vast stretches of land, although clearly, some areas will require such protection because of their high degree of sensitivity. A coherent biodiversity strategy will define levels of protection and management techniques that accommodate human uses without degrading the land or endangering species.

As part of our national biodiversity conservation strategy, we thought a book containing specific management guidelines and techniques for maintaining biodiversity within different landscapes would be useful for land managers and others committed to long-term resource conservation. Mostly, we recognized that a lack of information exists about how to manage land to conserve biodiversity, which contributes to the difficulty managers have in implementing effective conservation programs. To be sure, resource agencies at both the state and federal levels have come to appreciate the value of a more holistic approach to conservation. The Forest Service, for example, approved policy guidelines supporting ecosystem management in 1992. And the new Clinton administration, in its efforts to resolve the spotted owl controversy in the Pacific Northwest, has stated objectives that include more ecosystem-based approaches to land management. But appreciating and actually implementing ecosystem management are two very different things, and it remains to be seen how effectively the resource agencies take up the new standard.

We recognize, too, that it is not always easy to keep pace with the new scientific understandings of a constantly evolving field. For the past several years, we have helped to bring some of these new ideas to land managers through a series of training courses sponsored by the Bureau of Land Management's (BLM) Phoenix Training Center. As useful as the courses are, however, they reach only a small percentage of the nation's land managers. As an advocacy organization with an interest in promoting conservation education, Defenders wanted to carry the important messages of conservation biology to a wider audience, and so we developed the idea for this book. We wanted the book to be written by credible, independent scientists who have experience in government management systems, and who could, therefore, bring a certain empathy to the process. Both Reed Noss and Allen Cooperrider fit this bill. Prior to becoming independent consultants, Dr. Noss worked as a biodiversity project leader for the Environmental Protection Agency, and Dr. Cooperrider as a wildlife biologist for the Bureau of Land Management. Both scientists bring a wealth of knowledge and experience to this effort.

We also wanted the two authors to reflect Defenders' own philosophical position on the issue of biodiversity conservation, but at the same time feel free to express their cutting-edge ideas on conservation biology. While we may not see eye to eye on every issue, we respect both authors for their passion and vision. The cutting edge is often a lonely place; sometimes people come to appreciate such ideas and sometimes they never do. But if, as a society, we never provide a forum for such ideas, then we will be that much more impoverished, that much more deprived of the ideas that stimulate discussion and ultimately produce a course of action.

It is our sincerest hope that this book will not only offer land managers practical guidance for the complex issue of biodiversity conservation, but that it will also inspire respect for the kind of innovative thinking that traditionally has played such an important role in shaping the course of American conservation. Because conservation biology is an ever-evolving field, we recognize that this book will not be the final word on the subject. Nevertheless, with this book, Drs. Noss and Cooperrider make a significant contribution to the canon of conservation biology, and Defenders of Wildlife is proud to have sponsored this important educational effort.

Rodger Schlickeisen
President, Defenders of Wildlife

PREFACE

> It was the best of times, it was the worst of times, it was the age of wisdom, it was the age of foolishness, it was the epoch of belief, it was the epoch of incredulity, it was the season of light, it was the season of darkness, it was the spring of hope, it was the winter of despair, we had every thing before us, we had nothing before us . . .
>
> Charles Dickens (1859), *A Tale of Two Cities*

We sat in a dreary hotel conference room in Washington, D.C., in 1989, squirming through yet another round of debate about what biodiversity is and whether or not society should bother saving it. Our thoughts about the future of nature in the United States were equivocal. We were gathered as the Keystone National Policy Dialogue on Biological Diversity, a forum of agency representatives, Congressional staff, academic biologists, conservationists, and industry emissaries. But we did not know whether to be happy or sad, hopeful or hopeless. Our task was to develop a consensus report on how to conserve biodiversity in the United States. Opinions flew hotly about the room. People were there with missions, but those missions varied. Some were there to obfuscate issues and thwart progress. The future of biodiversity conservation seemed full of promise one minute, doomed the next. We were frustrated and tired. The coffee was stale.

Biodiversity has become a national and international issue, albeit a minor one in the minds of most elected officials. There is cause for guarded optimism. For the first time in our political history, we hear discussion of whole communities of species—indeed, *ecosystems*—in the marbled halls of Congress and in the offices of agency administrators. No longer is the talk just about big game, pretty places, or rare birds. Business as usual on our national forests and other public lands is not acceptable to a large and growing segment of the public. People who care about these lands want more than two-by-fours and fishing piers, more than high-tech recovery programs for a handful of glamour species. People want nature. And they want their government to save nature for them and their descendants while providing for all their material needs. As a people, Americans want it all and are accustomed to getting what they want. Yet, as we write this, the lands and waters of the United States and the world are being grossly impoverished. If present trends continue, future generations will inherit a depauperate Earth. Many nonhuman species have no future at all.

The Keystone National Policy Dialogue on Biological Diversity was convened "to provide a forum where a diverse cross section of groups and organizations actively concerned with the biological diversity issue could come together to (1) formulate consensus recommendations, (2) clarify areas of disagreement, and (3) produce a final report summarizing the group's deliberations" (Keystone Center 1991). The dialogue was funded by federal agencies and emphasized management of federal lands. Yet, after nearly two years of meetings, the Keystone Dialogue produced little besides yet another bureaucratic report. There were few explicit recommendations, no executive orders, no legislation—no teeth, no action. In the end, the status quo prevailed. Big trees keep falling, cows keep munching, bulldozers roar.

We write this book because we believe the status quo is unacceptable. Uncounted species going extinct each year, ruination of landscapes, impoverishment of the oceans, severe alteration of atmospheric chemistry—all these things are unacceptable to us, and we believe, to the American people. Consensus processes like the Keystone Dialogue, loaded with representatives from agriculture, timber, livestock, mining, and oil industries, cannot be expected to resolve conservation problems. Many people in these industries will reap short-term profits from environmental degradation. Why would they want the destruction to stop? If the people who profit from biotic impoverishment are given veto power over the future of life on earth, as can be the case in consensus processes and in the hidden operations of government, the public is not well served. Furthermore, we believe that the welfare of our nonhuman kin, usually ignored by the anthropocentric institutions of our society, must be considered when making decisions about how land should be managed. The voiceless must be given a vote.

This book provides an example of how conservation biology—science in the service of conservation—can be applied to land management issues. Our purpose is twofold: (1) to develop the outline of a broad land-based strategy for biodiversity conservation in the United States, and (2) to synthesize information and offer general guidelines for biodiversity conservation that can be adapted by land managers to specific situations. To a large extent, this book is "technology transfer." That is, we wish to translate the principles, techniques, and findings of conservation biology and other ecological sciences to policymakers, land managers, landowners, conservationists, and the public at large.

But we wish to do more than just transfer information. We wish to go where the Keystone Dialogue feared to tread and provide explicit advice on what needs to be done to protect and restore the full spectrum of native biodiversity in North America. We do not have all the answers—no one does or ever will—but we believe it is time for conservation biologists to stick their necks out and state openly what they do know about the science and art of conservation. They must offer guidance to society for conserving biodiversity in the face of uncertainty. This is a matter of professional responsibility, especially needed in a time of crisis. This book offers our advice as it stands for now, and we trust our approach will be refined, broadened, and improved on by others. We also hope most urgently that it will be applied.

We write with a sense of mission but also a sense of pragmatism. We have both worked for and with government agencies that control much of the land base in the United States. This experience has convinced us that internal changes in these agencies will be central to any advances in biodiversity conservation. Higher level changes can be facilitated politically, but the agencies must have intelligent, educated, and impassioned staffs at all levels. Those people making land management policy and those out there getting their hands dirty must be equally knowledgeable about and committed to the conservation of biodiversity.

Our intended audience for this book includes managers and staff of land-managing agencies, elected officials and their staffs, other policymakers, private landowners and land managers, environmentalists, and everyone else who cares about how land is managed and conserved. We focus on the United States as a case study because we are most familiar with the ecosystems and the institutions of this country. However, we discuss conservation issues in the United States within a context of global biodiversity and hope that the principles and guidelines offered here can be applied elsewhere. We focus on public lands because fully one-third of the United States is federally owned and because we believe that biodiversity, ecological integrity, wildness, and other broad public values provide the most legitimate basis for managing these lands.

Our emphasis is on managing higher levels of biological organization such as communities, ecosystems, and landscapes, rather than on conserving genes or populations. Nevertheless, we often discuss conservation needs of individual species, particularly those that are highly sensitive to human activities and those that serve as useful "flagships" for conservation efforts. We emphasize terrestrial and freshwater biodiversity and land-based management approaches, but we do so with full recognition that the oceans and the atmosphere must also be protected. More than many of our colleagues, we stress the central importance of reserves in an overall multiple-use management framework. At a time when politicians and managers are touting the virtues of ecosystem management (a revised concept of multiple use) as an alternative to wilderness and other reserve designations, someone needs to remind the public that most ecosystem management approaches are untested and that some species and habitats are extraordinarily sensitive to human activities. We do not provide details or even much advice for the political implementation of our strategy. We leave that task for others.

We have written a book about land management for biodiversity, arguably one of the most controversial enterprises anyone could undertake these days. The very word *biodiversity* raises either warm feelings or hackles, depending on one's views about nature. Although much of the controversy (such as that between the timber industry and old-growth protectionists) is ostensibly over facts and figures, it involves a more fundamental rift. Indeed, much of the conflict over land management results from differing value systems, philosophies, and associated aspirations. The polarization we see today runs deep and will be difficult to reconcile.

In the spirit of honesty and open communication, we include here a statement of our premises in writing this book—our ecological "bottom line." We do this so that those who disagree with our premises can recognize that their disagreement is philosophical, not technical. In essence, we believe that

- Conservation of biodiversity is among the most important missions of human society and should be a fundamental goal for management of both public and private lands.

- Management that harms native ecosystems, or management out of ignorance or lack of respect for biodiversity, is no longer excusable. Management must be ecologically informed and environmentally sensitive.

- Lands should be managed according to a principle of "no net loss" of native biodiversity or natural areas from the regional landscape.

- All native species and ecosystems must be adequately represented in protected areas in each region.

- Ecosystem conservation and species conservation are fundamentally compatible and complementary. Thus, all biologically endangered and threatened species should be managed to recover their populations to viable status.

- Society must protect species sensitive to human activities, for without protection these species may soon be extinct.

- Human population and resource consumption must be reduced to a level that can be supported sustainably without loss of biodiversity.

This book is organized to introduce and defend the concept of biodiversity, review some failures of past land management, offer an alternative vision for the future, and provide guidelines for achieving ambitious conservation goals. The first three chapters lay the groundwork for biodiversity conservation—what biodiversity is, why it is important, its status in North America, the strengths and limitations of past and current land management, and a framework for a new strategy. The next two chapters discuss methods for inventorying biodiversity and selecting areas for protection, and the design of regional and continental reserve networks. The following three chapters address ecological principles, threats, and management guidelines for forests, rangelands, and aquatic ecosystems, respectively. Finally, we suggest a monitoring program that must accompany new conservation efforts and conclude with a chapter on priorities for getting the job done.

Acknowledgments

Many people have taught us and influenced our views about biodiversity and conservation. Those who provided information, advice, or review of proposals, chapters, or technical papers related to this book include Peter Brussard, Blair Csuti, John Davis, Dave Foreman, Chris Frissell, Ed Grumbine, Larry Harris, Wendy Hudson, Malcolm Hunter, David Johns, Dennis Murphy, Dave Perry, Mike Scott, Michael Soulé, Fred Smeins, Sara Vickerman, Don Waller, David Wilcove, Jack Williams, and George Wuerthner. Those stalwart individuals who reviewed the entire book are Blair Csuti, Barbara Dean, Ed Grumbine, Wendy Hudson, Malcolm Hunter, Ken McGinty, Sara Vickerman, and Barbara Youngblood. Sara and Wendy deserve special recognition, as the idea for this book was theirs, and their diligence saw it through to completion. We also thank Mike Jennings for reminding us of the timely Dickens quote used in the Preface.

Finally, we gratefully acknowledge the financial support of Defenders of Wildlife, the National Fish and Wildlife Foundation, and the Department of Defense Legacy Resource Management Program. The final part of Reed Noss's work on this project was supported by the Pew Charitable Trusts.

SAVING NATURE'S LEGACY

CHAPTER ONE

BIODIVERSITY AND ITS VALUE

The earth never tires:
The earth is rude, silent, incomprehensible at first—Nature is
rude and incomprehensible at first;
Be not discouraged—keep on—there are divine things,
well enveloped;
I swear to you there are divine things more beautiful than words
can tell.

Walt Whitman (1856), *Leaves of Grass*

This book is an exercise in applied conservation biology. The fundamental question of conservation biology is a critical one: how can the variety of life be maintained in perpetuity? How can we help preserve "divine things more beautiful than words can tell"? No one has an answer to these questions. But scientists have learned a few things about how nature works and what kinds of human activities are compatible and incompatible with life on earth. In this chapter, we first define biodiversity and describe its major components, then discuss why diversity has become an issue in the United States. This leads into a discussion of the values of biodiversity and why management of biodiversity has become a regrettable necessity today.

What Is Biodiversity?

In little more than a decade, biodiversity progressed from a short-hand expression for species diversity into a powerful symbol for the full richness of life on earth. Biodiversity is now a major driving force behind efforts to reform land management and development practices worldwide and to establish a more harmonious relationship between people and nature.

Biodiversity. A symbol? An issue? A driving force? It would be easier if biodiversity could be measured by the quantity of bird species in a forest, wildflowers in a meadow, or beetles in a log. But simplicity is not one of the virtues of biodiversity. Ecosystems are more complex than we can imagine. Our most intricate machines—say, a space shuttle and all its ground-control computers—are simple toys compared to an old-growth forest, its myriad known and unknown species, and their intricate genetic codes and ecological interactions. Just identifying and counting species is difficult enough. The almost infinite complexity of nature defies our best efforts to classify, categorize, or even describe.

A common misconception is that biodiversity is equivalent to species diversity—the more species in an area, the greater its biodiversity. However, biodiversity is not just a numbers game. On a global scale, maintaining maximal species richness is a legitimate goal and requires keeping global extinction rates low enough that they are balanced or surpassed by speciation. When we consider species richness at any scale smaller than the biosphere, quality is more important than quantity. It is not so much the number of species that we are interested in, it is their identity. Fragmenting an old-growth forest with clearcuts, for example, would increase species richness at a local scale but would not contribute to species richness at a broader scale if sensitive species were lost from the landscape.

Diversification can all too easily become homogenization. The greatest cause of homogenization worldwide is the introduction of nonnative plants and animals, often called exotics. Exotics are species that have invaded new areas due to accidental or deliberate transport by humans. Although species naturally disperse and colonize new areas, so that floras and faunas change continually over long periods of time, human transport and habitat disturbance have greatly increased the rate and scale of invasions. Many regions have nearly as many exotic as native species today. Introductions of exotics may increase species richness locally or even regionally, but they contribute nothing positive to biodiversity. Rather, they pollute the integrity of regional floras and faunas and often alter fundamental ecological processes, such as fire frequency and intensity, and nutrient cycles. Thus, whole ecosystems are changed. Regions invaded by exotics lose their distinctive characters. Every place begins to look the same. The result is global impoverishment. For these reasons, we emphasize *native biodiversity,* not diversity per se.

The important task is not to define biodiversity, but rather to determine the components of biodiversity in a region, their distribution and interrelationships, what threatens them, how we measure and monitor them, and what can be done to conserve them. These topics are the subject of this book. But because working definitions are helpful to summarize what we are talking about, we propose the following modification of a definition developed by the Keystone Dialogue (Keystone Center 1991):

> Biodiversity is the variety of life and its processes. It includes the variety of living organisms, the genetic differences among them, the communities and ecosystems in which they occur, and the ecological and evolutionary processes that keep them functioning, yet ever changing and adapting.

This definition recognizes variety at several levels of biological organization. Four levels of organization commonly considered are genetic, population/species, community/ecosystem, and landscape or regional. Each of these levels can be further divided into compositional, structural, and functional components of a nested hierarchy (Noss 1990a). Composition includes the genetic constitution of populations, the identity and relative abundances of species in a natural community, and the kinds of habitats and communities distributed across the landscape. Structure includes the sequence of pools and riffles in a stream, down logs and snags in a forest, the dispersion and vertical layering of plants, and the horizontal patchiness of vegetation at many spatial scales. Function includes the climatic, geological, hydrological, ecological, and evolutionary processes that generate and maintain biodiversity in ever-changing patterns over time.

Why bother with this cumbersome classification? Because nature is infinitely complex. Unless we try to identify and classify the forms of this complexity, we are likely either to miss something or become hopelessly confused. If something falls through the cracks in our conservation programs, it may be lost forever. With each loss biodiversity is diminished. The earth becomes a less interesting place.

Conserving biodiversity, then, involves much more than saving species from extinction. As implied by our characterization of biodiversity, biotic impoverishment can take many forms and occur at several levels of biological organization. Hence, steps must be taken at multiple levels to counteract impoverishment. Below, we review some conservation issues, goals, and problems that can be addressed at each of four major levels of biological organization. We emphasize that a *comprehensive* conservation strategy must integrate concerns from all levels of the biological hierarchy.

Genetic Level

Genes, sequences of the DNA (deoxyribonucleic acid) molecule, are the functional units of heredity. Species differ from one another and individuals within species vary largely because they have unique combinations of genes. Gene frequencies and genotypes (individual organisms with a particular genetic make-up) within a population change over time as a consequence of both random and deterministic forces. Random changes include mutations that create new genes or sequences of genes, and loss of genes by chance in small populations (called sampling error or genetic drift). Deterministic changes include natural and artificial selection, where some genotypes are more successful reproducers than others. In the long run, genetic change leads to evolutionary change as individuals adapt to different situations and pass on their new traits to offspring. Genetic diversity is fundamental to the variety of life and is the raw material for evolution of new species. We will discuss evolution briefly in Chapter 2.

Conservation goals at the genetic level include maintaining genetic variation within and among populations of species, and assuring that processes such as genetic differentiation and gene flow continue at normal rates. Without genetic variation, populations are less adaptable and their extinction more probable, all else being equal. Small, isolated populations are more likely to diverge genetically, having fewer chances for genetic mixing with other populations. But at the same time small, isolated populations are more likely to suffer from inbreeding depression caused by mating between close relatives, which may result in reduced fertility and other problems (Frankel and Soulé 1981). Small, isolated populations also are subject to random loss of genes (genetic drift), which restricts their ability to adapt to a dynamic environment.

Conservationists talk much about saving the earth's genetic resources. But with the exception of some agricultural crops, commercial tree species, populations of rare vertebrates in zoos, and a handful of wild populations, we know very little about genetic diversity. Land managers seldom think about maintaining biodiversity at the genetic level. If our vision of conservation is long term, however, genetic variation must be better understood for all organisms.

Species Level

The species level of diversity is probably what most people think of when they hear the term *biodiversity*. Although in some ways species diversity is the best known aspect of biodiversity, we should bear in mind that the vast majority of species in the world are still unknown. Of an estimated 10 to 100 million species on Earth (Wilson 1992), we have named only about 1.8 million (Stork 1992). Known species are dominated by insects, half of them beetles (Fig. 1.1). But many invertebrates, bacteria, and other organisms remain to be discovered, even in the United States. Hundreds of invertebrate species can be found in one square meter of soil and litter in an old-growth temperate forest (Lattin 1990). Even more amazing, Norwegian microbiologists found between 4000 and 5000 species of bacteria in a single

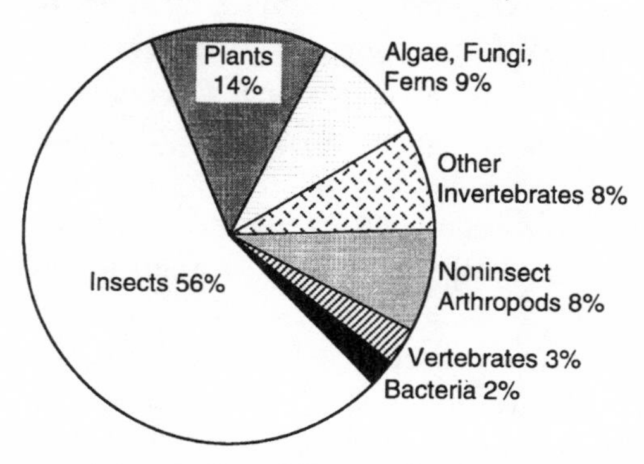

FIGURE 1.1 Taxonomic categories of species that have been named and described (adapted from Office of Technology Assessment 1987). Undescribed species, which outnumber described species by perhaps an order of magnitude, are probably mostly insects, other invertebrates, and bacteria.

gram of soil from a beech forest. About the same number of species, with little overlap, was found in a gram of sediment from off the coast of Norway (Wilson 1992). These findings raise the question of whether the tropical rainforests really are the most diverse habitats on Earth. We know too little about biodiversity to conclude much with certainty.

A population is a local occurrence of a species and is the unit that we usually manage. Conservation goals at the population/species level include maintaining viable populations of all native species in natural patterns of abundance and distribution. These goals grade into community-level goals of maintaining native species richness and composition, as discussed below.

Despite the problems and biases of single-species management, many species require individual attention, particularly when they have become so rare that heroic measures are needed to save them. In addition, certain kinds of species warrant management emphasis because their protection will conserve more than themselves. Especially important in this regard are keystone species, which play pivotal roles in their ecosystems and upon which a large part of the community depends (Noss 1991a). The importance of a keystone species is often disproportionate to its abundance. The beaver, for instance, creates habitats used by many species and also regulates hydrology and other ecosystem functions (Naiman *et al.* 1988). If we reduce beaver numbers through heavy trapping, then all else being equal, we impoverish the landscape. The beaver is not an endangered species, but it is greatly reduced or even absent from many regions where it was once abundant. Major declines of keystone species are more important ecologically than the loss of the last few individuals of rare species that play minor roles in their communities. This said, we must recognize that the term *keystone species* is poorly defined. Instead of a dichotomy of keystones and nonkeystones, communities may be better characterized by a wide range of interactions of variable strengths (Mills *et al.* 1993). Because we know so little about the ecological roles of species, each species must be considered important.

Some kinds of species have great pragmatic value for conservation, especially those we can characterize as "umbrellas" or "flagships" (Noss 1991a). To illustrate the umbrella concept, consider a carnivore (such as a grizzly bear or wolf) that requires millions of acres of land to maintain a viable population. If we secure enough wild habitat for these large predators, many other less-demanding species will be carried under the umbrella of protection. Umbrella species are often charismatic, so they also function as flagships or symbols for major conservation efforts. The grizzly bear, for instance, is a potent symbol for wilderness preservation in the northern Rocky Mountains. No umbrella is complete, however. Some endemic plant species have very small ranges—perhaps restricted to a single rock outcrop—that might not be protected in an ideal wilderness network established for grizzlies.

Animals and plants that are highly vulnerable to human activity often need to be managed individually, at least until their habitats can be protected by an ecosystem-level approach. Otherwise, biodiversity will continue to diminish with each extinction. Although we might accept the egalitarian notion that all species are ultimately equal, at any given place and time some species thrive on human activity and others suffer. Familiar examples of species that are extremely vulnerable to human activity are the northern spotted owl, threatened by logging of old-growth forests in the Pacific Northwest (Thomas *et al.* 1990); the red-cockaded woodpecker, endangered by logging of longleaf pine forests in the Southeastern Coastal Plain (Jackson 1986); and the desert tortoise, often shot or run over by motorized recreationists, forced to compete with livestock, collected for pets, and now ravaged by disease (U.S. Fish and Wildlife Service 1993). Species declines are signals that the environment is not healthy, but vulnerable species often require intensive care above and beyond immediate protection of their habitat.

Community or Ecosystem Level

In many cases, conservation is most efficient when focused directly on the community or ecosystem. A community is an interacting assemblage of species in an area. Terrestrial communities are usually defined by their dominant plants (for instance, the beech-maple forest), but functional or taxonomic groups of animals (for example, bird communities, lizard communities, herbivore communities) are also recognized. Functional groups of organisms (species that use a set of resources in similar ways, such as bark-gleaning birds) are often called *guilds.* Similarly, aquatic communities may be taxonomically or functionally defined, for example fish communities or littoral (shoreline) vegetation.

An ecosystem is a biotic community plus its abiotic environment. Ecosystems range in scale from microcosms, such as a vernal pool, to the entire biosphere. Many ecologists equate the terms *ecosystem* and *community,* except that ecosystem ecologists emphasize processes more than species and other entities. The Nature Conservancy defines natural communities by their most striking characteristics, whether biotic or abiotic. Thus, coastal plain pond, rich graminoid fen, black spruce-tamarack bog, and rich mesophytic forest are all described communities of New York State (Reschke 1990). These communities might also be called ecosystem "types." The variable spatial scale of ecosystems confuses the issue sometimes. Although scientists usually think of ecosystems as relatively discrete and existing at the same spatial scale as natural communities, conservationists often use the term *ecosystem* to encompass many different communities. For example, the Greater Yellowstone Ecosystem covers a diverse region of 14 to 19 million acres (see Chapter 5).

We consider conservation at the community or ecosystem level to complement, not replace, species-level management. The rationale for protecting ecosystems is compelling: if we can maintain intact, ecologically functional examples of each type of ecosystem in a region, then the species that live in these ecosystems will also persist. Representing all native ecosystems in a network of protected areas is the most basic conservation goal at the ecosystem level (see Chapter 4). Opportunities for adequate representation of ecosystems are being rapidly diminished as many of our native vegetation types are being reduced in area and degraded in quality (Noss *et al.* 1994).

Practicing conservation at the community or ecosystem level demands attention to ecological processes. Maintaining processes is not just a way to maintain species. Rather, processes are valuable for their own sake as part of the diversity of life. The processes that are most crucial for ecological health vary from ecosystem to ecosystem. In terrestrial communities some of the most important processes are fire and other natural disturbances, hydrological cycles, nutrient cycling, plant–herbivore interactions, predation, mycorrhizal interactions between tree and shrub roots and fungi, and soil building processes. All of these processes affect biodiversity at several levels and are included within our definition of biodiversity. They must be maintained within normal limits of variation if native biodiversity is to persist. Clearcutting and other intensive forest management may fail to conserve biodiversity because they disrupt nutrient retention and other ecological processes. Livestock grazing that interferes with basic ecological processes will also fail to conserve native biodiversity in rangelands.

Alpha, Beta, and Gamma Diversity

The variety of species in a defined area is one common measure of biodiversity at the ecosystem level. But to say that more diverse areas are better is misleading because measures of species richness or diversity neglect a most important consideration—the identity of species. One way to consider species diversity while paying close attention to species composition is to note the spatial scale of observation and how composition changes from one scale to another. The collection of species within an area of relatively homogeneous habitat is called *alpha* diversity or within-habitat diversity (Whittaker 1972, Karr 1976). Each site will have its own characteristic alpha diversity, although physically similar habitats in the same region can be expected to have similar species composition.

As we expand the scale of observation, we encounter variation in the underlying physical environment (environmental gradients). As we move along a gradient, say upslope, downslope, or from one soil type to another, we encounter new species adapted to these different conditions. The turnover in species along an environmental gradient is called *beta* diversity or between-habitat diversity. When we measure the diversity of species across several different habitats in a landscape, we are measuring beta diversity.

At a still broader scale, many environmental gradients are found and geographic replacements of species occur as range boundaries are crossed. Diversity at this regional scale is called *gamma* diversity. The alpha, beta, and gamma diversity concepts are useful for comparing biodiversity in different regions or in the same region under different management scenarios. Two regions of roughly equivalent gamma diversity may differ greatly in alpha and beta diversity. For example, Region A is mostly lowland forest with high alpha diversity but little habitat diversity. Thus, any site in the region is likely to contain roughly the same set of species. In contrast, Region B is mountainous, with tremendous differences in species composition between habitats but lower diversity within any single habitat. Generally speaking, a landscape in the eastern deciduous forest biome will have higher alpha diversity than many areas in the West. However, western landscapes characteristically have higher beta diversity due to the effects of aridity and topographic variation. Conservationists would have to consider very different scales and factors in planning reserve systems in the two regions. It is prudent to consider the maintenance of alpha and beta diversity within the broader context of gamma (and ultimately global) diversity.

Landscape and Regional Levels

If biodiversity occurs at multiple levels of organization, it is worth protecting at all levels. Forman and Godron (1986) defined a landscape as "a heterogeneous land area composed of a cluster of interacting ecosystems that is repeated in similar form throughout." Similarly, Urban *et al.* (1987) characterized a landscape as "a mosaic of heterogeneous land forms, vegetation types, and land uses." These definitions suggest that landscapes have a *pattern* and that this pattern consists of repeated habitat components that occur in various shapes, sizes, and spatial interrelationships. In many landscapes, this pattern consists of patches and corridors in a matrix, the matrix being the most common or interconnected habitat in the landscape (Forman and Godron 1986). Other landscapes are mosaics of many habitats, and the pattern is difficult to classify into discrete components.

We use the term *region* (also *bioregion* or *ecoregion*) to refer to large landscapes that can be distinguished from other regions on the basis of climate, physiography, soils, species composition patterns (biogeography), and other variables. Landscape or regional diversity is pattern diversity—the pattern of habitats and species assemblages across a land area of thousands to millions of acres—and can be considered a higher level expression of biodiversity. The pattern of species distributions and communities across a landscape has functional ramifications. Many animals, for example black bears, use more than one habitat type to meet their life history needs. We cannot protect these species by managing different communities in isolation. Bears and other wide-ranging animals are often important in dispersing seeds across a landscape. Disrupting bear movements by fragmenting the landscape may indirectly affect other species.

Adjacent habitats affect each other in many ways, including by microclimatic effects and transfer of nutrients, propagules, and disturbances across edges and ecotones. Because human activities often change landscape patterns, they have impacts on biodiversity that ripple through other levels of organization, affecting species composition and abundances, gene flow, and ecosystem processes. If a forest landscape is fragmented into small patches, those patches may experience a drier microclimate than the original forest, increased susceptibility to windthrow, loss of forest interior species, reduced genetic diversity within remaining populations, and invasion by weedy and exotic species (Burgess and Sharpe 1981, Harris 1984, Franklin and Forman 1987). These problems cannot be solved patch by patch, but only across all patches and their matrix. Hence, the regional landscape is an appropriate scale at which to identify important sites and patterns, and to manage and restore land for conservation purposes (Noss 1983, Turner 1989).

A primary conservation goal at the landscape or regional level is to maintain complete, unfragmented environmental gradients. This extends the representation goal beyond traditional ecosystem boundaries. Species richness and composition are known to vary along environmental gradients. The most commonly studied gradient is elevation. In the western Cascades of Oregon, the number of species of amphibians, reptiles, and mammals declines sharply with increasing elevation (Fig. 1.2). This presents a problem for conservation, because generally speaking, the low-elevation, high-diversity sites are private lands which are often heavily exploited and have few natural areas left. Mid-elevation sites are commodity-production public lands, and large protected areas (such as designated wilderness) occupy the high-elevation, lowest diversity sites. This biased pattern of habitat protection is common throughout the western United States (Davis 1988, Foreman and Wolke 1989, Noss 1990b). By contrast, in southeastern states with abundant wetlands, such as Florida, most wilderness areas are habitats too wet for commercial forestry. Preserving only the most species-rich sites or portions of environmental gradients is no solution, because different species occupy different portions. Alpine wildflowers are not found in the rich lowlands. Conservation programs must strive to maintain natural ecosystems and biodiversity across the full extent of environmental gradients.

The effects of natural disturbance on biodiversity often can be best appreciated at the landscape scale. Disturbances typically create patches in the landscape, of various sizes, that are used by different sets of species (Pickett and White 1985). Disturbance-recovery processes are complex. For example, most forest fires are mosaics of many different fire intensities, with some patches experiencing crown fires, other patches untouched, and a wide range of intensities in between. Recovery after fire varies with intensity (sometimes the soil may be sterilized and succession is slow), seed sources, weather, and other factors. Variability in disturbance regime is responsible for much of the habitat diversity found in natural landscapes. Generally speaking, the more diverse the habitat, the more species can coexist. However, diversification by disturbance has a limit. Actions that diversify habitat locally may reduce diversity at regional and global scales if disturbance-sensitive species are eliminated. Disturbance regimes will be discussed in more detail in Chapter 2.

In landscape ecology, context is just as important as content. Small reserves set aside for their content (for example, to represent plant community types or to protect a remnant population of a rare species) are heavily affected by their context when the surrounding landscape is altered. Eventually, a small reserve or a system of small, isolated reserves may fail to maintain the elements for which they were established. Natural fire regimes,

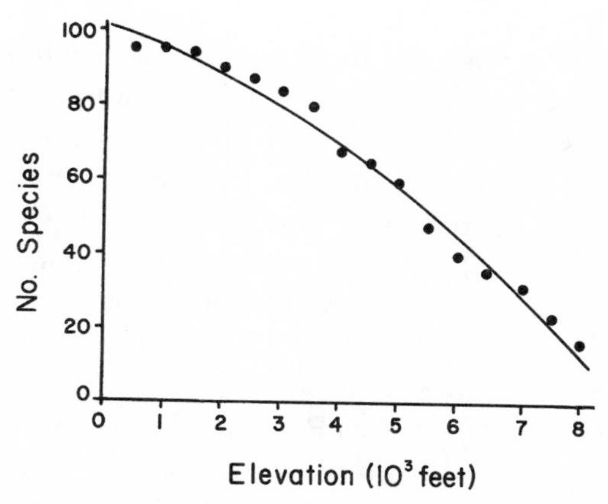

FIGURE 1.2 Relation between elevation and species richness of amphibians, reptiles, and mammals in western Oregon (from Harris 1984). The decline in species richness with increasing elevation is accompanied by a decline in density of individuals of all species combined. Used with permission of the author and University of Chicago Press.

migration of large animals, landform evolution, and hydrological cycles are ecological processes that can be perpetuated only by conservation at landscape and regional scales. Thus, a biodiversity conservation strategy is complete only when expanded to these scales.

Cultural or Social Diversity

Have we missed anything in our broad characterization of biodiversity? During the Keystone Dialogue on Biological Diversity and in similar forums, some participants insisted that human cultural or social diversity be included in the definition of biodiversity and in any strategy for its conservation. The Global Biodiversity Strategy (WRI, IUCN, UNEP 1992) makes this point strongly:

> Human *cultural diversity* could also be considered part of biodiversity. Like genetic or species diversity, some attributes of human cultures (say, nomadism or shifting cultivation) represent "solutions" to the problems of survival in particular environments. And, like other aspects of biodiversity, cultural diversity helps people adapt to changing conditions. Cultural diversity is manifested by diversity in language, religious beliefs, land-management practices, art, music, social structure, crop selection, diet, and any number of other attributes of human society.

On the face of it, inclusion of social diversity in a definition of biodiversity makes sense. We are fundamentally as much a part of Nature as any other species and share kinship and ecological interactions with all of life. But what would be the practical effect of including diversity of human languages, religious beliefs, behaviors, land management practices, etc., in a biodiversity definition and striving to promote this diversity in conservation strategy? We believe the effect would be to trivialize the concept and make it unworkable, even dangerous. As Kent Redford (personal communication) notes, "This definition allows Manhattan or Sao Paulo to be considered on equal footing with the Great Barrier Reef of Australia and makes impossible any coherent discussion of biodiversity conservation." We are not interested in maintaining social or cultural diversity if it means maintaining Nazis, slave owners, or those who enjoy using desert tortoises for target practice. This book is about how culture might adapt to nature. We want to conserve all cultural approaches that are compatible with conserving biodiversity. To combine cultural and biological diversity into one definition is to muddle the concept.

Why Has Biodiversity Become an Issue?

Why has biodiversity become an issue in the United States? Have conventional approaches to conservation failed? Consider the continent of North America 300 years ago. A description from an early explorer in Florida portrays the diversity of life in the Southeastern Coastal Plain, a richness paralleled in many different and glorious ways across the continent:

> We returned, viewing the Land on both sides of the River, and found as good tracts of land, dry, well wooded, pleasant and delightful as we have seen anywhere in the world, with great burthen of Grasse on it . . . the woods stor'd with abundance of Deer and Turkies every where . . . also Partridges great store, Cranes abundance, Conies . . . several Wolves howling in the woods, and saw where they had torn a Deer in pieces. Also in the River we saw great store of Ducks, Teile, Widgeon, and in the woods great flocks of Parrakeeto's . . . we measured many of the Oaks in several places, which we found to be in bignesse some two, some three, and others almost four fathoms; in height, before you come to boughs or limbs, forty, fifty, sixty foot, and some more. . . . Also a very tall large Tree of great bignesse, which some do call Cyprus. . . . (Hilton 1664, in Salley 1911)

The European explorers' impressions of a vast wilderness continent were accurate enough. All of this country was roadless, unpolluted, rich with wildlife, and incomparably beautiful. But this was not a wilderness "untrammeled by man," in the controversial language of our Wilderness Act of 1964 (Callicott 1991). The North American wilderness was a peopled wilderness, yet peopled sparsely and, for the most part, gently. An estimated 10 million native humans, 3 percent of the present human population, inhabited North America when the first white people arrived. In places the Indians modified their landscape considerably, especially through the use of fire (Day 1953, Pyne 1982). Hunting by their ancestors probably contributed to the extinction of large mammals near the close of the Pleistocene (Martin and Klein 1984). But native cultures occasionally enriched native biodiversity locally and perhaps regionally through diverse agricultural plantings (Nabhan 1982). Although the romantic notion of Indians as the original environmentalists is not entirely accurate (Callicott 1982), in general the native Americans seem to have lived in harmony with the rest of nature. Without such a relationship they would have had trouble persisting here for over 20,000 years. A culture that destroys its environment is suicidal.

The picture changed dramatically after the arrival of European settlers. The biological history of North America since then—a story seldom told in American history classes—has been one of profound impoverishment, particularly in the last 200 years. The slaughter of native Americans by early explorers and colonists is now well known, but the desire of Europeans for subjugation extended to other life forms as well. This subjugation continues today. In the words of Barry Lopez (1992):

> The assumption of an imperial right conferred by God, sanctioned by the state, and enforced by a militia; the assumption of unquestioned superiority over a resident people, based not on morality but on race and cultural comparison . . . the assumption that one is *due* wealth in North America, reverberates in the journals of people on the Oregon Trail, in the public speeches of nineteenth-century industrialists, and in twentieth-century politicians. You can hear it today in the rhetoric of timber barons . . . standing before the last of the old-growth forest, irritated that anyone is saying "*enough* . . . , it is enough."

European settlers saw the people, wildlife, and land of North America as something to be conquered, tamed, and subjected to their will. A concern for or even knowledge of what was being lost was altogether lacking.

As biologist Larry Harris describes it, "we swept across this continent so quickly . . . that we never really knew what was here" (quoted in Chadwick 1990). The great eastern deciduous forest exists today only as tattered remnants, growing slowly back in some regions, such as parts of the Northeast and southern Piedmont, but still being fragmented and subdivided in others. Cougar are gone from the East with the exception of a tiny population of Florida panthers on the verge of extinction in south Florida, and scattered but questionable reports farther north. The ivory-billed woodpecker, Carolina parakeet, Labrador duck, heath hen, great auk, and passenger pigeon (the most abundant landbird in the world when Europeans arrived) are gone from the earth, as are Merriam's elk, Audubon bighorn, the buffalo wolf, sea mink, and Caribbean monk seal. The estimated 40 million pronghorn that roamed the West before arrival of the pioneers were quickly reduced to fewer than 20,000. An estimated 60 million bison were reduced to fewer than a thousand by 1890 (Zeveloff 1988). The Boskowitz Hide Company of Chicago shipped more than 34,000 bison hides out of Montana and northern Wyoming in 1880. In 1884 they could get only 529 (Madson 1987).

These megafauna were only the most conspicuous losses. The Nature Conservancy estimates that over 200 full species of plants, plus many more varieties, and 71 species and subspecies of vertebrates have gone extinct in North America north of Mexico since European settlement (The Nature Conservancy 1992, Russell and Morse 1992). Over 750 species of plants and animals in the United States are federally listed as threatened or endangered, thus officially considered close to extinction. Another 3000-plus species are candidates for listing, yet at present rates of listing many of these candidates will be lost before receiving protection under the Endangered Species Act. Only five listed species have recovered enough to be removed from the list (Wilcove *et al.* 1993). The steady erosion of our native biodiversity is a direct consequence of the callous disregard we have shown for our environment and our evolutionary kin. This disregard continues today, ironically despite polls showing that 89 percent of the national public agrees with the statement "humans have an ethical obligation to protect plant and animal species" (Shindler *et al.* 1993).

Opponents of conservation often point out that extinction is natural and not worth worrying about. However, with the exception of a few mass extinction events in ancient geological history, the rate at which new species are created has exceeded the rate of extinction. Therefore, the number of species on Earth seems to have slowly but, with a few punctuations, steadily increased over time. That trend is being reversed today. As explained by Wilson (1985):

> No comfort should be drawn from the spurious belief that because extinction is a natural process, humans are merely another Darwinian agent. The rate of extinction is now about 400 times that recorded through recent geological time and is accelerating rapidly. Under the best of conditions, the reduction of diversity seems destined to approach that of the great natural catastrophes at the end of Paleozoic and Mesozoic Eras, in other words, the most extreme for 65 million years. And in at least one respect, this human-made hecatomb is worse than at any time in the geological past. In the earlier mass extinctions, possibly caused by large meteoritic strikes, most of the plant diversity survived; now, for the first time, it is being mostly destroyed. (Knoll 1984)

Recent extinctions and the ever-expanding list of endangered species in North America show that the biodiversity crisis is not just a tropical problem. Although current extinction rates in the tropics, estimated at somewhere between 10,000 and 150,000 species lost per year over the next few decades (Wilson 1988, Diamond 1990), are higher than in the temperate zone due to the apparently greater diversity of tropical systems, some North American ecosystems are more endangered than tropical rainforests and stand to lose as great a proportion of their species. Examples from the United States of natural communities being destroyed faster than tropical rainforests include freshwater habitats in California (Moyle and Williams 1990) and old-growth forests of the Pacific Northwest (Norse 1990). In 1992, newspapers throughout the country reported a NASA study showing that the loss of old-growth forests in the Pacific Northwest surpassed in rate and extent the clearing of the Amazon basin. Satellite photographs vividly compared the damage (Fig. 1.3). Earlier, a *National Geographic* article portrayed the loss of virgin forests in the 48 conterminous states (Findley 1990), which amounts to at least 95 percent of the forests that greeted the first European settlers. Ancient forests have finally captured the public's attention, but they are not the only ecosystems disappearing. Estimates of ecosystem decline throughout the United States are shockingly high (Noss *et al.* 1994). We have already lost far more than most Americans realize.

Why Are We Concerned About Biodiversity?

Aldo Leopold (1953) observed: "The last word in ignorance is the man who says of an animal or plant: 'What good is it?' If the land mechanism as a whole is good, then every part is good, whether we understand it or not. If

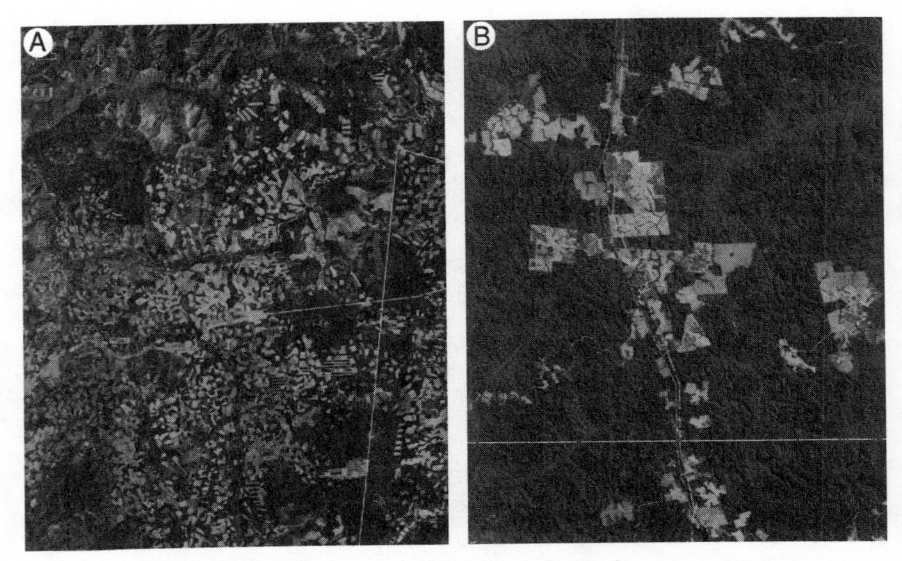

FIGURE 1.3 Satellite images from NASA show forest fragmentation more severe in the Mt. Hood National Forest of Oregon (A) than in the Amazon basin of Brazil (B). Source: NASA/GSFC (1992).

the biota, in the course of aeons, has built something we like but do not understand, then who but a fool would discard seemingly useless parts? To keep every cog and wheel is the first precaution of intelligent tinkering."

Do people care if species go extinct and natural areas are converted to shopping malls? Public opinion polls in the United States show that Americans are concerned about endangered species. For example, a recent poll of 1000 registered voters spanning the demographic, political, and geographic spectrums of the U.S. showed that 66 percent support the Endangered Species Act and only 11 percent opposed protecting endangered species (Stolzenburg 1992). Another poll showed that 78 percent of the national public believes that greater protection should be given to fish and wildlife habitats on federal forest lands; 65 percent disagreed with the statement that endangered species laws should be set aside to preserve timber jobs (Shindler *et al.* 1993). Although these polls asked mostly about species, not ecosystems, we can expect that many people will oppose destruction of natural areas, especially areas near and dear to them.

Are there more fundamental reasons for protecting species and ecosystems, besides public support for such actions? Many texts have examined the values of biodiversity (e.g., Ehrlich and Ehrlich 1981, Prescott-Allen and Prescott-Allen 1983, 1986, Norton 1986, Wilson 1988, World Wildlife Fund 1991). The value of biodiversity is our fundamental assumption. If we did not believe in it, we would not be writing this book. However, it is worth reviewing briefly the types of value that humans ascribe to nature. Often arguments about what is proper management of natural resources can be put in perspective, if not totally resolved, by understanding how people value nature in different ways. The limitations of different value justifications for saving nature also need to be understood. Although we prefer to think that nature has essentially one value—with no necessary distinction between utilitarian and intrinsic—we partition this value below for purposes of discussion.

Direct Utilitarian Values

The kind of value easiest to appreciate, for many people, is the utilitarian or instrumental value of a species or other natural resource. That the "what good is it" question is so often asked suggests that many people value things largely for their direct utility for humans. Though incomplete as a justification for saving biodiversity, such values are real.

The medicinal value of certain plants and invertebrates provides a powerful argument for conservation, as does the value of wild gene pools for agriculture and wild populations for food. Wild species provide an estimated 4.5 percent of the Gross Domestic Product of the United States, worth $87 billion annually in the late 1970s (Prescott-Allen and Prescott-Allen 1986). Fisheries contributed 100 million tons of food to people worldwide in 1988 (FAO 1988). One-fourth of all prescription drugs in the United States contain active ingredients extracted from plants, and nearly 3000 antibiotics are derived from microorganisms (WRI, IUCN, UNEP 1992). These statistics suggest that it is in our best interest to prevent extinctions of species that are potentially useful to us. What if the Pacific yew, until recently considered a trash tree and destroyed during clearcutting in the Pacific Northwest, were extinguished before we discovered that it contained taxol, a valuable new drug for treating several forms of cancer? This question is of more than academic interest. By one estimate, only about 5000 (2 percent) of the 250,000 described species of vascular plants have been screened for their chemical compounds (World Wildlife Fund 1991). We are driving species to extinction without even trying to learn what they might contribute to human society.

Arguments based on utility are limited, however. Leopold (1949) observed that "one basic weakness in a conservation system based wholly on economic motives is that most members of the land community have no economic value." Similarly, Ehrenfeld (1988) lamented, "what biologist is willing to find a value—conventional or ecological—for all 600,000-plus species of beetles?" What happens if we thoroughly screen a plant for medicinal compounds and conclude that it has none? Do we then say it is permissible to extinguish that species? Conservationists often fall into the trap of justifying species preservation for utilitarian purposes, thereby sanctioning the humanistic attitude that is responsible for the biodiversity crisis (Ehrenfeld 1978, 1988). The attitude implied by economic valuations is that the worth of a species depends on its direct ultility to humans. If a species does not benefit us, it is worthless.

At best, the utilitarian argument for biodiversity conservation is a double-edged sword. Under some circumstances it might help gain public support for protecting species and ecosystems, but in other cases it can be used to justify destruction of seemingly worthless forms. In all cases, it encourages disrespect for species in and of themselves. Thus, we are troubled that current arguments for maintaining international biodiversity, such as those expressed in the Global Biodiversity Strategy produced by the World Resources Institute (WRI), World Conservation Union (IUCN), and United Nations Environment Programme (UNEP) (1992), are thoroughly utilitarian; they hinge almost entirely on presumed benefits to humans. The sustainable development theme of the Global Biodiversity Strategy and related international conservation programs is potentially dangerous. Sustainable development could do more harm than good to biodiversity if strict protection of sensitive areas is not part of the program (Robinson 1993).

Indirect Utilitarian Values

Natural ecosystems and biodiversity also provide benefits to humans that are indirect, yet essential. Paul and Anne Ehrlich (1981) call these benefits "ecosystem services." Every habitat on Earth, including urban and agricultural environments, is an ecosystem that receives and transforms energy, produces and recycles wastes, and relies on complex interactions among species to carry out these functions. But urban and agricultural ecosystems are dependent on natural ecosystems for their sustenance. Solar energy is the basis of virtually all food chains (rare exceptions include chemically based communities in deep-sea vents) and is converted to chemical energy by photosynthetic plants. Plants, including crops, often depend on animals to pollinate their flowers and disperse their seeds, on nitrogen-fixing bacteria to convert molecular nitrogen to a form that can be assembled into proteins, and on microorganisms to convert complex organic compounds into inorganic nutrients that can be taken up by their roots. Animals, fungi, and microbes in an ecosystem have comparable interdependencies. Thus, an ecosystem is a richly interconnected web of relationships greater than the sum of its parts.

But how does a natural ecosystem benefit humans, besides providing pharmaceuticals and other products? An entire book could be written on this subject. Ehrlich and Ehrlich (1981) describe ecosystem services upon which human civilization is entirely dependent, including: (1) maintaining atmospheric quality by regulating gas ratios and filtering dust and pollutants; (2) controlling and ameliorating climate through the carbon cycle and effects of vegetation in stimulating local and regional rainfall; (3) regulating freshwater supplies and controlling flooding (wetlands, for example, can act as giant sponges to soak up moisture during rainy periods and release water slowly during dry periods); (4) generating and maintaining soils through the decomposition of organic matter and the relationships between plant roots and mycorrhizal fungi; (5) disposing of wastes, including domestic sewage and wastes produced by industry and agriculture, and cycling of nutrients; (6) controlling pests and diseases, for example through predation and parasitism on herbivorous insects; and (7) pollinating crops and useful wild plant species by insects, bats, hummingbirds, and other pollinators.

The public-service functions of ecosystems remain little known to most people, perhaps because we need to understand ecology in order to appreciate the functional relationships that underlie these services. Our society, by any measure, is ecologically ignorant. The role of biodiversity in supporting ecosystem services is striking, but we cannot easily predict how many or what kinds of species can be lost before ecosystems break down. Because all ecosystems contain some functional redundancy (with different species playing similar roles), we might impoverish an ecosystem substantially before impairing basic ecological processes such as nutrient cycling. From a ultilitarian perspective, we don't *need* every species. Some simplified, human-created ecosystems may perform virtually all of the public service functions reviewed above. The danger is that natural ecosystems have evolved their functional relationships over thousands or millions of years, whereas our experiments in manipulating ecosystems are comparatively brief. Who knows when we may lose a species or set of relationships critical to ecosystem function?

Recreational and Esthetic Values

Probably most people who care about the environment are motivated primarily by their personal appreciation of nature's beauty. John Muir, founder of the Sierra Club and a leading force in the creation of the U.S. national park system, firmly believed that exposure of ordinary people to wild places would foster an attitude to save these places (Fox 1981). Leopold (1949), too, noted that people will behave ethically only toward something they can experience and have faith in. Recreational and esthetic enjoyment of nature often leads directly to appreciation of nature for its own sake, that is, to a spiritual or ethical appreciation of biodiversity. Without people motived by their experiences of wild places, we would arguably have fewer wild areas remaining and the status of biodiversity in North America would be even more precarious.

Despite the critical role of these kinds of human experience in promoting conservation, areas set aside to fulfill recreational or esthetic objectives do not necessarily meet biodiversity conservation goals. Many national parks, wilderness areas, and other large reserves selected on the basis of esthetic criteria are relatively depauperate biologically. The Forest Service evaluates the "need" for wilderness designation on the basis of expected recreational visitor days, not on biological criteria. As a result, most wilderness areas are rock and ice, or other such scenic but not particularly diverse lands.

Many managers of national forests and other public lands, forced to reduce commodity production because such uses were unsustainable and in violation of environmental laws, are turning to recreation as an alternative use of these lands. This trend can be risky to the extent that it emphasizes motorized recreation. Managers justify road-building, leaving logging roads open to the public, and allowing use of off-road vehicles by arguing that the public needs access to these lands. Motorized recreation is almost always destructive of biodiversity. Furthermore, it encourages an attitude of dominance over nature. As Leopold (1949) put it: "It is the expansion of transport without a corresponding growth of perception that threatens us with qualitative bankruptcy of the recreational process. Recreational development is a job not of building roads into lovely country, but of building receptivity into the still unlovely human mind."

Thus, conservation arguments based on promoting human enjoyment are incomplete, at best. A deeper reason for protecting nature must be found.

Intrinsic, Spiritual, and Ethical Values

The limitations of utilitarian arguments for conserving biodiversity leave an alternative: the appreciation of wild creatures and wild places for themselves. We believe that nature and biodiversity possess *all* the kinds of value reviewed above, but that intrinsic values (or the spiritual and ethical appreciation of nature for its own sake) offer the least biased and ultimately most secure arguments for conservation. Virtually all religious traditions recognize the value of a human being—for example, a newborn baby—as at least partially independent of what that person might do for us. Why shouldn't we feel the same way about other creatures? The acknowledgment that natural objects and processes are valuable in themselves reflects a basic intuition of many people. Science cannot prove or disprove intrinsic value. Yet as scientists, we see no objective reason for believing that humans are fundamentally superior to any other organism. If we have value, then all natural things have value.

The ethical basis for respecting and protecting nature was expressed eloquently by Aldo Leopold (1949) in his famous essay on the land ethic. "A thing is right," Leopold wrote, "when it tends to preserve the integrity, stability, and beauty of the biotic community. It is wrong when it tends otherwise." Ecologists might quibble with Leopold's choice of such imprecise terms as integrity and stability, noting that Nature is instead dynamic and unpredictable (Botkin 1990), and they might wonder how we can objectively measure beauty. But few challenge his primary message that ethical obligations must encompass more than our fellow human beings (Nash 1989, Noss 1992a). Many philosophers have joined Leopold in calling for consideration of all life in our decisions about what is morally right or wrong (Devall and Sessions 1985, Rolston 1988, Naess 1989). To do so is to expand our circle of ethical concern beyond the individual self and ultimately to the ecological self—the land as a whole (Fig. 1.4).

Without moral consideration of the needs of other creatures, policies for protecting biodiversity remain on shaky ground. Thus, we urge a reaffirmation of the World Charter for Nature, adopted by the United Nations General Assembly in 1982, which stated: "Every form of life is unique, warranting respect regardless of its worth to man, and, to accord other organisms such recognition, man must be guided by a moral code of action."

Where Have We Failed?

The United States has a long and venerable conservation history, with many accomplishments worthy of pride. Our laws, including the National Park Organic Act of 1916, the Wilderness Act of 1964, the Endangered Species Act of 1969 and 1973, and the National Environmental Policy Act of 1969 and 1973, are held up as models for the world to follow. Yet the biodiversity crisis continues to worsen. Our air and water get dirtier, the ozone layer thins, and most significantly for biodiversity, more and more natural habitat is destroyed in the name of progress and jobs. Why have the American people allowed this to happen?

Let us briefly examine the failure of conventional conservation practices and institutions to address the "big picture" (this topic will be explored in

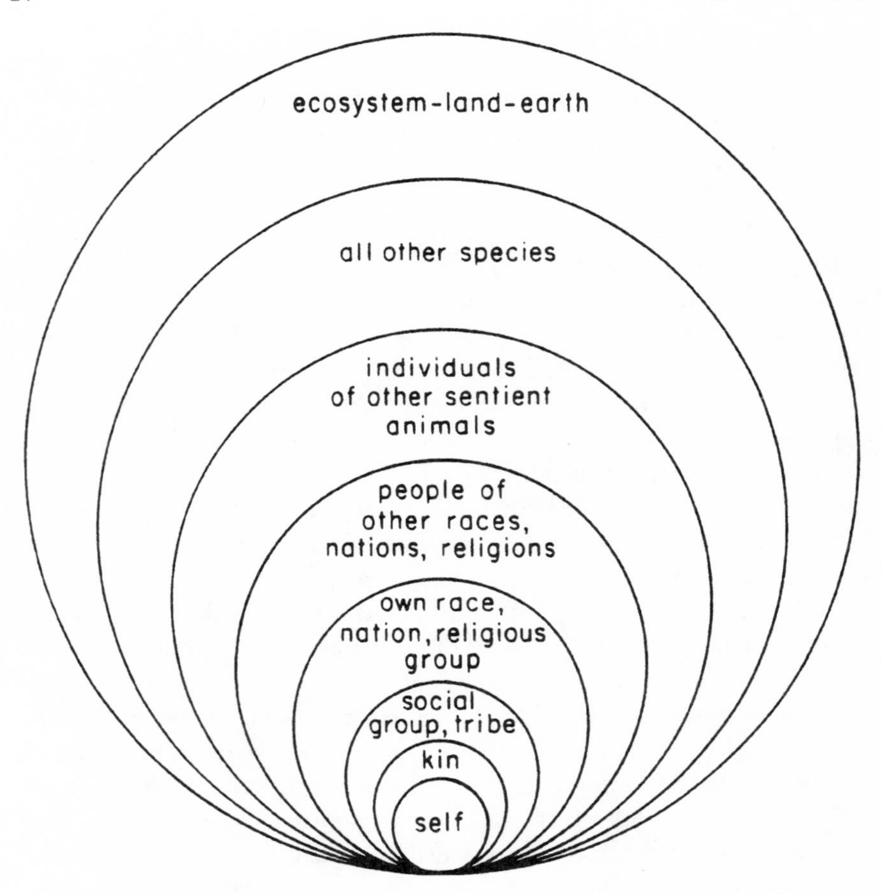

FIGURE 1.4 An ethical sequence, showing circles of moral concern and obligation increasing from the self and immediate family to all other species and the earth as a whole (the ecological self). Concern for higher levels is an extension, not a replacement, of traditional ethical concern for human beings. From Noss (1992a). Used with permission of Routledge, Chapman & Hall.

more depth in Chapter 3). We have laws, regulations, and agencies set up to protect aspects of the environment, yet none of them has measured up to the challenge of protecting biodiversity as a whole. We have trained four generations of professionals in forestry, wildlife management, fisheries, and range conservation, yet these disciplines remain fragmented and are often narrow. The traditional approach to conservation has been piecemeal: species by species, resource by resource, project by project, threat by threat.

The first laws in the United States oriented toward natural resources were game protection laws. Market hunting was big business in the late nineteenth century. After George Perkins Marsh (1864) published *Man and*

Nature, wildlife declines became widely recognized. Sportsmen led the fight to enact more laws protecting wildlife, including a bill outlawing bison killing in 1874, pocket-vetoed by President Grant for anti-Indian reasons; the Lacey Act (1900), authorizing the federal government to prohibit interstate transport of illegally taken game and wildlife parts; and the Migratory Bird Treaty with Canada (1916) and Mexico (1936), which protected nongame birds. Meanwhile, the Boone and Crockett Club was established in 1887, largely to promote sport over market hunting, and The Audubon Society was formed in 1886 to combat the fashion of wearing the feathers and skins of birds.

About the same time, Americans began formally protecting places where wildlife and scenery could be enjoyed. The Yellowstone Park Protection Act passed in 1892, twenty years after actual creation of the Park. Yellowstone and other national parks were not established primarily to protect wildlife, but rather "to conserve the scenery and the natural and historic objects and the wild life" within parks, and "to provide for the enjoyment of the same in such manner and by such means as will leave them unimpaired for the enjoyment of future generations" (National Park Organic Act of 1916). Also in 1892, the Sierra Club was founded by John Muir and other mountain hikers (Fox 1981). The first federal wildlife refuge, Pelican Island, Florida, was established by Theodore Roosevelt in 1903 to protect egrets, herons, and brown pelicans from plume hunters. Before leaving office in 1909, Roosevelt created 51 refuges for birds and mammals (Zaslowsky 1986), but he also escalated the war against wolves, bears, puma, and other predators (Worster 1977).

Conservation history in the United States is largely a series of responses to urgent threats against popular species and scenic wonders. Game management rose as a formal science in the United States in the 1930s, with the publication of *The Bobwhite Quail* by Herbert Stoddard (1931), and *Game Management* by Aldo Leopold (1933), and with the founding of The Wildlife Society in 1937. Most of wildlife management since then has been concerned with promoting population surpluses of favored species. Some single-species projects have worked remarkably well. The intended beneficiaries, game species like the wild turkey and endangered species like the peregrine falcon, have made remarkable comebacks as a result of reintroductions and intensive management. But such management has cost millions of dollars per species. Efforts this expensive can hardly be applied to all species. Even if such enormous funds were available, the conflicting needs of different species would make management inefficient and conflict-ridden.

Many biologists now acknowledge that, despite notable successes, single-species management has caused many problems. For instance, game managers traditionally advocated forest harvest patterns that maximized fragmentation of the landscape because many game species thrive where forage and cover areas are interspersed and edges between habitat types are abundant (Leopold 1933). But we now know that while fragmentation may benefit deer and other game, it does not contribute to native biodiversity (Noss 1983). Forest fragmentation has benefited deer so much in the upper Midwest that several conifer species are not regenerating in many areas due to overbrowsing (Alverson *et al.* 1988). Intense herbivory from white-tailed deer threatens many rare plant species throughout the eastern and midwestern United States (Miller *et al.* 1992). In addition to manipulative habitat management, control of predators for the benefit of game animals and domestic livestock has led to many ecological problems (Dunlap 1988). Amazingly, costly and destructive predator control continues on public lands today.

When wildlife managers expanded their concern beyond game species and began considering nongame and endangered species, the circle of conservation broadened. Leopold (1933) recognized a historical and progressive sequence of wildlife management: (1) restriction of hunting, (2) predator control, (3) reservation of game lands, (4) artificial replenishment, and (5) environmental controls. Environmental control is habitat management and is based on ecological principles and empiricism. As acknowledged by Leopold, this approach has considerable potential for expansion to nonhunted species. But the spectrum of species considered by most conservation programs today remains limited.

Endangered species efforts, by definition, do not worry about life forms until they teeter on the brink of extinction or appear to be rapidly approaching that brink. A recent study found that 39 plant species in the United States were listed only when 10 or fewer individuals were known to exist; a freshwater mussel, *Quadrula fragosa,* was not listed until a single nonreproducing population remained (Wilcove *et al.* 1993). Because we wait so long to list species, saving them is bound to be expensive and the chances of success are poor. Most of the funding for recovery of listed species has been devoted to a few popular vertebrates (Kohm 1991), and even then recovery targets are usually less than what it would take to restore population viability (Tear *et al.* 1993).

Endangered species programs, although critical as a safety net to catch imperiled species where other actions fail, are obviously reactive rather than proactive. They make no attempt to identify potentially vulnerable species before they begin the slide toward extinction. They usually fail to recognize opportunities for protecting suites of species, such as those associated with an endangered ecosystem type, in a cost-effective way. They pay no attention to levels of organization beyond species. Although the first stated purpose of the Endangered Species Act of 1973 is "to protect the ecosystems upon which endangered species and threatened species depend," the agencies have never taken this ecosystem protection mandate seriously, and Congress has never told them how they might do so.

Nongame species programs have the potential to be proactive and taxonomically broad, but with very few exceptions (states such as Missouri, which has a portion of its sales tax allocated to conservation) they have received little funding. Furthermore, their focus has usually been on vertebrate groups popular with the public, such as birds, and they have made little attempt to determine the conservation needs of such groups as salamanders, mites, liverworts, or fungi. Because there has been no effective national nongame program, state programs operate without full knowledge of what is going on around them and have little incentive or opportunity to coordinate programs with other states and nations.

The continually expanding list of endangered species and ongoing degradation of entire ecosystems is proof enough that current approaches to conservation are flawed. In the case of the Endangered Species Act, grossly inadequate funding and political interference with listing and recovery actions are much to blame (Bean 1991). But today the Endangered Species Act and other conservation laws are being asked to do more than what they were designed to do. Knowledge of conservation problems and techniques has expanded greatly since these pieces of legislation were enacted. The interdisciplinary field of conservation biology is growing at a phenomenal pace. And, hopefully, our esthetic and ethical appreciation for life has deepened since the early days of American conservation.

Many conservationists now insist that we move beyond game species, endangered species, and other popular organisms, and start inventorying and protecting whole assemblages of species, habitats, and ecosystems before they decline further. What's more, the grassroots members of the "New Conservation Movement" (Foreman 1991) are urging ecological restoration at a massive scale, including removal of roads and developments in many areas and reintroduction of large predators. Dissatisfied with a biologically simplified America, they want back much of what has been lost.

And why not? At first glance, a vision of North America with regained wildness and biodiversity seems unrealistic, even utopian. But when we consider that restoration at this scale is a process requiring decades or even centuries, it begins to make sense (Soulé 1992). Perhaps recovery is inevitable. The human population cannot grow forever, and must either plateau, decline gradually, or crash. In any case, repairing the damage our culture has done and giving other creatures a fair chance for life is the job of enlightened management today and in the future. This is our chance to pay retribution.

Why Management?

Throughout this book we use the term *management.* All land management is biodiversity management, whether intended or not. All land-use decisions—including a decision to designate a reserve, put a fence around it, and leave it alone—are land management decisions with significant consequences for biodiversity. It is much better to manage biodiversity by design rather than by default. As the most powerful species on earth, we can alter biodiversity worldwide and are expressing that capability in a frightening way. Accepting responsibility for our actions means not only that we carefully consider the effects of management on biodiversity, but also that our management programs be designed explicitly to protect and restore native biodiversity.

In some respects, management is an arrogant concept. How can we presume to manage nature if we can't even manage ourselves? What right do we have to manage and manipulate landscapes for human ends, be they conservation or development? Why not let landscapes manage themselves and let organisms fulfill their evolutionary destinies with little human interference? We are sympathetic to these concerns but believe that, for better or worse, humans are now responsible for the continuing or ending of 3.5 billion years of organic evolution. Such power must be wielded carefully and wisely.

Land managers often become overly enthusiastic and manipulate habitats that should be left alone. On the other hand, stopping land management would in many cases be an ecological disaster. In landscapes already fragmented by human activity, the remaining natural areas are vulnerable to all sorts of threats from outside their boundaries, including edge effects. For example, the brown-headed cowbird, a parasite that lays its eggs in the nests of other birds, may need to be controlled if other bird species are to survive in remaining patches of forest. Management of the endangered Kirtland's warbler and black-capped vireo includes trapping and removing cowbirds. Without such intrusive management, these endangered birds cannot reproduce successfully. Exotic plants, such as Japanese and bush honeysuckles in eastern forests and kudzu in the South, often seriously compete with native species and need to be controlled. Without management, invasive exotics overrun many landscapes and eliminate some native species. In all such cases, we have created a dangerous situation—an environment favorable to proliferation of weedy and exotic species—that we must now correct through management or risk losing more biodiversity as sensitive species decline.

Vegetation types that require frequent fire to maintain their native species composition and structure offer some of the strongest arguments for ecologically informed management (see Chapter 2). A small, isolated patch of prairie in a fragmented landscape is not likely to receive lightning strikes often enough to burn regularly and sustain its natural structure. If we do not manage these fire-dependent systems by use of prescribed fire, they will lose their native biodiversity.

The role of management in conservation strategy can be denied only at great risk. But perhaps management is optimally a set of interim measures that will help ecosystems recover their natural values. The ideal future may be one where management is no longer needed because ecosystems are wild and healthy enough to take care of themselves. Even if one does not accept this long-term goal, it cannot be denied that large, essentially self-managing wilderness areas are among the most important reservoirs of native biodiversity today. No one really knows how to manage a natural ecosystem. Thus, "we should at least keep our minds open to the proposition that nature—if given a chance—can still manage land better than we can" (Noss 1991b).

We fear, unfortunately, that letting things be is not a safe option for much of our landscape in the near future, however valuable it may be as a guiding principle for large wilderness areas and possibly for some distant future worldwide. But concerning management, one thing is clear: Traditional approaches have failed to protect biodiversity. A new approach to land conservation, built on what we feel is the best of past and current approaches, is developed later in this book. But first, we examine in more detail the forces responsible for evolution of biodiversity in North America and the limitations of past approaches to managing this incredible diversity.

CHAPTER TWO

BIODIVERSITY: CREATION AND DESTRUCTION

> Our place is part of what we are. Yet even a "place" has a kind of fluidity; it passes through space and time...A place will have been grasslands, then conifers, then beech and elm. It will have been half riverbed, it will have been scratched and plowed by ice. And then it will be cultivated, paved, sprayed, dammed, graded, built up. But each is only for a while, and that will be just another set of lines on the palimpsest. The whole earth is a great tablet holding the multiple overlaid new and ancient traces of the swirl of forces. Each place is its own place, forever (eventually) wild.
>
> Gary Snyder (1992), *The Practice of the Wild*

During a recent interagency course on biodiversity, we posed a set of questions to land managers about the habitats and species they were charged with managing. Our questions included "When were the present plant communities formed in geological time?"; "What were climatic conditions like then?"; "What species were present then that are absent now?"; and "What ecological processes are responsible for maintaining the biotic community?" Referring to particular species in the habitats that course participants were managing, we asked "Where did the species originate or evolve?"; "What were environmental conditions like there?"; and "How do current conditions compare with conditions under which the species evolved?" Answers given by managers to both sets of questions were often less than satisfactory. Perhaps these questions seem academic and of little relevance for on-the-ground managers, who must deal with present conditions and immediate "brush fires." However, failure to address such questions has led to some regrettable resource management decisions. It is comparable to practicing medicine without taking patient histories or understanding anatomy or physiology.

To devise practical methods to conserve biodiversity, biologists and managers need to understand (1) the geological and evolutionary forces responsible for generating biodiversity, (2) the ecological processes that maintain biodiversity, and (3) the forces that threaten biodiversity. This understanding is needed whether dealing with the gene, species, ecosystem, or landscape level, and whether management focuses primarily on biodiversity conservation or on something else, such as wildlife, timber, or livestock production.

Land managers have made mistakes because they did not understand or consider biogeographic and ecological factors. For example, giant sequoias found only in the southern Sierra Nevada of California are geological and successional relicts dependent on recurring fire for persistence (Harvey *et al.* 1980). Silvicultural practices developed for expanding, widespread, or late-successional tree species have not worked for sequoias. Similarly, range managers have applied management principles from the Great Plains, where plant communities evolved in the presence of large herding herbivores, to Great Basin and southwestern desert areas, which had no such fauna for thousands of years.

Managers also need to understand better the processes that are threatening biodiversity. In many cases, knowledge of factors that shaped or maintained a biotic community will provide clues to current threats. If some process critical to the life histories of native species is disrupted, biodiversity can be expected to decline. For example, trout and salmon (Family Salmonidae) need clean gravel beds for spawning. These beds are maintained by periodic high water in streams, termed "flushing flows." Dams, water diversions, and excessive soil erosion from logging or road building can reduce the frequency and effectiveness of flushing flows, and cause a decline in salmonid populations.

In this chapter we review some major forces that have created, shaped, and maintained the biodiversity of North America and some other forces that now threaten it. Though brief and general, our review highlights some types of knowledge required to develop effective conservation plans and programs.

Generation of Biodiversity

Before outlining the forces responsible for shaping biodiversity in North America, we must discuss how biological variation develops. This discussion assumes some knowledge of evolutionary theory and is not intended to be a primer; readers lacking such background may wish to consult a general biological text or one of the many popular books that deal with the subject, such as Ehrlich (1986).

Evolution

Ever since Darwin's exposition of the theory of evolution by natural selection in *The Origin of Species*, biologists have devoted much effort to elucidating the finer points of how this process works. Despite continuing disagreement about aspects of evolution, such as whether speciation proceeds at an even pace or in bursts, modern evolutionary theory is adequate to explain the general processes by which variation among organisms has developed. Since we have stated that biodiversity can be conveniently recognized at four levels (gene, species, community, landscape), let's consider how variation develops at each of these levels.

Variation within a species results ultimately from mutation and recombination of genes. Since certain genes confer a competitive advantage in a given environment, individuals having these genes will more successfully survive and reproduce (i.e., natural selection for these genes has occurred). If a species has colonized many different areas, it may develop genetic differences among populations due to selection for genes best adapted to each particular habitat. For example, populations living in drier habitats may develop physiological or behavioral adaptations (such as concentrated urine or nocturnal activity) to conserve water. Thus, variation among populations of the same species is often a function of the diversity of habitats that the species occupies across its range.

Of the countless examples of natural selection, some patterns are common enough to have been codified into rules of thumb. For example, Bergman's rule states that individuals in northern populations of endothermic ("warm-blooded") species are larger than those in southern populations, presumably because larger bodies have lower surface–volume ratios and thus better conserve heat. Hence, moose in Alaska are predictably larger than moose in Yellowstone National Park. Similarly, populations of many plants and animals vary in shape and color across their range. Sometimes the adaptive value of a trait can be explained scientifically and sometimes it cannot. For example, the northern flicker, a common woodpecker, has yellow under its wings in the eastern United States but salmon color in most of the West. Why? No one knows for certain. Some traits may not be adaptive; instead, they may be genetically linked (e.g., close together on the same chromosome) to traits that are adaptive, or they may represent examples of neutral evolution or random changes in gene frequencies (genetic drift) that often occur in small, isolated populations.

New species arise when populations diverge genetically (and often, physically) and are no longer able to interbreed, at least in the wild. Reproductive isolation can evolve in several ways, and usually follows the physical isolation of populations by geographic barriers such as rivers, glaciers, or mountain ranges. For example, mountain sheep of the genus *Ovis* were separated by Pleistocene (Ice Age) glaciation and evolved into the thin-horned Dall sheep in Alaska and Canada and the bighorn sheep to the south (Geist 1971). Several bird species, including warblers, also may have arisen when glaciers separated ancestral populations. Similarly, the eastern and western diamondback rattlesnakes were apparently the same species until some time during the Pleistocene when populations in southeastern and southwestern refugia became isolated and differentiated into distinct species (Futuyma 1979). Any physical or topographic feature that prevents movement by plants or animals can isolate populations and allow new species to evolve. Alternately, isolation can lead to extinction if the isolated populations are too small.

Island Biogeography

Islands are inherently surrounded by barriers and many islands have floras and faunas that differ drastically from the continental areas with which they are associated. For example, the Channel Islands off the coast of California have approximately 100 species of endemic plants (Davis 1990). Endemic species or races are ones that are native to a particular place and found only there. Beginning with Charles Darwin's trip to the Galapagos Islands, much of our understanding of evolutionary processes has come from the study of islands.

The species–area relationship, first noted with respect to islands, is one of the great empirical generalizations of biogeography. As the area of any habitat declines, so does the number of species. Johann Reinhold Forster, naturalist on Captain Cook's second tour of the Southern Hemisphere in 1772–1775, noted that "islands only produce a greater or lesser number of species, as their circumference is more or less extensive" (cited in Browne 1983). Later studies confirmed the species–area effect for oceanic islands and extended the effect to habitat islands such as caves, isolated wetlands, alpine grasslands, and forest fragments. Typically, a tenfold decrease in habitat area cuts the number of species by half (Diamond 1975).

What are the causes of the species–area relationship? Scientists have long argued over this question. The most straightforward explanation in many cases is habitat diversity. As area increases, so does the diversity of physical habitats and resources, which in turn support a larger number of species (Lack 1976). Imagine a large oceanic island containing desert, montane forests, streams, and marshes. A volcano erupts and annihilates all living space on the island except part of the desert, which contains few of the species that lived in the other habitats; diversity has been reduced accordingly.

Abundant evidence supports the generalization that larger populations have a smaller chance of going extinct. A small island or nature reserve may not contain enough area for a single home range of a large animal. For example, a grizzly bear will not find enough to eat for long in a reserve of a few thousand acres. Other species, for reasons not entirely understood, avoid settling in small tracts of forest or other seemingly suitable habitat. Studies in the eastern United States have confirmed that many songbirds are area-sensitive and breed only in large tracts of forest, even though their individual territories consist of only a couple acres (Whitcomb *et al.* 1981, Robbins *et al.* 1989). Random factors may be partially responsible for the species–area relationship; the more individual plants or animals sampled by a researcher, the more species that sample will contain just by chance (Connor and McCoy 1979). The odds of finding species with low population densities increase as more area is sampled.

The most famous and controversial explanation for the species–area relationship is the equilibrium theory of island biogeography (MacArthur and Wilson 1963, 1967). MacArthur and Wilson proposed that the number of species on an island represents a balance between immigration (or colonization) and extinction (Fig. 2.1). Over time, species on an island continually go extinct, but other species immigrate to the island from the mainland or other islands. Islands near the mainland experience higher rates of immigration than remote islands because the dispersal distance is shorter. Large islands contain larger populations and consequently suffer lower rates of extinction. Island size may affect immigration rates as well, as larger patches will be easier for dispersing individuals to locate. Islands close to an immigration source may also have lower extinction rates, as small populations can be augmented by immigrants of the same species, a so-called "rescue effect" (Brown and Kodric-Brown 1977). Therefore, equilibrium theory predicts that large, close islands will contain the most species, all else being equal.

MacArthur and Wilson (1963, 1967) elaborated many mechanisms related to island biogeography, and later studies have tested and refined these postulates. Much of this work may seem academic and without practical application, but many of the mechanisms proposed have turned out to be useful in understanding the process of extinction and how to prevent it.

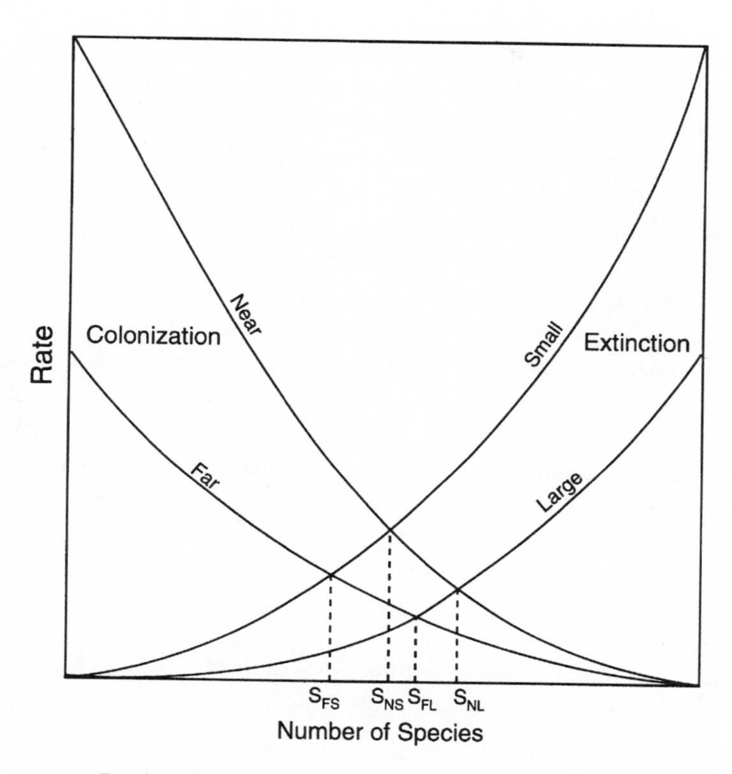

FIGURE 2.1 Predicted species richness on an island, represented as a balance between rate of colonization (immigration) to the island and rate of extinction, according to the equilibrium theory of island biogeography (MacArthur and Wilson 1967). In this model, colonization is affected mainly by island distance from the mainland (near or far); extinction is affected by island size. Species richness corresponds to the intersection of the colonization and extinction curves. The greatest number of species is predicted to occur on islands that are near and large (S_{NL}). From Wilcox (1980). Used with permission of Sinauer Associates.

INTERRELATIONSHIPS

As plants and animals colonize and adapt to new environments, relationships may develop in which two or more species interact to their mutual benefit. Certain insects have evolved to pollinate only a specific plant. Plants have evolved that depend on only one or a few species of animals to disperse their seeds. Some microorganisms have adapted to life in the gut of animals, where they digest cellulose to the mutual benefit of themselves and their host. These relationships can become quite intricate. For example, biologist Chris Maser (1988) describes the complex relationship between the northern flying squirrel and mycorrhizal fungi and their overall role in maintaining the health of the forest ecosystem in which they live. The squirrels feed on the fruiting bodies (mushrooms and truffles) of mycorrhizal fungi that trees depend on for obtaining soil nutrients through their roots. When the squirrels defecate, the fungal spores are dispersed from tree to tree. Many biologists make their living trying to sort out and understand these interdependencies. However, the documented relationships represent only a few well-studied cases. Millions of such relationships remain unknown.

The existence of mutualistic relationships and other interdependencies has led some ecologists to suggest that the biotic community is a holistic organism that is more than just the sum of its parts. However, we now know that the parts are interchangeable to a degree. Communities change in composition over space and time as species respond individually to climate and other changes. This is not to say that relationships are unimportant, but only that their nature is forever dynamic.

The Landscape Perspective

Finally, we can extend our view of how biodiversity is generated to a landscape level. A regional landscape usually consists of many different plant communities with associated animals. But these communities shift in their position on the landscape over time. Consider the northern Great Plains, which is dominated by tall grasses like big bluestem and short grasses like blue grama. For thousands of years, millions of bison traversed this region in large herds, grazing and trampling down areas to bare dirt well mixed with manure. Areas that were periodically denuded would be quickly colonized by invader plants such as fringed sagebrush that can effectively use the nutrients released into the soil. Prairie dogs would also move into these areas. Large animals such as pronghorn antelope, attracted to disturbed areas to forage, helped disperse seeds of invader plants into these areas. Over time, the flora and fauna of a disturbed site would change in a more or less predictable pattern until eventually tall grasses dominated again. This pattern of systematic replacement of biotic communities, called succession, is familiar to all foresters, range managers, and wildlife biologists. However, the end point of succession is not as predictable as once thought, particularly in times of rapid climate change. Succession also involves an element of chance, especially with regard to which species get to a disturbed site first.

The Great Plains example provides a few lessons for land managers about landscape ecology. First, disturbance and the subsequent pattern of plant and animal succession is a natural process that helps maintain long-term ecological health by recycling or importing nutrients and invigorating soils. Second, disturbed sites depend upon surrounding areas for recovery. When a patch of grassland is denuded, adjacent areas with early successional species provide the seed sources that allow colonization by new plants. Later, adjacent mature tallgrass communities provide the supply of seeds and animals to return the area to a tallgrass stand.

Thus, community diversity within a regional ecosystem is not only a function of topographic or landform diversity but also a function of natural disturbance cycles. These disturbance cycles may be generated by the activities of animals or by climatic or geological events such as fire, flooding, landslides, or volcanic activity. In any case, they are unending.

Geologic and Biogeographic Forces

At an even broader scale, environmental change is eternally creating new expressions of biodiversity. Six major forces have had a profound effect on the current patterns of biodiversity in North America: plate tectonics/continental drift, uplift/erosion, inundation, catastrophes, climate change/glaciation, and biological invasions.

Plate Tectonics/Continental Drift

Approximately 250 million years ago all the present continents were joined together into a single giant continent, called Pangaea. Then about 200 million years ago, in the Cretaceous period, the continents began to separate into Laurasia in the north and Gondwanaland in the south. Eventually North America split off from Eurasia and formed a separate continent. Florida, however, seems to have been a part of Africa that later became attached to North America.

Continental drift isolated many life forms from each other and led to the development of distinct floras and faunas on each of the continents. Geological forces related to the shifting of continental plates also helped create a variety of habitats on each continent, including marine shelves, mountain ranges, freshwater rivers, and lakes.

Uplift/Erosion

The drift of continents has been accompanied by geological uplift, both from pressure ridges created when plates are pushed against each other and from volcanism. Pressure ridges were the major factors that uplifted most mountain ranges of the continent—the Cascade/Sierra Nevada, Rocky Mountains, and Appalachians. Uplift in each case was followed by erosion, creating the pattern of mountain ridges and valleys, and lakes and rivers in endless variety, depending on the erosiveness and penetrability of the substrate and the pattern of rain and snowfall.

This uplift and erosion, which continues today, created many new habitats. Plants and animals adapted to a range of elevational microclimates and to riparian and aquatic habitats associated with river and lake systems. Volcanic eruptions also created distinct habitats with steep elevational gradients and often with substrates rich in minerals.

Inundation

Accompanying continental movement and uplift were periods of inundation. About 110 million years ago much of the central plains of North America was covered by seas. Similarly, a large portion of the Great Basin was covered by a huge inland lake, Lake Bonneville, during much of the Pleistocene (between 10,000 and 2 million years ago). Most of the rivers running out of the mountains within or surrounding the Great Basin ran into this lake. Since that time the climate has become more arid, and many of these streams no longer have perennial flows. The Great Salt Lake in Utah is a remnant of Lake Bonneville, and the biodiversity of the Great Basin has been markedly influenced by fluctuations in the size and salinity of this lake and its tributaries. Similarly, Florida was covered by ocean for much of its geologic history until 25 million years ago; its highest ridges were formed originally as coastal dunes (Webb 1990). Because these ridges were often isolated from one another by seawater, they developed distinct endemic floras and some endemic animals, most evident today in the endangered Florida scrub of the Lake Wales Ridge.

Catastrophes

Scientists have found much evidence of periodic events that caused mass extinctions of species (particularly animals) across the planet. One of the most significant of these events marks the boundary of the Mesozoic and Cenozoic eras (approximately 63–65 million years ago). This event was characterized by a massive die-off of reptiles, especially the dinosaurs, and was followed by the rise of mammals. Increasing evidence suggests that this mass extinction was caused by a meteorite or comet as large as 6 miles in diameter striking the earth. According to this theory, the collision released a tremendous cloud of dust into the atmosphere, dramatically cooling the earth. Most plants persisted through this extinction event, apparently because of their seed banks. Animals, such as small mammals, that could feed on these seeds and were not too bothered by cold presumably also survived. Those without these adaptations perished.

Several mass extinctions are apparent in the fossil record, and probably also relate to ancient cataclysms. Five major events are recorded, and for each event 10–20 million years were needed for the diversity of species on earth to return to predisturbance levels. These extinctions and subsequent diversifications in new evolutionary directions helped shape the biodiversity of North America. Catastrophes on a global scale are extremely rare and have not occurred for 63 million years. However, the current wave of extinctions resulting from human activity is eliminating an estimated 27,000 species each year, making it the sixth great mass extinction in the earth's history (Wilson 1992).

Climate Change/Glaciation

Climate change has had a major effect on the structure of biotic communities. Climate has changed continuously, at one rate or another, throughout the history of the earth. Plants and animals have shifted their distributions in response to these changes. For example, during wetter times, redwoods covered large parts of the West from Colorado to the Pacific Coast. Climatic changes isolated and fragmented redwood populations and eventually led to the development of distinct species. Two relict species of redwoods remain, the giant sequoias of the western Sierra Nevada and the coastal redwoods along the north coast of California.

One of the most important climate-induced forces shaping the biodiversity of North America has been the periodic advances and retreats of continental glaciers in relatively recent times from the Pleistocene to the present. At least 22 separate cycles of glaciation have occurred in the Northern Hemisphere over the last million years (Graham 1986). Although many of these glaciations covered most of what is now Canada, Alaska, and as far south as the present-day Ohio River Valley, unglaciated areas (refugia) existed even in northern latitudes. Animals and plants that survived in these areas often evolved into distinct species, subspecies, or varieties. For example, there are several pairs of closely related bird species (northern shrike and loggerhead shrike; Bohemian waxwing and cedar waxwing, and so on) in which one member of the pair has a more northern but overlapping breeding range. In each case these pairs are believed to have descended from a common ancestor and then separated during a Pleistocene glaciation. After the ice receded, the southern species spread northward, and the northern species spread south and eastward (Pielou 1991).

Glaciation not only isolated species populations, but the advance and retreat of glaciers created many unique landforms and habitats such as glacial cirques so common in the Rocky Mountains and Sierra Nevada, moraines, eskers, and kettle holes. Furthermore, the soils in formerly glaciated areas often differ greatly from those in unglaciated zones, resulting in distinct plant and animal communities.

Climates have usually changed slowly in the past, but humans now appear to be changing the climate rapidly by increasing concentrations of carbon dioxide and other greenhouse gases in the atmosphere, as we will discuss later.

Biological Invasions

Ever since the continents separated, plants and animals have periodically invaded North America. A few seeds arrived on the wind, drifted across oceans, or were carried in mud on the feet of birds. Some small animals may have rafted across on logs or other debris. Spiders ballooned. At times major land bridges were available to aid colonization of terrestrial species. During glacial periods over the last 60,000 years, North America was connected to Asia near Alaska by an area called Beringia. This land bridge was once as wide as 200 miles and at times had a temperate climate more typical of areas further south today. Many species, including large mammals such as bison, bighorn sheep, elk, and moose, are believed to have reached North America by crossing Beringia during one of these periods of lower sea levels. All of these species have close relatives in the Old World. For example, the red deer in Europe is the counterpart of the larger American elk. These two animals are considered to be closely related subspecies.

Ultimately, the most influential of the invaders of North America were the humans who entered the continent from Asia during the Pleistocene. When and by which routes these people arrived in North America is still controversial, but the Beringian route and a coastal route through the same region appear most likely (Hoffecker *et al.* 1993). Humans probably arrived in North America at least 20,000 years ago, and probably in several waves thereafter, although the timing of these colonizations is uncertain.

More controversial is the effect of these early human invasions on biodiversity. Martin (1967, 1973) and others have postulated that aboriginal people had a major if not dominant role in exterminating many of the large mammals on the continent about 10 to 12 thousand years ago. This theory is not universally accepted, but has considerable support (Brown and Gibson 1983). Although humans occupied perhaps the entire continent at that time, they lived in relatively low densities. Nevertheless, evidence from around the world suggests that every major human colonization of a new continent or island has been accompanied by a wave of extinctions, especially of large mammals and flightless birds (Diamond 1982, 1984, Martin and Klein 1984).

The human invasion from Eurasia and Africa beginning around 1500 has had the most pronounced effect on the biodiversity of North America. Not only did these people develop population densities far higher than the continent had ever experienced, but their per capita use of resources was much higher and typically quite inefficient. Furthermore, they brought with them many organisms from Europe and Africa that became naturalized here, often with devastating effects. These exotic organisms include horses introduced into the Southwest by the Spaniards, carp introduced by German immigrants into our waterways, and the many deliberate introductions of birds such as starlings and house sparrows. Introduced species also include plants such as tamarisk, introduced as an ornamental tree from the Middle East and now spread throughout riparian areas of the Southwest; and cheatgrass, which was accidentally introduced from contaminated grain and has spread throughout most of the West. Many of these introductions have disrupted ecological processes and caused displacement of native species. In other cases, introduced species hybridized with closely related varieties or species, thereby polluting native gene pools.

Fish provide a good example of the damaging effects of exotic species on native biodiversity. Of the 1033 species of freshwater fish in North America, 27 (or 3 percent) have become extinct within the past 100 years and another 265 (or 26 percent) are vulnerable to extinction (Miller *et al.* 1989). Displacement by introduced species has been implicated as a cause of decline in 68 percent of these species, topped only by physical habitat destruction at 73 percent (the percentages add up to more than 100 because more than one cause affects most species). In addition, hybridization with other species and subspecies is listed as a cause of decline in 38 percent of species.

This pattern of displacement and genetic pollution is well documented in fishes, but is by no means unique to them. Exotic species are one of the greatest threats to native species and to human-disturbed ecosystems in the world (Reid and Miller 1989).

The introduction of exotic diseases to which native species had little or no resistance has sometimes had devastating effects. For example, domestic sheep carried diseases that when transferred to the closely related bighorn sheep of North America caused mortality resulting in local extirpation, a pattern that continues today except when domestic sheep are carefully controlled within bighorn ranges. Similarly, canine distemper, introduced in North America by domestic animals, particularly dogs, is fatal to the endangered black-footed ferret and may be partly responsible for the historic reduction in ferret numbers (Schroeder 1987).

Ecological Processes Maintaining Biodiversity

Over thousands or millions of years, biotic communities change dramatically, for instance from prairie to forest and back again. However, over shorter periods, measured in decades or centuries, many communities remain remarkably stable. In this section, we outline the processes responsible for maintaining the integrity of communities, ecosystems, and landscapes. Conserving ecological processes is essential to conserving biodiversity. Six interrelated categories of ecological processes that biologists and managers must understand in order to effectively conserve biodiversity are: (1) energy flows, (2) nutrient cycles, (3) hydrologic cycles, (4) disturbance regimes, (5) equilibrium processes, and (6) feedback effects.

Energy Flows

The flow of energy through an ecosystem is fundamental to maintaining its function. Our sun is the source of virtually all energy useful to organisms, and energy is lost continuously as it moves through an ecosystem. The most basic way in which life captures energy is through photosynthesis. Plants form the basis for food chains through which energy is passed to higher trophic levels. Any reduction in plant or leaf biomass will reduce the amount of energy flowing through an ecosystem.

Energy is captured and expended in ways other than photosynthesis. For example, solar energy evaporates water, which returns as precipitation. Upon reaching the ground, precipitation is the prime mover of physical objects in terrestrial ecosystems. Surfaces of the earth such as rock faces absorb radiant energy, and the energy from such surfaces creates a unique microsite used by many plants and animals.

Nutrient Cycles

Nutrient cycles are the processes by which elements such as nitrogen or carbon move through the biotic and abiotic components of an ecosystem. Ecosystems function by cycling and recycling; otherwise a system would eventually become depleted of essential elements or nutrients. For example, phosphorus is a nutrient that limits plant growth in many ecosystems. Phosphorus is found in certain phosphate rock formations in the earth's crust. Slowly, through weathering and erosion, phosphorus moves into rivers and to the ocean. There it forms insoluble deposits on the bottom of shallow areas near the coast, where it can eventually be uplifted. Other phosphorus settles in deep marine deposits, which for all practical purposes do not cycle back through land. Because geological uplifting is so slow, phosphorus is being washed into the sea faster than it returns to land (Miller 1982). Fish catches return 54 million kg (60,000 tons) of phosphorus each year, and the phosphorus-rich guano of fish-eating birds, such as pelicans, gannets, and cormorants, returns another 3100 million kg (350,000 tons). But these amounts are small compared to the larger amounts of phosphorus that erode from the land to the oceans each year.

The cutting of forests and other land clearing accelerates natural erosion losses.

Conversely, elements or compounds that cannot be cycled will accumulate somewhere in the ecosystem and may cause toxicity. Pesticides such as DDT, which are not natural components of an ecosystem, are neither metabolized nor detoxified as they pass through the ecosystem and tend to accumulate in toxic proportions in the fat of animals high in the food chain. DDT is proven to have caused declines in animal populations. DDT and other chlorinated-hydrocarbon pesticides concentrate in tissues of peregrine falcons and other predatory and fish-eating birds and can cause eggshell thinning, behavioral abnormalities, and other problems resulting in reproductive failure. The decline of peregrine falcons in North America and their subsequent recovery are closely correlated with prevalence of DDT in the environment and in the bird tissues (Craig 1986). Populations and northern subspecies that were not exposed to pesticides did not decline as did more southern populations.

Of particular importance in nutrient cycles is the role of decomposers. Ecologists often focus on the food chains of plant–herbivore–carnivore since this is the most visible part of a nutrient cycle to us. However, without the invertebrates and microorganisms that decompose and recycle dead material at each stage, the world would soon be without soil nutrients to feed the plants that capture the sun's energy. And there would be lots of dead bodies lying around.

Hydrologic Cycles

Another process critical for life is the hydrologic cycle, the process through which water moves from ocean to atmosphere to land and back to the ocean. Water is a finite resource, renewed and regulated in complex ways. Water is necessary for all life from the molecular level through the ecosystem level. Furthermore, it is the major vehicle through which materials, both abiotic and biotic, are transferred through an ecosystem. Hence, water is a key factor in the occurrence and distribution of organisms. For example, vernal pools, containing water for only a short period of the year, often support plant species and associations not found on adjacent higher ground. Proper water cycling is critical to maintaining the biodiversity and functioning of biotic communities.

Disturbance Regimes

As noted earlier, most ecosystems are subject to regularly or sporadically recurring events such as fires, windstorms, landslides, or floods. These events are often called catastrophic, but historical records show that they have always occurred and that ecosystems have complex responses to them. What appears to devastate a natural community at a local scale or in the short term by causing death and destruction may actually be essential to rejuvenation and persistence at a broader spatiotemporal scale.

Many plant and animal species are not only adapted to disturbances, but depend on them for survival. A well-studied example is the Kirtland's warbler, which requires homogeneous thickets of five- and six-year-old jack pines interspersed with grassy clearings for breeding. This kind of habitat is created and maintained by intense fires (Lowe *et al.* 1990). If fires are suppressed, as is usually the case in managed forests, habitat for Kirtland's warbler would disappear. Restoring and maintaining a regular pattern of fires has helped conserve this endangered species.

Prairies, other grasslands, oak savannas, and ponderosa and longleaf pine forests often depend on frequent, low-intensity ground fires. These fires recurred historically at intervals of 1 to 25 years, depending on the particular community and site conditions. The life histories of the dominant species in these communities have been shaped evolutionarily by fire (Mutch 1970, Platt *et al.* 1988). Without fire, these communities gradually change to other types that may be less diverse or healthy (e.g., Walker and Peet 1983, Anderson *et al.* 1987, Habeck 1990). Most fire-dependent communities evolved in areas with frequent thunderstorms and lightning ignitions, although in some cases, burning by native Americans greatly increased the extent of the community over thousands of years (Pyne 1982). In all such cases, fire must be considered a normal and necessary part of the ecosystem.

Fire suppression, both active and passive (the latter occurs when firebreaks such as roads, clearcuts, agricultural fields, and developments prevent the natural spread of fire), has harmed fire-dependent communities. If a fire-dependent community goes too long without fire, it may accumulate so much woody fuel that a fire would be catastrophic, killing adult trees and perhaps even sterilizing the soil. Prescribed fire, designed to mimic the natural fire regime, seems to offer the best hope of maintaining natural diversity in these communities.

A knowledge of natural disturbance regimes is essential to conservation of biodiversity. Ecologists have developed theory and some evidence to suggest that species diversity will be highest at some intermediate frequency or intensity of disturbance (Fig. 2.2; Connell 1978, Huston 1979, Pickett and White 1985, Petraitis *et al.* 1989). Intermediate should not be interpreted as median. An optimal level of disturbance will need to be determined experimentally or through extensive observation for any given community. Determining what that intermediate intensity or frequency is remains a difficult problem for any community. At the landscape level, the influence of distur-

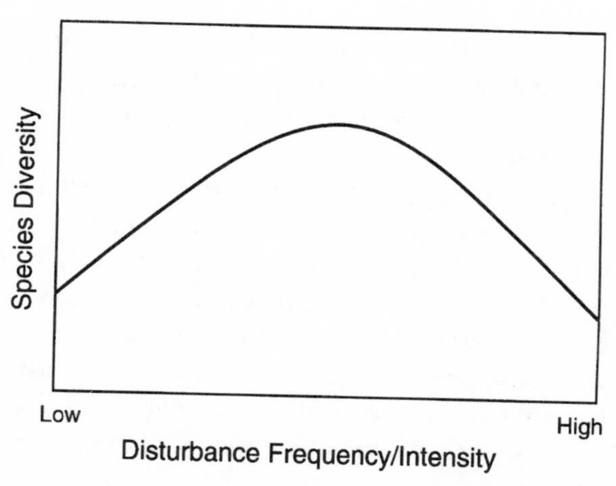

FIGURE 2.2 The intermediate disturbance hypothesis, which indicates that species diversity within a given patch should be highest at intermediate frequencies or intensities of disturbance (from Hobbs and Huenneke 1992, after Connell 1978). Used with permission of the Society for Conservation Biology and Blackwell Scientific Publications, Inc.

bance on species diversity is even more complex because many habitats and species assemblages are involved.

Perhaps more instructive for land managers is the realization that species in any region have adapted, through evolution, to a particular disturbance regime. If we radically alter that regime, many species will be unable to cope with the change and will be eliminated. For example, the patchwork created by clearcutting differs from the mosaic created by fire or windthrow in fundamental ways. A landscape of dispersed clearcuts and tree plantations—a common pattern across much of the United States—has less dead wood and other structure within patches, greater contrast between patches, and more pronounced edge effects than a naturally disturbed landscape (see Chapter 6). It also has roads and vehicles, which lead to other problems.

Human actions that dampen or eliminate natural disturbances are likely to be a threat to biodiversity in many kinds of environments. For example, many riparian plant species such as cottonwoods become established after floods, which create new deposits of bare silt and gravel where seedlings can establish. Eliminating periodic flooding by building dams may prevent regeneration of many species and drastically alter riparian plant communities.

Although disturbances are fundamentally natural, they can become unnatural and truly catastrophic when their frequency, severity, or other qualities are modified by human action. Thus, although many streams are subject to overbank flooding on the order of 100 years, an increase in frequency to every 10 years by upstream logging and grazing must be considered unnatural and potentially destructive to biodiversity. Either increases or decreases in the frequency of fire can have dramatic effects on habitat structure and species composition.

Human actions that mimic natural disturbances are much less likely to interfere with ecosystem function and threaten biodiversity than human actions that impose novel disturbance regimes on an ecosystem. Consider, for instance, livestock grazing in the northern Great Plains, which once supported bison. Restoring free-ranging bison would be the ideal management strategy from a biodiversity perspective. Short of this ideal, a livestock grazing system that mimics bison grazing patterns is more likely to be compatible with the region's remaining biodiversity than some grazing pattern based on English pasture management.

While many species and communities are well adapted to frequent disturbance, seemingly natural disturbances can still have harmful effects. In particular, they may leave communities vulnerable to invasion by exotic species, and all the inherent problems exotics cause. In an extensive review of scientific literature on disturbances, particularly on grasslands, Hobbs and Huenneke (1992) described how changing the frequency and intensity of disturbance may affect both natural diversity and susceptibility to invasion by exotics (Fig. 2.3).

Equilibrium Processes

An ecological concept of historical importance to conservation but disputed today is equilibrium. An equilibrium condition is one in which two opposite forces exist in a balanced state. For example, in many mule deer populations, birth rates roughly equal death rates, resulting in relatively stable numbers over a period of, say, 20 years. If birth rates dramatically increase without a corresponding increase in death rates, then the population would explode. If the opposite happened, the population would die out.

Equilibrium processes could potentially operate at all levels of organization: gene, species, ecosystem, and landscape. The examples shown in Table 2.1 represent only a few of many. These processes are useful to recognize because they tend to confer stability upon an ecosystem. However, not all ecosystems are stable; probably most are stable only within a limited range of environmental conditions. Modern ecological theory holds that equilibrium conditions are often fleeting and can be recognized at some spatial scales but not at others (Botkin 1990, Noss 1992a).

Feedback Systems/Homeostasis

Conventional ecological theory and resource management principles in North America have emphasized the homeostatic or negative feedback

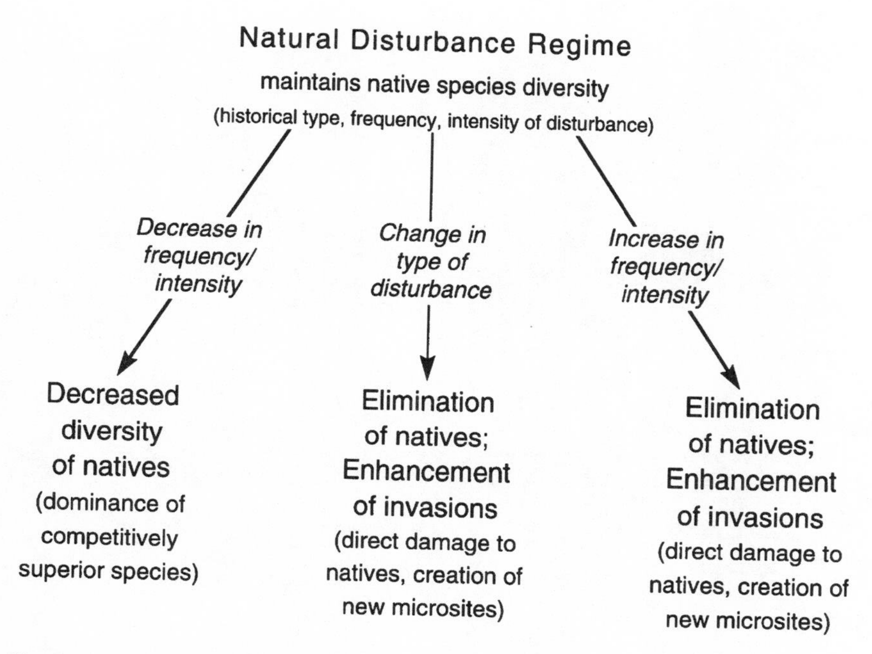

FIGURE 2.3 Any change in the historical disturbance regime of an ecosystem may alter species composition by reducing the importance of native species, by creating opportunities for invasive species, or both (from Hobbs and Huenneke 1992). Used with permission of the Society for Conservation Biology and Blackwell Scientific Publications, Inc.

processes of ecosystems. Early plant ecologists described plant succession as an orderly process by which one plant community replaces another in a predictable sequence that eventually reverts to a "climax" community. Ecosystems do exhibit some homeostatic behaviors. They tend, for instance, to respond to minor disturbances by moving back to the predisturbance state, an example of negative feedback. Controls on population size such as predation, food supply, or social behavior are also negative feedback or density-dependent mechanisms.

Negative feedback confers stability upon a system because it moves the system back toward the "original" state. For example, eastern deciduous forests that have been cleared for farming and then abandoned will revert to forests that at least superficially resemble the original ones, though they may be missing many components.

The most familiar example of a negative feedback system is a thermostat controlling the temperature of a house. When the temperature drops below the setting, the thermostat turns on the heat until the set temperature is reached. Conversely, if the house gets too hot, the thermostat will

TABLE 2.1 Potential Equilibrium Processes and Results of Disequilibrium

Level	Example	Process	Relationship	Consequences
Gene	Genetic diversity	Background mutation rate (BMR) versus rate of gene loss (RGL)	BMR > RGL	Diversification
			BMR = RGL	Genetic stability
			BMR < RGL	Loss of genetic diversity
Individual organism	Perennial plant carbohydrate reserve	Carbohydrate storage (CS) versus carbohydrate drawdown (CD)	CS > CD	Increased plant vigor
			CS = CD	Plant maintenance
			CS < CD	Decreased plant vigor; plant death
Population	Population dynamics	Birth rate (BR) versus death rate (DR)	BR > DR	Population increase
			BR = DR	Population stability
			BR < DR	Population decrease; extinction
Community	Species richness	Species immigration rate (SIR) versus species extinction rate (SER)	SIR > SER	Increased species richness
			SIR = SER	Stable species richness
			SIR < SER	Decreased species richness
Ecosystem	Nutrient balance	Nutrient import (NI) versus nutrient export (NE)	NI > NE	Ecosystem enrichment
			NI = NE	Ecosystem stability
			NI < NE	Ecosystem degradation
	Hydrologic balance	Groundwater recharge (GWR) versus groundwater discharge (GWD)	GWR > GWD	Net aquifer recharge
			GWR = GWD	Aquifer stability
			GWR < GWD	Aquifer drawdown; desertification
	Soil dynamics	Soil development (SD) versus soil loss (SL)	SD > SL	Soil buildup
			SD = SL	Soil stability
			SD < SL	Soil erosion; soil depletion
	Forest dynamics	Biomass removal (BR) versus forest growth (FG)	FG > BR	Forest succession/forest regeneration
			FG = BR	Sustained yield forestry
			FG < BR	Forest depletion; desertification

trigger the air conditioning system, which runs until the temperature drops to the desired setting. Negative feedbacks are common in nature and include the ability of mammals and birds to regulate their body temperature within a narrow range physiologically. Many ectotherms, such as reptiles, also regulate their body temperature within a narrow range, but they do so behaviorally by seeking suitable microhabitats.

Biologists and particularly resource managers tend to think of biological systems as responding to disturbance with negative feedback, thereby leading toward stability. However, recent evidence from both forests and rangelands (Archer and Smeins 1992, Niering 1987, Perry *et al.* 1989, Schlesinger *et al.* 1990) suggests that positive feedback processes that destabilize ecosystems are also important, and that ecosystems may be quite vulnerable to unusual disturbances. The implication is that if a new disturbance is strong enough or recurs with high enough frequency, the system may lose the ability to return to its original state. As a case in point, cheatgrass invaded many intermountain rangelands of the West following the introduction of livestock and overgrazing of the native bunchgrasses. Cheatgrass, an annual grass, carries fires well and can outcompete native grasses after fire. The shift from native grasses to cheatgrass has been accompanied by an increase in fire frequency, and now even with removal of the original disturbance (overgrazing), the community will not revert to perennial grass. Cheatgrass-dominated communities have become the steady state in many areas of the Intermountain West. Most ecosystems will eventually reach some relatively stable state, but in terms of biodiversity it may be a highly impoverished state.

Recognizing the potential for positive feedback in ecosystems is important, since so much theory of natural resource management (wildlife management, forestry, range management) is based on the assumption that ecosystems behave as negative feedback or self-perpetuating systems. Foresters assume, for example, that a clearcut will always return to the climax forest. Similarly, range managers have assumed that relaxation of grazing pressure will allow a system to return to climax grassland. Experience has shown that this does not always happen. Hence, we need to be more cautious in our treatment of the land.

Threats to Biodiversity

In this section we examine in more detail some major forces that threaten biodiversity. The threat to biodiversity extends well beyond the chainsaw, cows, or bulldozer that cause the immediate and visible destruction. The threat ultimately involves the fundamental tendency of our species to reproduce excessively, use resources profligately and selfishly, discount the future, and not worry about the needs of other people or (even less) other species (Soulé 1991).

ULTIMATE THREATS

The driving force behind loss of biodiversity is an increasing human population and consumption of resources. Figure 2.4 shows how overutilization of resources can lead to loss of biodiversity and extinction. The amount of resource use and magnitude of impact depends on three factors—human population size, per capita consumption, and efficiency of use:

$$\text{Resource Use} = \frac{\text{Population Size} \times \text{Per Capita Consumption}}{\text{Efficiency of Use}}$$

This relationship suggests that when aboriginal populations existed in North America in low numbers, and where individuals used few resources, the threat to biodiversity was minimal.

We are now facing an ominous situation where human populations virtually everywhere in the world are increasing. Per capita consumption is high and is being deliberately encouraged by advertising and governmental economic incentives (Durning 1992). Furthermore, efficiency of use is low because we are extremely wasteful. The result is that over 40 percent of the world's net terrestrial primary productivity is now used by humans, and the proportion is ever increasing (Vitousek *et al.* 1986).

INTERMEDIATE THREATS

The intermediate-level threats to biodiversity revolve around the way we use resources and the consequences of that use. Returning to Figure 2.4, we see that resource use or exploitation can threaten biodiversity in several ways. A complete review of intermediate threats is not necessary here; many of them, such as habitat loss and deterioration, are well documented elsewhere as they have been studied by applied ecologists for years. Others, such as the impact of roads, have been overlooked or underpublicized in many cases. Threats of importance in particular ecosystems such as forests or rangelands will be discussed in detail later. We focus now on three threats of widespread and general significance: habitat fragmentation, roads, and global warming.

Habitat fragmentation. Our understanding of the insidiousness of habitat fragmentation has increased dramatically in the last 20 years. This new understanding contradicts much of the conventional wisdom of wildlife and forest management.

Habitat fragmentation is one of the greatest threats to biodiversity worldwide (Burgess and Sharpe 1981, Noss 1983, 1987a, Harris 1984, Wilcox and Murphy 1985). Fragmentation is often considered to have two components: (1) decrease in some habitat type or perhaps all natural habitat in a landscape; and (2) apportionment of the remaining habitat into smaller, more isolated pieces (Wilcove *et al.* 1986). Although the latter component is fragmentation per se, it usually occurs with deforestation or other massive habitat reduction (Harris 1984). An almost inevitable consequence of

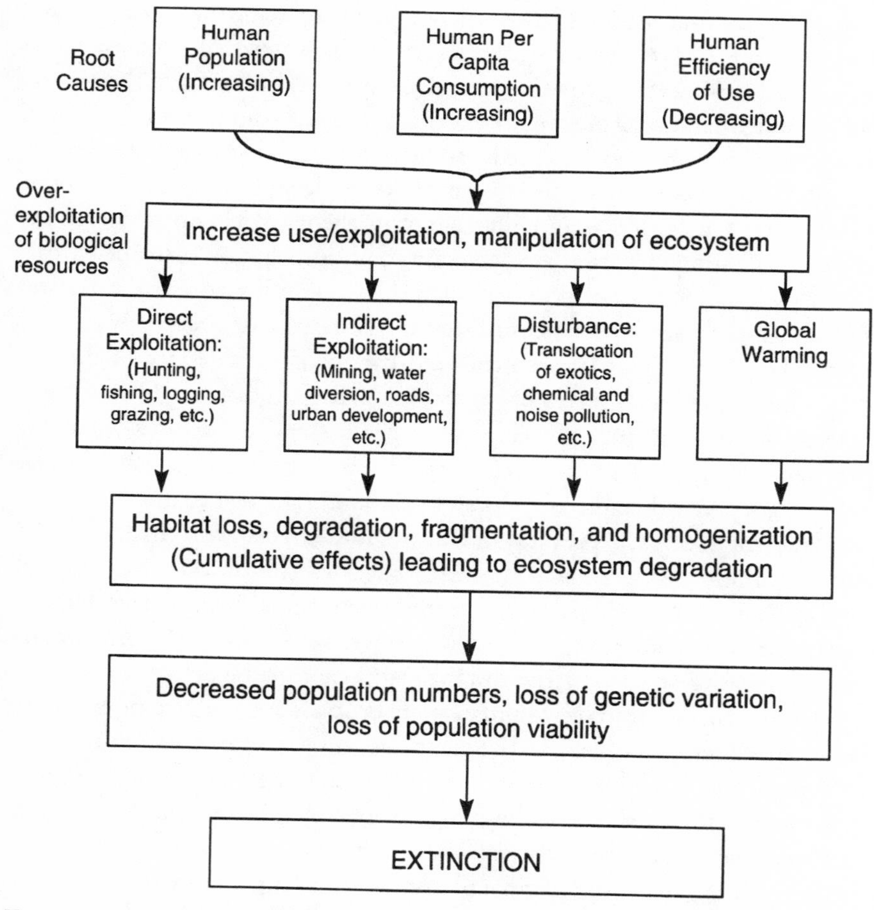

FIGURE 2.4 Relationship between root causes, overexploitation of resources, and loss of biodiversity. The final result is extinction.

human settlement and resource extraction in a landscape is a patchwork of small, isolated natural areas in a sea of altered land. A case study of fragmentation in Wisconsin (Fig. 2.5) shows a gradual reduction in forest area, accompanied by trends toward smaller and more isolated patches, as a forested landscape was converted to agriculture.

As usually happens in science, as ecologists learned more about fragmentation, the process turned out to be much more complex than once thought. Early fragmentation studies viewed the process as a species–area problem analogous to the formation of land-bridge islands as sea levels rose since the Pleistocene. Hence, island biogeographic theory was invoked to explain losses of species as the area of habitats declined and their isolation increased. Certainly, there are good analogies between real islands and caves, lakes, prairies in a forested landscape, or pieces of remnant forest in agricultural land. But there are differences, too. The water that surrounds real islands provides habitat for few terrestrial species. In contrast, the matrix surrounding habitat islands may be a rich source of colonists to the island, many of which are invasive weeds or predators on species inhabiting the island. Thus, species richness does not always decline on isolated habitat patches, as predicted by island biogeographic theory. Richness may even increase (at least temporarily) as species invade from adjacent disturbed areas. In such a case, species composition often shifts toward weedy, opportunistic species while sensitive species of habitat interiors are lost (Noss 1983, Lynch 1987). The matrix in a fragmented landscape is also in a state of flux, as crops are planted and harvested, as tree plantations go through their rotations, as farming or silvicultural methods change, and as human settlements grow and decline. Thus the external environment of a habitat patch is not as constant or predictable as the water surrounding a real island.

Fragmentation is a process and ecological effects will change as the process unfolds (Fig. 2.6). In the early stages of the process, the original landscape is perforated by human-created openings of various sizes, but the matrix remains natural habitat. At this stage, we would expect the abundance of native species of the original landscape to be affected little, although the access created by human trails or roads may reduce or extirpate large carnivores, furbearers, and other species subject to human exploitation or persecution. Such losses are well documented historically. Also, a narrow endemic species whose sole habitat just happened to be in an area converted to human land use would also be lost. As human activity increases in the landscape, the gaps in the original matrix become larger, more numerous, or both, until eventually they occupy more than half of the landscape and therefore become the matrix. A highly fragmented landscape may consist of a few remnant patches of natural habitat in a sea of converted land. Many landscapes around the world have followed this pattern of change.

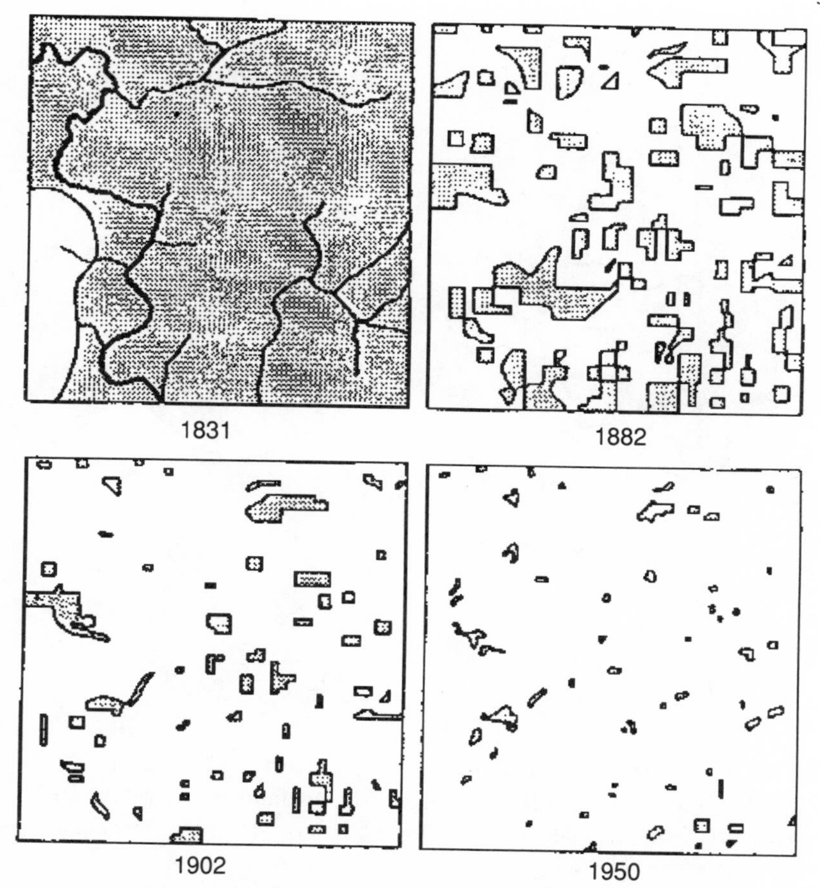

FIGURE 2.5 Changes in wooded area of Cadiz Township, Green County, Wisconsin, during the period of European settlement. Shaded area represents the amount of land in forest in 1831, 1882, 1902, and 1950. From Curtis (1956). Used with permission of the University of Chicago Press.

Fragmentation does not necessarily spell extinction. A species might persist in a highly fragmented landscape in three ways. First, it might be able to survive or even thrive in the matrix of human land use. A number of weedy plants, insects, fungi, microbes, and vertebrates such as European starlings and house mice fit this description. Second, it might be able to maintain viable populations within individual habitat fragments; this is an option only for plants, microbes, and small-bodied animals with modest area requirements. Or third, it might be highly mobile. A mobile species could integrate a number of habitat patches, either into individual home ranges or into an interbreeding population. Pileated woodpeckers, for example, have learned to fly among a number of small woodlots to forage in landscapes that were formerly continuous forest (Whitcomb *et al.* 1981,

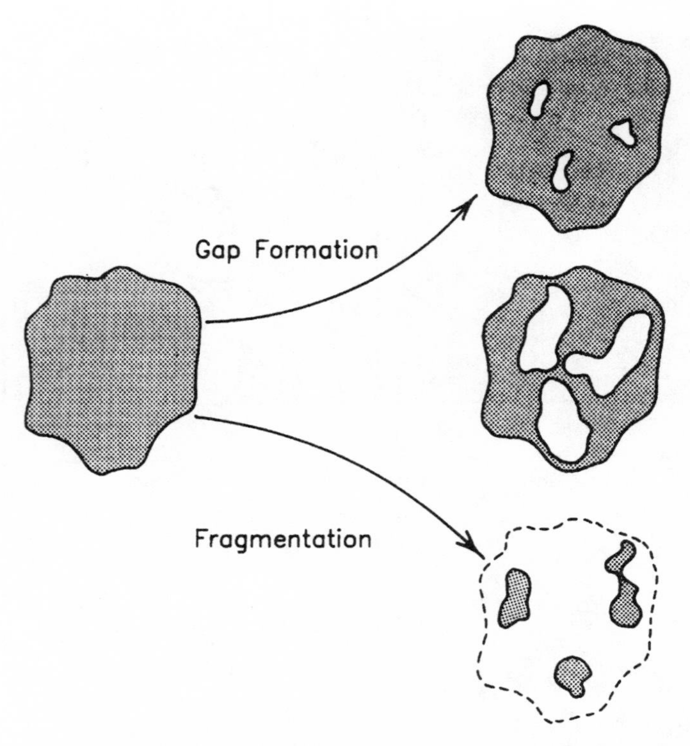

FIGURE 2.6 A fragmentation sequence begins with gap formation or perforation of the landscape. Gaps become bigger or more numerous until the landscape matrix shifts from natural to anthropogenic habitat. From Wiens (1989). Used with permission of the author and Cambridge University Press.

Merriam 1991). A species incapable of pursuing one or more of these three options is bound for eventual extinction in a fragmented landscape.

Besides the problem of small populations in small habitat patches being more likely to go extinct, small patches are also greatly affected by their surroundings. Sun, wind, rain, and other physical factors create an environment near the edges of a habitat patch different from that in the interior, particularly for forests with relatively closed canopies. Predators, competitors, and parasites may also thrive in the disturbed habitat near an edge and penetrate some distance into the patch.

Roads. Roads are increasingly being recognized as a severe threat to sensitive wildlife and natural ecosystems. Wilderness advocates have long fought road-building proposals, largely on recreational or scenic grounds. However, biology also provides evidence of detrimental effects of roads. Landscape ecologists have found that roads block movement of small animals.

Wildlife biologists recognize problems with open roads that expose large mammals to heavy hunting pressure, poaching, and harassment. Fisheries biologists worry about sedimentation of streams as a result of slope failures and erosion from roads. Humane societies worry about roadkills. And conservationists are concerned about the role of roads in stimulating more development and resulting habitat destruction. Only recently have biologists recognized that the cumulative effect of all these factors constitutes a leading threat to biodiversity.

Roads are movement barriers to some species of small vertebrates and invertebrates. To the extent that animals hesitate to cross roads, roads fragment populations into smaller demographic units that are more vulnerable to extinction from any number of causes. A study in southeastern Ontario and Quebec found that several species of small mammals rarely ventured onto road surfaces when the road clearance (distance between road margins) exceeded 20 m (Oxley *et al.* 1974). The mammals whose movements were inhibited by roads included eastern chipmunks, gray squirrels, and white-footed mice. In Germany, several species of carabid beetles and two species of forest rodents where shown to rarely or never cross two-lane roads. Even a narrow, unpaved forest road closed to public traffic served as a barrier (Mader 1984). In an Oregon study, dusky-footed woodrats and red-backed voles were found at all distances from an interstate highway but never in the highway right-of-way, suggesting that these rodents did not cross the highway (Adams and Geis 1983). Road clearances can also be barriers in more open habitats. In a study of the effects of a highway on rodents in the Mojave Desert (Garland and Bradley 1984), only one white-tailed antelope squirrel, out of 612 individuals of eight species captured and 387 individuals recaptured, was recorded as having crossed the road. A nine-year study in a Kansas grassland found that very few prairie voles and cotton rats ever crossed a dirt track 3 m wide that bisected a trapping grid (Swihart and Slade 1984). Many other studies have documented barrier effects of roads (Bennett 1991), even for animals as large as black bears (Brody and Pelton 1989).

Paved roads often affect mortality rates more directly; more than one million vertebrates are killed each day on roads in the United States (Lalo 1987). In Florida, roadkill is the leading known cause of death for all large and medium-sized mammals, with the single exception of the white-tailed deer (Harris and Gallagher 1989). Roads through national forests, such as the Ocala National Forest in Florida, are significant mortality sites for threatened vertebrates. In the Big Cypress National Preserve, expensive underpasses and fencing have been constructed to channel movement of Florida panthers under Interstate 75. Panthers and other animals have used these underpasses, but holes in the highway fenceline limit their effectiveness and point to a need for frequent maintenance (Foster and Humphrey 1991).

Another impact of roads stems from the access they provide to legal and illegal hunters. Open road density has been found to be a good predictor of habitat suitability for large mammals, with habitat "effectiveness" and population viability declining as road density increases. The sensitivity of hunted elk populations to open road density is well established. According to research in the Northwest and the northern Rockies, a road density of one mile per square mile of habitat can decrease habitat effectiveness for elk by 40 percent, compared to roadless watersheds. As road density increases to six miles per square mile, elk habitat use falls to zero (Lyon 1983, Wisdom *et al.* 1986). Mountain lions may also avoid roads. In Arizona and Utah, existing populations of cougar are concentrated mostly in areas of low road density, but road avoidance was found to be limited to paved and improved dirt roads (Van Dyke *et al.* 1986).

Studies in northern Wisconsin and Minnesota have found that wolves cannot maintain populations where road density exceeds about 0.9 miles per square mile (Thiel 1985, Mech *et al.* 1988). Wolves generally do not avoid roads and often follow them as convenient travelways. But following roads brings them into contact with people who shoot them. Similarly, most grizzly bears die near roads, and, in many areas, grizzly habitat use near roads is significantly reduced (McLellan and Mace 1985, McLellan and Shackleton 1988). Mattson *et al.* (1987) found that grizzlies in Yellowstone National Park are also harmed by roads. Adult females and subadult males displaced by dominant bears into areas near roads and developments have a greater chance of being killed. Mattson and Knight (1991) concluded that, given uncertainty about the viability of the Yellowstone grizzly population, "we cannot afford to increase the area impacted by secondary roads and major developments."

Black bears can also be vulnerable to road access and may not be able to maintain populations in the southern Appalachians where road density exceeds 0.8 miles per square mile (Brody 1984). The vulnerability of black bears to road access in the Appalachians results from unusually high poaching pressure. In the Great Dismal Swamp of Virginia and North Carolina, roads may not be such a threat. A protected bear population spent more time near roads than expected and used roads as travel corridors (Hellgren *et al.* 1991). In much of the West, black bear populations persist in areas with fairly high road densities, perhaps because poaching is not yet intense. This persistence could be temporary, however. With increased value of bear gall bladders and other body parts in the Asian marketplace, survival of black bears in any roaded landscape cannot be assured.

Although inhibiting dispersal of some species by blocking movement, roads also help disperse species that thrive in disturbed roadside habitats or travel by way of vehicles. In the Northwest, Port Orford cedar root rot fungus, black-stain root disease fungus, spotted knapweed, and the gypsy moth are all known to disperse and invade natural habitats via roads and vehicles (Schowalter 1988). Other deleterious effects of roads, too numerous to elaborate here, include soil and water pollution, erosion, sedimentation of streams and decline of fisheries, edge effects, overcollecting of rare plants and animals (for example, cacti and king snakes), elimination of snags (upon which many cavity-nesting birds and mammals depend) for firewood or road safety, and a number of indirect and cumulative effects (Diamondback 1990, Bennett 1991).

Many land managers close roads to public use at least seasonally to protect wildlife such as elk or grizzly bears. However, the barriers erected may be inadequate, and many driveable roads are not included in agency road inventories. A detailed study on the Flathead National Forest in Montana found that the Forest Service failed to include in its inventory 70 percent of the short-term and temporary roads. Some 80 percent of "obliterated" roads inventoried by the Forest Service were driveable by ordinary passenger vehicles, and 38 percent of the barriers erected for road closures were ineffective (Hammer 1986, 1988, 1990). A study by the Oregon Department of Fish and Wildlife found that roads planned for closure after timber sales on the Siuslaw National Forest were actually closed only 21.4 percent of the time (Ingram 1991). Even roads truly closed to public use, unless fully revegetated, still function as movement barriers to small animals, as sources of sediment to streams, and as trails along which poachers may walk.

Thus, road construction, road use, and the physical road itself may have a number of negative effects. A decision to build a new road or upgrade an existing road must be considered carefully. The need for a critical assessment of potential road impacts is especially strong where habitat of threatened or endangered species is involved. In many such cases, not only is it desirable to prevent new roads from being built; it may also be advisable to close and obliterate existing roads.

Global warming. A looming threat to biodiversity is global warming (Peters and Lovejoy 1992). The predicted warming is a result of increasing levels of carbon compounds in the atmosphere, especially carbon dioxide (CO_2) but also methane and others. Atmospheric CO_2 comes from a variety of sources, the primary ones being combustion of fossil fuels (oil and coal) and the cutting and burning of forests worldwide. This increase in CO_2 creates a "greenhouse effect" that warms the earth's surface. Current estimates are that atmospheric CO_2 is increasing at a rate of 1 to 2 percent per year, resulting in a doubling of atmospheric CO_2 in about 50 years, with resulting temperature rises of from 1 to 5°C.

A warming of 3°C (the median prediction) would result in a world warmer than it has been for the past 100,000 years and would cause massive changes in species distributions and ecological processes. Sea levels would rise, patterns of rainfall would change, and storms and wildfires would increase in many regions. An increased temperature of 3° is equivalent to a latitudinal shift of about 250 km (156 miles) or an altitudinal shift of 500 m (0.3 miles). Species already confined to the tops of mountains could be extirpated as their habitats are "pinched off" (Fig. 2.7). Other species could presumably move north, although the rate of change expected may be 10–40 times more rapid than previous changes in climate, at least since the Pleistocene, making it difficult for species to migrate quickly enough. Further-

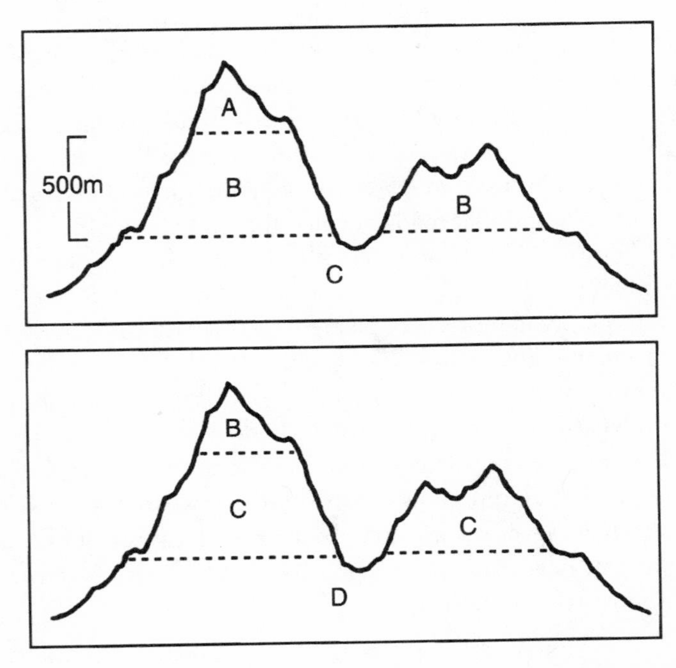

FIGURE 2.7 Response of species to global warming by shifting altitudinal distribution. (*Top*) Initial altitudinal distribution of three species, A, B, and C. (*Bottom*) Species distribution after a 500-m shift in altitude in response to a 3°C rise in temperature. Species A becomes locally extinct. Species B shifts upward, and the total area it occupies decreases. Species C becomes fragmented and restricted to a smaller area, while species D successfully colonizes the lowest altitude habitats (from Peters 1988).

more, changes in climate would be taking place concurrently with habitat fragmentation and other human stresses on biodiversity, resulting in many cumulative impacts (Fig. 2.8).

Consequences

As the intermediate impacts on ecosystems intensify and accumulate, populations of many species decline. Decline is usually accompanied by loss of genetic variation, uneven sex ratios, and other threats to the viability of

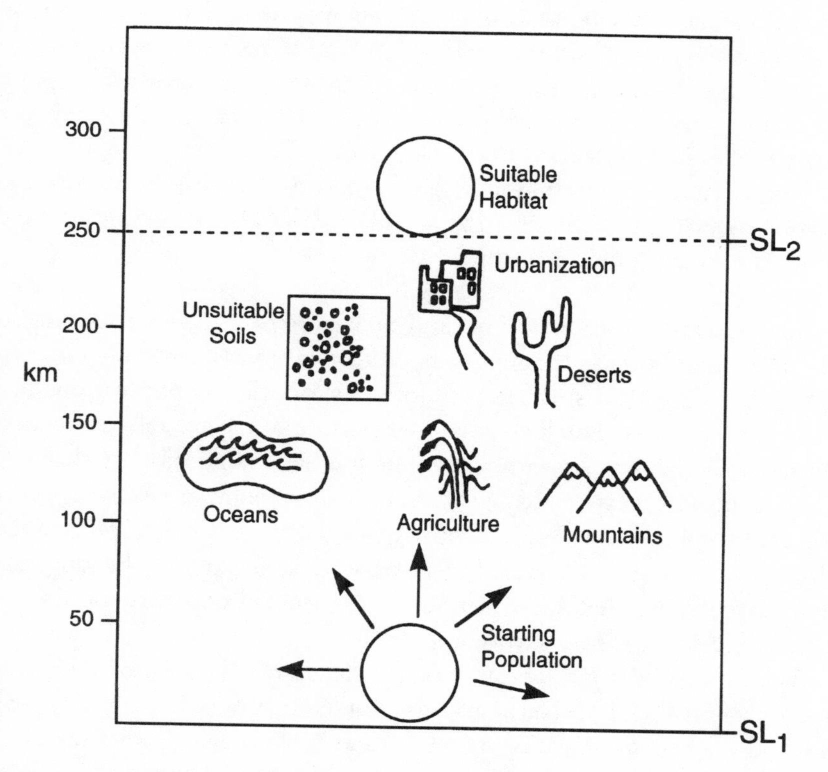

FIGURE 2.8 Response of species to global warming by shifting latitudinal distribution. The figure illustrates a plant species consisting of a single population with a distribution determined solely by temperature. After a 3° C rise in temperature the population must shift 250 km to the north (from SL_1 to SL_2) to survive. Shifting might occur by simultaneous range contraction from the south and expansion by dispersal and colonization to the north, which depends on propagules being able to find suitable habitat to mature and in turn produce propagules that can colonize more habitat to the north. Propagules must pass through or around natural and artificial obstacles like mountains, lakes, cities, and farm fields (from Peters 1988).

populations. These proximate factors are what directly lead populations down a virtually inescapable vortex to extinction (Gilpin and Soulé 1986).

The most basic rule about population viability is that small populations are more vulnerable to extinction than larger populations. Analyses of population viability attempt to determine the probability that a particular population will persist for a given period of time, and ask something like: "What is the size of the population that has a 95 percent chance of surviving 100 years?" With few exceptions, populations on the order of thousands of individuals appear necessary for long-term persistence (Soulé 1987, Thomas 1990). Population viability analysis also allows one to ask questions about the probability of persistence under different management scenarios. For example, how much will the probability of persistence increase for a population of grizzly bears if we reduce mortality rates by reducing road density from 3 miles to less than 0.5 miles per square mile?

Several factors, both random and deterministic, predispose small populations to extinction (Shaffer 1981, Soulé 1987). The most important of these factors are reviewed below.

Environmental variation and natural catastrophes. Bad winters, droughts, hurricanes, and other random or unpredictable environmental phenomena affect both large and small populations. As a result, all populations fluctuate through time. Small populations are simply more likely to fluctuate down to zero. A species with a large geographical range has less chance of all of its members experiencing the same environmental variation, especially for catastrophic events such as hurricanes, but also for less pronounced changes in habitat conditions. In a given year, some parts of the range may have favorable weather and other parts experience drought. A species with a restricted range has no such spatial buffer.

When a population of one species fluctuates, populations of other species with which it interacts may also fluctuate. A poor mast year for oaks is a bad year for acorn-eating mammals and birds. A rare species can be especially vulnerable to population increases among its predators, competitors, parasites, and diseases (Shaffer 1981). If an exotic predator invades a small habitat island, it can quickly extirpate suitable prey species. Introduced rats, cats, and other predators have been a primary source of extinctions on oceanic islands (Vitousek 1988).

Predictable environmental change can also threaten small populations, especially in fragmented landscapes. Natural succession may eliminate meadows and other early seral habitats upon which certain species depend. As trees invade, sun-loving wildflowers and butterflies may perish. In small islands, the chances of natural disturbances occurring and creating earlier successional habitats are reduced (Pickett and Thompson 1978, Gilbert 1980).

Demographic stochasticity. Chance variation in age and sex ratios and other population parameters, termed *demographic stochasticity*, can change birth and death rates. In small populations, extinction occurs quickly when deaths outpace births (Lande 1988). Some species have a threshold density or number of individuals below which the population cannot recover. This *Allee effect*, named after the animal ecologist W.C. Allee, is likely when organisms modify their environment physically or chemically to encourage their survival, when group defense against predators is necessary, or when social interactions and mating success depend upon some critical population density. In the United States, extreme sociality may have contributed to the extinctions of the Carolina parakeet, passenger pigeon, and heath hen (Soulé 1983). The decline of these populations to some threshold level may have stifled reproductive behavior and led to rapid extinction.

Genetic deterioration. If we consider the fate of populations over a fairly long time span, genetic integrity becomes virtually as important as demographic stability to persistence. Small, isolated populations on habitat islands are prone to two kinds of genetic problems: inbreeding depression (caused by mating between close relatives) and genetic drift (a random change in gene frequencies). Populations with low genetic diversity may show reduced fertility and survivorship, especially when inbred, and in the long run will be less able to respond adaptively to environmental change (Frankel and Soulé 1981, Schonewald-Cox *et al.* 1983). The Florida panther, for example, has been restricted to south Florida for many decades due to habitat destruction and persecution across the rest of its original range in the Southeast. The remaining population of fewer than 50 individuals is suffering effects of inbreeding, including reduced fertility (Ballou *et al.* 1989). Ultimately, small populations on habitat islands, if they survive at all, may lose their evolutionary potential unless enriched by gene flow from other populations.

Metapopulation dynamics. Many species are apparently distributed as metapopulations, that is, as systems of local populations linked by occasional dispersal that wards against demographic or genetic deterioration (Levins 1970, Gilpin and Hanski 1991). Marshes and other wetlands, for example, are often distributed as patches corresponding to potholes or other topographic depressions. Species restricted to these wetlands have populations scattered about the landscape. It is the fate of these local populations to wink off and on over time. Dispersal from one wetland to another allows for recolonization of vacated habitats, or genetic enrichment of an existing population, so the metapopulation as a whole persists.

Suppose that urban development occurs in the upland matrix in which the wetlands are found. Frogs, water snakes, turtles, muskrats, and other aquatic animals that once dispersed through the forest between wetlands now face significant barriers to movement: roads, buildings, parking lots, and the like. Meanwhile, development may drain or pollute many of the wetlands. The wetlands may be visited by children who like to collect small animals for pets. When a local population is extirpated in this scenario, it has little chance for reestablishment. If local populations go extinct but are not reestablished, the entire metapopulation (and eventually, perhaps, the species) gradually dies off. Because increases in local extinction rates and declines in immigration rates are cumulative effects, even common species are not immune to the effects of widespread habitat alteration and fragmentation.

Current Status of Biodiversity in North America

Although most people now realize that biodiversity is declining, no comprehensive statistics are available to document all aspects of this loss. People have simply not been keeping good track of their natural heritage. Nevertheless, ample evidence suggests that substantial impoverishment has occurred in North America and that the rate of loss is increasing. Some general aspects of this impoverishment were reviewed in Chapter 1. Here, we present additional information.

Species Loss

At least 71 vertebrate species and subspecies have gone extinct in North America, north of Mexico, over the last 500 years (Nature Conservancy 1992), plus perhaps 217 full species of plants (Russell and Morse 1992). Undoubtedly many more invertebrates, nonvascular plants, fungi, and bacteria have been lost, with most of these extinctions going unnoticed because many of the species were not even known to science. The United States has never had a systematic biological inventory, so the total number of extinctions can only be guessed.

One obvious indicator of biotic impoverishment is the number of species listed under the Endangered Species Act (Table 2.2). In the United States alone, some 796 species were listed as threatened or endangered as of May 1993, and thousands more are candidates for listing. As noted in Chapter 1, very few species listed under the Endangered Species Act have shown any evidence of recovery. More alarming, the implementation of the Act by federal agencies virtually precludes recovery. An analysis of U.S. Fish and Wildlife Service recovery plans found that 28 percent of recovery goals were set at or below the size of the species population at the time the recovery plans were written; 60 percent of vertebrates had recovery goals that would keep them biologically endangered (Tear *et al.* 1993).

Furthermore, many species such as bison have been extirpated from most of their original range in North America but are not listed as endangered or threatened because the remaining populations appear stable. Many more species have been gradually fragmented into small populations that may be slowly on their way to ultimate collapse. Indeed, population viability theory (as reviewed above) suggests that the present diversity of species in landscapes across the continent is temporary. Indeed, we may be seeing the "living dead."

Ecosystem Degradation

Species extinction is only the last and most obvious stage of biotic impoverishment. Of greater long-term concern is the degradation of ecosystems and landscapes. Measures of ecosystem loss or dysfunction are not as straightforward as species extinctions, in part because ecosystems are much less easy to classify. No accepted national classification of vegetation exists for the United States. Nevertheless, statistics on broad landscape categories such as deserts, wetlands, riparian areas, forests, and rangelands (and on more specific associations in many regions) suggest that habitat loss and degradation are widespread in North America (Noss *et al.* 1994).

Loss of forests in the United States after European settlement proceeded as rapidly as in any region of the world. By 1920, some 96 percent of the virgin forests of the northeastern and central states had been logged, with other regions not far behind (Reynolds and Pierson 1923). By 1980, some 85 percent of the virgin forests in the United States had been destroyed, with losses estimated as 95 to 98 percent for the conterminous 48 states (Postel and Ryan 1991). Some forest types today remain in only a fraction of their former abundance. Old-growth forests of the Pacific Northwest are a well-known example, with losses approaching 90 percent (Norse 1990). Longleaf pine, which once dominated the uplands of the southeastern coastal plain, has been reduced by over 98 percent since settlement and is our most endangered major forest type (Noss 1989). Not all of the loss of longleaf pine can be blamed on logging; some is due to fire suppression, which results in invasion of stands by hardwoods and loss of native species dependent on

TABLE 2.2 Number of Species Listed as Endangered or Threatened in the United States as of November 30, 1992[a]

Category	Endangered	Threatened	Total
Mammals	56	9	65
Birds	73	13	86
Reptiles	16	18	34
Amphibians	6	5	11
Fishes	55	36	91
Snails	7	6	13
Clams	40	2	42
Crustaceans	9	2	11
Insects	14	9	23
Arachnids	3	0	3
Plants	298	72	370
TOTAL	577	172	749

[a] From USDI, Fish and Wildlife Service 1992.

open, sunny conditions. Many other forest types in the United States can be considered endangered or threatened (Noss *et al.* 1994).

Rangelands in the United States have been described as deteriorated in a series of reports beginning in 1936 (U.S. Department of Agriculture 1936; USDI Bureau of Land Management 1962, 1975a, 1975b; U.S. Comptroller General 1977; U.S. General Accounting Office 1988a, 1988b, 1991; Wald and Alberswerth 1985, 1989). The Bureau of Land Management has recently claimed that public rangelands are in improved condition (U.S. Bureau of Land Management 1990), but has presented no credible evidence to support this contention. In fact, their own estimates suggest that at least 52 percent of the public ranges are in lower seral stages [mid-seral = fair (36 percent); early seral = poor (16 percent)]. One could argue that a certain proportion of rangelands would be in this category naturally, but the percentages would be quite different and seral stages would shift across the landscape over time (see Chapter 7). Very few areas in the West have not been grazed at some time by livestock. For sagebrush steppe, a major vegetation type in the Intermountain West, less than 1 percent has never been grazed and 30 percent has been heavily grazed, with dominance concentrated in a few woody plants (West 1994).

Grasslands and savannas are generally the most endangered terrestrial ecosystems in the United States, with most losses due to conversion to agriculture and, secondarily, to fire suppression or overgrazing and subsequent invasion by exotics. Among the types that have declined in extent by over 98 percent since European settlement are longleaf pine savannas and *Arundi-*

naria gigantea canebrakes in the Southeastern Coastal Plain; tallgrass prairies east of the Missouri River and on mesic sites everywhere; bluegrass savanna–woodland and prairies in Kentucky; Black Belt prairies in Alabama and Mississippi and Jackson Prairie in Mississippi; oak savanna in the Midwest; wet and mesic coastal prairies in Louisiana; lakeplain wet prairie in Michigan; sedge meadows in Wisconsin; Hempstead Plains grasslands on Long Island (New York); prairies and oak savannas in the Willamette Valley of Oregon; the Palouse Prairie of Oregon, Washington, and Idaho, plus similar communities in Montana; and native grasslands of all types in California (Noss *et al.* 1994).

The degradation of wetlands is similarly severe and is better documented than most ecosystem declines because of the long-standing interest of hunters and conservationists in wetlands. Nationally, 50 percent of animals and 33 percent of plant species listed under the Endangered Species Act (as of 1989) are dependent on wetlands (Nelson 1989). Yet because of draining and filling, many wetlands have been totally converted to nonwetland habitats such as agricultural fields and home sites. Estimates are that only 45 percent of original wetlands and 10 to 30 percent of original riparian vegetation remain in the United States. In some regions, the losses have been much greater, with many midwestern states and the Central Valley of California having less than 10 percent of their original wetlands (Dahl 1990, Noss *et al.* 1994). Some 98.5 percent of the riparian vegetation of the Sacramento River in California has been destroyed (U.S. Congress, Office of Technology Assessment 1967).

Finally, statistics about water are equally alarming. Water quality and availability have deteriorated in virtually every region of the United States. Only 3.9 percent of the nation's streams are considered to have "maximum ability to support populations of sport fish and species of special concern" (Judy *et al.* 1984) and less than 2 percent of streams in the 48 conterminous states are of high enough quality to be worthy of federal designation as wild or scenic rivers (Benke 1990). Astoundingly, almost all of the water in the Colorado River is diverted before it reaches the Gulf of California (*High Country News* 1987). We can be sure that the "water wars" of the West will intensify if current population and consumption trends continue, and that aquatic habitats nationwide will be increasingly endangered (see Chapter 8).

Conclusion

Human civilization, particularly the European variety, has not been kind to North America. The geological, physical, ecological, and biological forces that naturally generate biodiversity have not ceased operating; they will go on long after human civilization as we know it has crumbled. But natural forces have been overwhelmed over the last few centuries by the plow, bulldozer, chainsaw, dredge, dam, and other tools and artifacts of civilization, and by simply too many people consuming too much of everything. Biodiversity at every level is critically imperiled. Can we "wise up" and begin turning degenerative trends around? The following chapters will consider this question in some depth.

CHAPTER THREE

CONSERVATION STRATEGIES—PAST, PRESENT, AND FUTURE

> O! what men dare do! what men may do! what men daily do, not knowing what they do!
>
> William Shakespeare

Can any effort be more important than the battle to maintain biodiversity? Many special interests vie for use of the land—developers, farmers, loggers, miners, oilmen, cattlemen, recreationists of various stripes—but there is only one general interest: life on earth. To protect life will require a grand effort with leadership from the very highest levels, massive public education about the severity of the crisis, and a commensurate level of funding. Conservation today is truly a last-chance campaign to defend the sanctity of life.

If biodiversity is so important, why have we let it become so threatened? The precarious status of life on earth today suggests that past approaches to conservation, while they may have been as good as could be expected for the times, simply have not worked. Despite polls showing widespread public interest in the environment and despite legions of trained conservation professionals in the field and bureaus, the fabric of life wears thinner each day. The status quo approach to conservation has failed. We need a new strategy based on ecological understanding, conservation experience, and a basic respect for life.

Most people in the United States know that their environment is being degraded. All they have to do is look out their windows. Some people are deeply concerned about this degradation and have worked hard for environmental causes such as recycling and cleanup of hazardous wastes. We will need to harness the energy of these people—and many more—to the new cause of biodiversity if we are to achieve broader conservation goals. Moreover, people need to understand that many past conservation practices no longer make sense in light of our current understanding of ecological systems. Some knowledge of conservation history also will help us comprehend the limitations of past and current approaches to conservation and appreciate why something dramatically different is needed.

The conservation movement has not been static over the years. Rather, the movement has changed in response to what were perceived as the critical problems of the day and in response to new scientific understanding of natural systems. We have learned a few things. By appreciating how our thinking has evolved over the last century, we may gain a better idea of what institutional changes are needed now to correspond to our new understanding. In this chapter, we examine the conservation strategies of the past and present, then propose a new strategy for protecting and restoring biodiversity in the United States.

Attitudes about Conservation

We will concentrate here on the conservation movement as it developed in North America over the last 500 years. But to understand the roots of conservation movements we must look back much farther. The native Americans had a time-tested conservation ethic well before the first Europeans arrived on the continent. Although some native cultures lived more sustainably with their environment than others, many writers have noted that native Americans had a culture, religion, and lifestyle that emphasized strong ties to the land and respect for all life forms. The Europeans who settled here brought with them strikingly different attitudes about the environment. Few of the European invaders paused in their conquest long enough to learn from the native Americans. The dominant attitudes of the European invaders included a belief in "man the omnipotent," a reductionist view of the world, the idea that natural resources are virtually limitless, and a naive faith that technology could solve all problems. The idea that humans are all-powerful dates back at least to the Renaissance. The new settlers to this continent believed that "man" (the male of the species) was superior to all other creatures of the world and thus should naturally and rightfully dominate and subjugate other life forms. The related reductionist/mechanical view of the world is based on the faith that things can be understood rationally by observing them in isolation. Reductionism is the opposite of holism, which holds that things do not operate in isolation, but that, as John Muir stated, "when we examine anything in the Universe we find that it is hitched to everything else." Holistic understanding also involves emotion and intuition, rather than just rational cognition. Muir was an immigrant from Scotland; thus, not all people of European descent are rigid reductionists!

At the time of Columbus's voyage to America, the oceans seemed vast and were largely uncharted, and the North American continent appeared to the new settlers as an almost limitless source of land, timber, grass, and wildlife. The frontier was endless, or so it seemed. The European proclivity to invade these bountiful lands and populate them was greatly facilitated by the technology they brought with them—metalwork and gunpowder, to name two. This technology allowed the European settlers to survive and populate rapidly in an environment they little understood. Later, as new technologies such as the steam engine and railroads emerged, the Europeans and their descendants naturally applied them to further expansion and exploitation of the continent. Given this history, it is not surprising that North American society, to this day, turns to technology to solve its problems.

Central to the European attitude is an anthropocentric (human centered) view of the world. Humans were created in God's image. The imperfect world of nature can be improved by human technology and human wisdom. These attitudes have molded and shaped North American civilization over the last 500 years. They have resulted in vast destruction of biodiversity and natural resources. Yet, these same attitudes have largely shaped the response to such destruction—the conservation movement.

Past and Present Conservation Strategies

Conservation strategies in North America since European settlement can be grouped into three broad categories: (1) species protection; (2) land protection or preservation; and (3) natural resource management. *Conservation* is a relatively new word, but each of these strategies eventually came to be associated in some degree with the conservation movement or with the term *conservation.* Many agencies, organizations, and individuals have focused on only one approach, but others have pursued broader strategies that contained elements of two or more approaches.

Species Protection

Legislation to protect species in North America dates back to at least the 1620s when the government of Bermuda issued a proclamation protecting the cahow and the green turtle (Matthiessen 1959). Rhode Island established a May–November closed season on deer in 1639, and Massachusetts followed with a similar law in 1694. Ironically, one of the next pieces of legislation was the authorization of a bounty of one penny per wolf in Massachusetts. These actions exemplify the single-species approach that continues to this day. Certain desirable species are to be protected because of their value to humankind, while other "undesirable" species need to be eliminated.

An overview of species protection legislation shows that during the early days of the European settlement the concern was mostly over the larger, usually edible, birds, mammals, and fishes. However, over the years, legislation gradually expanded to include a wider range of species until eventually (with the Endangered Species Act of 1973) it included plants and invertebrates. At each step in this progression, the driving force for legislation was a recognition that some species with perceived value to humans had become scarce enough to limit our use of them. Usually the concern was not that a species might become extinct, but that it was becoming too uncommon to continue exploiting profitably for meat, eggs, fur, feathers, or whatever. On the other hand, those species considered of little value or harmful could be ignored or killed. Thus bison, wolves, and many other "varmints" were shot, trapped, and poisoned to near extinction. That they persist at all is testimony not to our wisdom but to our inefficiency in exterminating them.

The species protection movement evolved into the twentieth century field of game and wildlife management. However, the linkage is not as close as one would think. By the time that game management emerged as a science, most market hunting for birds and mammals was no longer legal (although trapping, a kind of market hunting, still was). Rather, the new emphasis was on killing animals for sport. Sport hunters typically abhorred market hunters; other people in the humane movement, which also began to develop in the early twentieth century, hated both (Dunlap 1988).

Although single-species approaches have been reasonably effective in conserving particular animals of interest to people, they have proven of limited value as an overall conservation strategy. Species conservation, until very recently, was never pursued for the purpose of conserving biodiversity. At least four principal limitations of the species approach can be identified. First, until recently the approach has concentrated only upon a few species—the so-called "charismatic megavertebrates." Thus, while there has been concern about deer, elk, and condors, species such as the rough-leaved loosestrife or the desert slender salamander have received little attention.

Second, species protection represents an inefficient piecemeal approach to conservation. More than 4400 species of vertebrates inhabit North America. When plants, invertebrates, and microbes are added to the total, the numbers become phenomenal but are not precisely known. If we add the biodiversity represented by distinct races, subspecies, and varieties, the numbers again increase substantially. The needs of so many species cannot be considered individually.

A third and major limitation of species protection efforts is that they have focused on direct threats to species (mainly animals) from such human activities as hunting and trapping. This protectionist approach made sense when direct killing was the principal threat to species. But, as reviewed in the preceding chapter, in recent years species have usually become rare due to destruction, deterioration, or fragmentation of their habitats. Very few of the early species protection efforts gave any consideration to habitat. To this day, shooting a bluebird is illegal; but you can cut down the snag it nests in without penalty, even though cutting down the snag harms the species far more in the long run. For many years shooting ducks out of season would bring a big fine, but draining the marsh was not only legal but subsidized by the government. And, of course, plants were rarely considered under the single-species approach.

Finally, most species protection efforts have been reactive, being concerned with species only when they had become rare rather than when they were abundant but perhaps declining. Conservationists now understand that the best time to save species is while they are still common.

Land Protection and Preservation

Another significant conservation force in North America was the movement to protect certain lands as parks, wilderness areas, wildlife refuges, and other forms of preserves. The national parks have become the most obvious symbols of this movement, beginning with the creation of Yellowstone National Park in 1872. Actions to preserve special places provide clear evidence that the people of the United States and Canada recognized that something of importance was being destroyed by human activity. Legislation to protect such areas has a history that goes back many years. However, designation of a park or preserve has often proven inadequate to conserve the area's biodiversity and was in fact not intended to do so.

Parks by themselves are inadequate to conserve biodiversity for at least five reasons (Cooperrider 1993). First, there are not nearly enough parks in the right areas. Statistics only show part of this complex issue, but a comparison of potential natural vegetation (PNV) types in North America with National Park Service administered lands showed that 33 percent of the PNV types were not represented in any National Park lands (Crumpacker *et al.* 1988).

Second, the existing parks and preserves are not large enough. The oldest and largest national park in the conterminous United States, Yellowstone, is not large enough to support viable populations of many species (Clark and Zaunbrecher 1987). There is evidence that loss of species from parks is inversely correlated with their size (Newmark 1985). In addition, many parks and preserves are still suffering from damage done before they were established or during their early years as parks.

Another major limitation of the park strategy is that incompatible uses within parks are threatening their internal biodiversity. Few if any parks or preserves have been set aside for the sole or even major purpose of conserving biodiversity. In fact, typically little thought went into determining the purposes of parks, much less into the ideal size and shape to accomplish these purposes. Many national and state parks have been managed more for recreation than for preserving natural values. Similarly, most wilderness areas have been fought for and designated because of their value for primitive recreation, not for biodiversity.

Finally, parks and preserves are rarely buffered from outside influences that threaten internal or regional biodiversity. Many parks, especially in the West, are logged, mined, and grazed by livestock right up to their borders. Problems with the reserve strategy, as presently pursued, will be explored in detail in Chapter 5.

Natural Resource Management

The third major faction of the conservation movement in North America was that associated with natural resources management (forestry, range management, game management, fisheries management). Although this movement first coined the term *conservation*, the driving force behind it was resource development and use. One of the founders of this movement and the first Chief of the U.S. Forest Service, Gifford Pinchot (1910) wrote, "The first principle of conservation is development, the use of the natural resources now existing on this continent for the benefit of the people who live here now." Pinchot is also widely known for his statement that there are only two things in this world: people and natural resources. By this philosophy, if a species is not a resource, it is worthless.

Despite its limitations, natural resources management has contributed in some ways to the conservation of biodiversity. Most programs have developed in response to a perception of loss or degradation of some element of biodiversity (forests, forage species, game species) and have helped slow or even reverse losses. Most programs have generated useful information about the ecology of forests, rangelands, and wild plants and animals, although the information was often piecemeal. Moreover, some natural resources professionals have recognized the limitations of their fields as usually practiced. Aldo Leopold, a trained forester and the virtual father of wildlife management, became one of the most eloquent spokespersons for a land ethic that included all elements of biodiversity (Leopold 1949). In his later life, Leopold worked tirelessly to change and broaden both management programs and the training of natural resource professionals.

Nevertheless, natural resource management in North America shows a strong influence of its European cultural heritage. The biodiversity of the continent has been viewed by most managers solely as a resource to be used, an attitude that Grumbine (1992) has labeled "resourcism." Conservation became an extension of business, a necessary process of the industry and commerce of nations. Nature was viewed as an imperfect manager of these resources—inefficient, sometimes out of control, and often chock full of "undesirable" plants and animals, "decadent" forests, and other symptoms of managerial neglect. But humans, with their wisdom and superior technology, could improve upon the imperfections of nature. Forests, rangelands, and watersheds could be made more efficient and more productive. Undesirable plants and animals could be controlled, and newer or improved organisms added or increased in abundance. These attitudes remain so ingrained in the colleges, agencies, and individual managers, that it is worthwhile to look at them in more detail.

Forestry. Forestry was already developed as a profession in Europe before being imported to North America. In traditional European forest management, the forester had virtually dictatorial powers over what happened in the forests, which were intensively managed compared to those in North America. Gifford Pinchot, arguably America's most influential forester during the late nineteenth and early twentieth centuries, was trained in European forestry and adopted many European attitudes. According to this view, foresters could manage forests not only much better than nature, but much better in isolation from public or private pressures (Behan 1966). To this day, the holier-than-thou attitude is well entrenched in the forestry profession and in agencies such as the U.S. Forest Service.

Although early actions for forest conservation, such as establishing the U.S. Forest Service, were motivated by a perception that forests were being destroyed and watersheds threatened, later evolution of the profession shifted the emphasis from protectors of the forest and of the public interest to facilitators of forest extraction. Indeed, the forestry profession largely became an extension of the timber industry as the focus of forest management moved toward production of a narrow range of forest products—sawlogs and pulp. Forestry schools receive large grants from timber interests, and with some notable exceptions, few faculty are vociferous critics of corporate timber interests or their forest practices.

European and American forestry has always advocated the need for sustainability, but the definition and application of the concept in North

America has been quite limited. Sustainability was defined in terms of sustained yield of products, primarily timber. Although lip service was paid to other components of the forest such as soil, water, and wildlife, no real effort was made to manage the forest as an ecosystem. Forest management practices evolved toward those that increased wood production (Table 3.1). The ideal forest became the plantation or tree farm, where all aspects of growing tree boles (the harvestable stem of the tree) could be planned and controlled. The forest became a factory.

The limitation of the resourcism attitude is nicely illustrated by a statement from forester Henry H. Carey (Carey 1989 cited in Behan 1990):

> People often ask me, "Why aren't our forests sustainable? Aren't we replanting after we cut?" My response is that a forest is not a plantation. A forest is a mosaic, including ancient trees, venerable giants providing shelter for wildlife. A forest is clear, pure water, rivers alive with trout and salmon. A forest is wildness, silence. A forest is tall, straight trees for house beams and logs. A forest is a wellspring of value for rural people. A forest is the deep, rich soil soaking and storing water and nutrients.
>
> The fact that we can grow young trees where all this once stood does not mean we have perpetuated the forest. The task of protecting and restoring the forest is more complex. It demands all the skill, imagination, and dedication of foresters. This is the art of forestry.

Range management. While the change of forestry into a service profession for the timber industry is regrettable, foresters are not the only natural resource professionals to have moved in this direction. Range management may be viewed as the counterpart of forestry for lands without trees. And like forestry on this continent, range management developed in response to a perception that rangelands were deteriorating—from overgrazing. The history of the profession has followed a similar path of resourcism, with livestock production becoming the primary focus.

Range management has been defined in one of its early and standard textbooks as "the science and art of optimizing the returns from rangelands in those combinations most desired by and suitable to society through the manipulation of range ecosystems" (Stoddart *et al.* 1975). As with forestry, the emphasis is on products and production (returns), on improvement over nature (optimizing), and on technology. A list of common range management practices reinforces our assertion that range management has focused on production and improvement of nature (Table 3.2). Chapter titles from

TABLE 3.1 Common Forestry Practices and Their Effects on Biodiversity

Forestry practice	Purpose	Effect on biodiversity
Planting of exotics (nonnative species) or genetically "improved" tree species	Improved yield of commercial tree species	Replacement of native species; loss of genetic purity or genotypes of native stock
Pesticide spraying	Protection of forest tree species of commercial value	Reduction in vast numbers of nontarget insect species; secondary effects on nontarget vertebrates and other organisms; major disruption of ecosystem
Clearcutting/reforestation	Maximum utilization of existing tree biomass/maximum speed of new forest growth	Artificial cycle of disturbance; truncated succession; loss of species richness; loss of structural and functional diversity
Clearcutting/even-aged management	"Efficient" regulation of the forest; maximum profit from growing trees	Shortened successional cycle; loss of forest structural diversity; many known and unknown effects on forest ecosystem
Slash-burning	Site preparation for new forest; esthetics (neatening up the forest)	Major loss of structure, biomass, and nutrients from forest ecosystem
Tree thinning	Increased growth of commercial tree species	Reduced structural diversity in forest
Brush removal/herbicide spraying	Removal of species believed to be delaying reestablishment of commercial tree species	Truncated succession with loss of important successional processes (soil redevelopment, soil inoculation, etc.); loss of species richness and diversity of successional stages
	THE BOTTOM LINE	
Maximum production of commercial sawtimber and pulpwood	Maximum forest output; maximum profit	Reduced structural and species diversity of forest; deterioration of forest ecosystem

TABLE 3.2 Common Range Management Practices and Their Impact on Biodiversity

Range management practice	Purpose	Effect on biodiversity
Reseeding with exotics (nonnative species)	Increased forage production for livestock	Displacement of native species
Brush or tree removal (chaining/burning/herbicide application)	Increased forage production for livestock	Replacement of woody species with grasses or other herbaceous species; loss of structural and species diversity
Water development	Increased water for livestock; better livestock distribution	Usurpation of water required by wildlife species; spreading of livestock impacts into previously ungrazed or lightly grazed portions of the landscape
Fencing	Control of livestock movement	Impairment of movement of some wildlife species
Predator control	Protection of livestock from native predatory animals	Extirpation or reduced abundance of predators; reduced abundance or extirpation of nontarget species
Salting (placing of salt on range to affect livestock distribution)	Better livestock distribution	Spreading of livestock impacts into previously ungrazed or lightly grazed portions of the landscape
Pesticide spraying	Control of insects harmful to livestock forage species	Reduction in vast numbers of nontarget insect species; secondary effects on nontarget vertebrates and other organisms; major disruption of ecosystem
Herbicide spraying	Removal of "noxious" plants competing with livestock forage species	Reduced plant species diversity
	THE BOTTOM LINE	
Maximizing livestock utilization of forage	Maximum livestock production; maximum profit	Decreased forage availability for wildlife species; decreased species diversity; degradation of ecosystem

a standard textbook on "Range Developments and Improvements" include noxious plant invasion and plant control, mechanical plant control, herbicidal plant control, range seeding, range fertilization, and rodent and insect control (Vallentine 1975).

Like forestry, the range conservation profession has developed close ties to industry. Several years ago, a conference speaker asked a group of professional range conservationists in the Bureau of Land Management two questions. The speaker first asked, "How many of you consider that you are in the business of managing the vegetative resources of the public lands?" Two or three tentatively put up their hands while most of the others looked rather puzzled. The second question was, "How many of you consider that you are in the business of facilitating livestock production from the public lands?," to which everyone immediately raised their hand. "This is the problem," the speaker dryly noted.

Wildlife management. Wildlife managers have often been viewed, at least by some environmentalists, as the white hats among the natural resource professionals. Yet wildlife management has a remarkably parallel history to that of forestry and range management. We use the term *wildlife management* here to include both fisheries and wildlife management.

Aldo Leopold (1933), defined game management (as it was called then) as "the art of making land produce sustained annual crops of wild game for recreational use." Again, we see the emphasis on human desires and production ("making land produce"), rather than on nature as a whole or for its own sake. Leopold was sensitive to the limits of technology and broadened his views considerably during the 1930s and 1940s, but many others in the wildlife field experienced no such metamorphosis. A 1986 Forest Service publication entitled "Wildlife and Fisheries Habitat Improvement Handbook" contains an encyclopedic listing of everything from stream habitat improvement to upland habitat improvement (Payne and Copes 1986). As with range improvement there are sections on vegetation control, fertilizing lakes, fertilizing streams, artificial nests, and so on.

Perhaps the height of what has been termed "techno-arrogance" can be seen in our fish hatchery programs. In the Pacific Northwest, humans have virtually destroyed salmon and steelhead fisheries through dams, logging, and other habitat destruction, and then tried with huge expenditures, but limited success, to raise and release salmon in hatcheries (Meffe 1992). Although many wildlife managers abhor the thought of artificial propagation through game farms or hatcheries, many states still raise and release game birds and hatcheries have been a mainstay of fisheries programs for years. Other practices, such as winter feeding of elk and deer and introducing exotics for sport, continue, suggesting that the wildlife and fisheries profession is far from untouched by the desire for human domination and control of nature.

Like foresters and range managers, wildlife and fisheries managers have focused on the production of a single species or group of species such as deer, ducks, and pheasants for hunting and bass and trout for fishing. For many species, early emphasis was placed on determining an allowable harvest through the study of population dynamics, while paying little attention to the habitat or ecosystem that supported the animals. A listing of common wildlife management practices (Table 3.3) reveals similar purposes and effects on biodiversity as those for forestry and range management.

Like range and forestry, wildlife and fisheries management has had a strong tie to the user group—in this case sport hunters and fishers and manufacturers of hunting and fishing equipment. Most state fish and game agencies to this day are funded mainly from sales of hunting and fishing licenses. Similarly, much fisheries and wildlife research is supported by taxes on sporting arms and ammunition and on sport fishing equipment. It is not surprising that most research and management have focused on species of recreational or economic value.

From Resourcism to Disciplinarianism

As each of these movements and disciplines evolved, they developed professional societies and academic departments in colleges and universities to further their cause, funding, and professional standing. In most cases, these entities developed independently, so we have today many different professional societies (Society of American Foresters, Society for Range Management, American Fisheries Society, The Wildlife Society), each with its own journal, annual meeting, and so forth. Similarly, most colleges of natural resources still have separate departments of forestry, range management, wildlife management, and sometimes fisheries. Professors and students in one department tend to interact very little with their counterparts in other departments, precluding meaningful interdisciplinary ventures. Often departments are openly competitive for funds, students, and recognition.

Government agencies have also organized along disciplinary lines. Initially whole agencies were devoted to one type of use—the Forest Service for forests, the Grazing Service (the predecessor of the Bureau of Land Management) for rangelands, the Bureau of Reclamation for water, and so on. As these agencies matured and saw the need for more diversified approaches, they still tended to organize along the same lines. For instance, the Bureau of Land Management has separate programs for wildlife, range,

TABLE 3.3 Common Wildlife and Fisheries Management Practices and Their Effects on Biodiversity

Wildlife and fisheries management practice	Purpose	Effect on biodiversity
Artificial stocking	Increased recreational opportunities and/or harvest	Displacement of native populations; genetic deterioration of native species; ecosystem disruption
Artificial feeding	Maximum productivity of existing game species	Disease and other problems in native species; development of dependence upon artificial feeding; unknown ecosystem effects
Predator control	Maximum productivity of existing game species	Reduced abundance or extirpation of predators; reduced abundance or extirpation of nontarget species
Introduction of exotics	Increased recreational hunting and fishing	Displacement of native species; disruption of ecosystem function
Vegetation manipulation (herbicide spraying, planting)	Improved habitat for target wildlife species	Reduced plant species diversity or richness
Harvest regulation to maximize yield	Maximum harvest of game species	Reduced within population diversity (sex and age class diversity); unknown effects on ecosystem
Use of structures (nest boxes, "guzzlers," etc.)	Maximum production of target species	Increased reliance on artificial as opposed to natural habitats; unknown effects on ecosystem
Water development	Maximum production of target species	Disruption of hydrologic cycle; draining of small aquifers; other unknown effects on ecosystem
	THE BOTTOM LINE	
Maximization of yield of game species	Maximum recreational opportunities; maximum profit from recreational equipment; maximum income to fish and game agencies from license sales	Ecosystem deterioration from focusing on a few species to the detriment of others and the ecosystem as a whole

forestry, soil and water, and other resources, each with separate funding and staff leadership.

Disciplinarianism has resulted in a fragmented and inefficient pattern of natural resource management. Individuals trained in one discipline work on problems in isolation from other specialists, even within the same agency. Agency land use plans are often written as if there are separate landscapes to provide for timber, wildlife, livestock forage, clean water, and recreation.

Thus, the natural resource disciplines, despite their substantial contributions to modern conservation, have some severe limitations. In particular, the disciplines are:

1. Narrowly focused on a particular segment of the ecosystem (tree boles, livestock forage, a wildlife species, irrigation water) rather than on the ecosystem as a whole;
2. Oriented toward products (sawtimber, livestock, recreational opportunities, agricultural crops) rather than toward ecological processes that sustain the commodities and amenities;
3. Closely tied to the industries that profit from the products (forest products industry, livestock industry, recreation, sporting arms and tackle manufacturers and retailers, agribusiness) rather than to society as a whole; and
4. Based upon the belief that humans can improve upon nature, particularly through technology.

Emerging Conservation Movements

Beginning in the late 1960s the conservation movement began to change dramatically. Heretofore "preservationists" (those concerned with natural things for some intrinsic value or purpose other than consumption) had been a fairly small, even elite group. But, they did manage to stir up public sentiment for grand causes, as did John Muir for national parks at the turn of the century (Fox 1981). Similarly, resource management had been mostly concerned with rural areas and with people who worked or played in such areas, and was conducted by government agencies charged with managing these areas and uses. In the meantime, with the flight of people from farms to urban and suburban areas, the population of North America had become less rural. Fewer people directly depended on natural resource extraction to earn a living.

Ecology, the science that deals with living organisms and their relationships with the environment, was largely an academic pursuit before the late

1960s. Few people knew what the word *ecology* meant (perhaps few do even today). But the environmental movement that emerged in the 1970s and 1980s in North America illuminated serious problems in ecosystems at all scales, and challenged many of the prevailing assumptions about resource management, ecological problems, and human responses to such problems. During the past 25 years, the conservation movement has changed substantially. Three overlapping movements or trends stand out: (1) the birth of environmentalism; (2) a tendency toward conservation by legislation and litigation; and (3) the emergence of conservation biology.

Environmentalism

The demarcation of a new era in conservation is arbitrary. Just as the U.S. Constitution did not arise spontaneously, the conservation movement of the 1970s and 1980s in North America was built on the thought and actions of many people and organizations over many decades. Yet two events of 1969 and 1970 in the United States, the passage of the National Environmental Policy Act of 1969 (NEPA) and the first Earth Day on April 22, 1970, stand out as powerful symbols of a new era.

NEPA is best known for having required the preparation of environmental impact statements before major federal actions could proceed. But a NEPA provision with even broader reach was one declaring that the policy of the Federal Government is to use all practicable means to administer federal programs in the most environmentally sound fashion. If this policy were ever fully implemented, the consequences could be revolutionary.

Shortly after NEPA was passed, Americans celebrated the first Earth Day. Demonstrations and protests against environmental abuses were held at more than 1,500 colleges and 10,000 elementary and secondary schools in 50 states. In a single day, extensive television and newspaper coverage boosted awareness of environmental problems to millions of citizens. Almost overnight, *ecology* became a household word.

Thus 1970 marked in many ways the beginning of a new era: Stuart Udall, former Secretary of the Interior and long-time conservationist, wrote:

> It [Earth Day] exposed so many Americans to the subject of ecology for the first time that some latecomers concluded that the environmental movement began in this country on Earth Day. But that was bad history (like asserting that the U.S. civil rights movement began at the Lincoln Memorial on the day Martin Luther King, Jr., gave his magnificent "I had a dream" speech), for it ignored the nurturing process that had brought ecological truths

> into the mainstream of American thought. But Earth Day was indeed a red-letter day, a coming of age for the American environmental movement. (Udall 1988)

Overall, Earth Day symbolized a rejection or at least a questioning of many of the attitudes that had permeated American thinking about nature and natural resources since the first Europeans landed on this continent. Many citizens came to realize that resources were limited, and came to question the role of technology in causing and solving environmental problems. Unfortunately, subsequent Earth Days have proven to be more media events than substantive calls for change.

Conservation by Legislation and Lawsuits

For 10 years, beginning with NEPA in 1969, some of the strongest environmental legislation in American history was passed. At the federal level, new laws included the Clean Air Amendments of 1970 (the "Clean Air Act"), the Clean Water Act of 1973, the Endangered Species Act of 1973, the Federal Land Policy and Management Act of 1976, and the National Forest Management Act of 1976.

This environmental legislation forced government and citizens to begin to address major environmental problems, albeit in a piecemeal way. Many states passed similar laws, such as state endangered species acts and NEPA-like legislation that required environmental impact statements for state-sponsored projects. Two important and interrelated results of this environmental legislation stand out: an increased citizen awareness and a mobilized environmental activism characterized by protests and lawsuits.

NEPA, together with the Freedom of Information Act, allowed the public to become much more aware of the activities of the Federal Government and of how these activities affect the environment. Large dams, nuclear power plants, new roads into wilderness, and many other projects were now subject to public scrutiny. Such scrutiny, together with better ecological understanding, led to increased citizen concern and activism. Tactics learned earlier from the anti-Vietnam War movement, including demonstrations and protests, were used to capture media attention and influence the general public and those in power. Later the product boycott, popularized by Cesar Chavez in the farm labor movement, was used to pressure businesses to forgo legal but environmentally damaging practices such as killing dolphins while fishing for tuna.

Legislation also provided a basis for lawsuits against agencies that failed to comply with the provisions of environmental laws. A plethora of lawsuits against agencies resulted, creating a new branch of law, "environmental law."

[A review of 155 cases cited in a current environmental law summary shows that all except 12 were cases since 1970 (Findley and Farber 1992).] Two major organizations, the Environmental Defense Fund and the Natural Resources Defense Council, were formed to contest environmental damage in court, using not only the law but independent scientific testimony. These organizations flourished, and older conservation organizations such as the Sierra Club, National Audubon Society, National Wildlife Federation, The Wilderness Society, and Defenders of Wildlife strengthened their legal efforts.

But environmental legislation and subsequent lawsuits had some aspects that were not so positive. As in other areas of law, environmental lawsuits tended to be divisive, resulting in polarized attitudes and solutions that satisfied neither party. Furthermore, the courtroom has never been a particularly good forum for explaining or debating science. Thus many court challenges tended to fall back upon procedural challenges ("the agency failed to comply with EIS requirements of NEPA . . . ") rather than deal with substantive issues ("the agency is taking an action harmful to the environment . . . ").

One of the biggest weaknesses of environmental laws and their enforcement is the fragmented way in which they deal with issues. While clean water, clean air, wilderness, and a full complement of native species are all parts of a fully functioning and healthy regional ecosystem, each of these issues has been dealt with separately. A piecemeal approach may be more manageable, administratively and legally, but is not especially effective in preventing continuing ecosystem degradation. Furthermore, it is not consistent with the ecological understanding that all things are interconnected. Even the Endangered Species Act of 1973, the strongest law in the United States (and perhaps the world) for protecting biodiversity, has not reduced the threat of species extinction (Bean 1988), although arguably much of the problem with the Act is due to insufficient funding, political intervention, and lack of agency willpower.

Perhaps the greatest weakness of environmental management by legislation, lawsuit, and protest is that it is reactive rather than proactive. We are trying to substitute laws and enforcement for an environmental ethic. In the long run, the earth and its biodiversity can be maintained only through a fundamental human conviction that any action that harms the environment is ethically wrong rather than merely against the law. Leaders of business, government, and other organizations must reach a state of mind where they would not want to be involved in destructive practices because of personal ethics, rather than because it is legally questionable. Citizens need to fight to protect their environment, not only for their immediate families, but for all living things now and in the future. And they need to be active participants in the conservation and restoration of species and habitats.

Conservation Biology

The limitations of traditional conservation have stimulated the emergence of a new discipline and movement—conservation biology. Conservation biology is described by several of its leading spokespersons as follows: "Conserving biology is an emerging synthetic discipline that deals with the basic issue of eroding biological diversity . . . it derives its theoretical basis from the pure sciences, such as population genetics, demography, biogeography and community ecology. However, it uses the resulting principles to address applied problems in conservation, such as the loss of genetic diversity, loss of species diversity and loss of diversity of ecosystems. Conservation biology now joins forestry, wildlife management and fisheries management as the newest player in the area of applied ecology" (Temple *et al.* 1988).

Scientists have been leaders in the movement to conserve species and natural areas throughout most of Western history (Grove 1992). With conservation biology this noble tradition continues, but now with greater urgency. Conservation biology is science in the service of conservation. It is the body of knowledge (albeit a body always expanding and shifting in new directions) that allows a book like ours to be written.

Conservation biology is not a typical science. Although it is fundamentally ecological and relies on principles of ecology, it is also cross-disciplinary and depends on the interaction of many different fields. Geography, geology, sociology, education, philosophy, law, economics, and political science are just as important to the successful practice of conservation biology as are wildlife biology, forestry, ecology, zoology, botany, genetics, and other biological sciences. A conservation biologist, even if mainly a plant geneticist, must be conversant with the other disciplines that compose conservation biology to apply findings to real-world conservation problems. Thus, conservation biology is a "metadiscipline," a level of knowledge that transcends the individual disciplines that compose it by leading to insights not directly deducible from any discipline alone (Jacobson 1990).

Conservation biology is mission oriented (Soulé and Wilcox 1980, Soulé 1985). It has a job to do, and that job must be accomplished quickly. The values underlying conservation biology run the gamut from utilitarian to purely ethical. But because it is concerned with all kinds of life, conservation biology differs from traditional natural resources disciplines, which focus on products for human consumption (Table 3.4). The fundamental belief of conservation biology is that biodiversity is *good* and should be conserved. The mission of conservation biology, then, is to conserve as much

Table 3.4 Comparison of Traditional Resource Management with Conservation Biology

Parameter	Traditional resource management (forestry, wildlife management, range management, etc.)	Conservation biology
Time scale	Short-term perspective	Long-term perspective
Orientation	Oriented toward use of resources	Ecologically oriented
Attitude	Confidence in human knowledge about resources and human effects on nature	Uncertainty about human knowledge and human effects on nature
Role of humans	Human domination/control of ecosystems/landscape	Humans living within ecosystem limits
Role of ecosystem	Emphasis on "improving" ecosystems	Emphasis on maintaining natural ecosystems
Use of technology	High-tech management	Management with emphasis on minimal or appropriate technology
Risk	Management with high risk to ecosystem/biodiversity (linear management)	Management to minimize risk to ecosystem/biodiversity (adaptive management)
Species emphasized	Emphasis on "improved" or introduced species	Emphasis on native species
Intraagency programs	Fragmented/disciplinary	Integrated
Interdisciplinary interaction	Competition	Cooperation
Interagency interaction	Competition	Cooperation
Responsiveness	Management responsive to bureaucracy/commodity users	Management responsive to long-term needs of earth and local people

of global biodiversity as possible and to allow evolution to continue generating biodiversity. Because biodiversity is being lost so rapidly, conservation biology can also be described as a crisis discipline.

Conservation biology is growing rapidly in popularity. It has more public appeal than perhaps any biological science except health and medicine. The Society for Conservation Biology, founded in 1985 and now with nearly 5000 members, has experienced more rapid growth than any professional scientific society in American history. *Conservation Biology,* the Society's journal, has become the most widely cited (and, presumably, most widely read) periodical in applied ecology worldwide. This growth, although in many ways encouraging, can be seen as a direct response to environmental deterioration. As biodiversity declines, interest in conservation biology expands.

What is lacking in conservation biology are tested strategies and tactics for conserving biodiversity at various scales of space and time. Managers, biologists, and citizens—convinced of the biodiversity crisis—are unsure of what to do. After all, nothing in recent history has worked. We cannot turn to model regions or societies—except for those of the past about which we know very little and of which little of their culture remains—and say "This is the way to conserve biodiversity within a sustainable society." We are groping in the dark.

The following section introduces a national strategy for managing land and people to achieve the goal of no net loss of biodiversity in North America. Our strategy is based on, as far as we can determine, the best available knowledge, scientific and otherwise. It is also based on a land ethic and on our personal, emotional attachment to the land. We are convinced that the strategy we offer, despite its preliminary nature and other limitations, is more likely to conserve biodiversity than current strategies.

Forging a National Strategy

> The ethical imperative should therefore be, first of all, prudence. We should judge every scrap of biodiversity as priceless while we learn to use it and come to understand what it means to humanity. We should not knowingly allow any species or race to go extinct. And let us go beyond mere salvage to begin the restoration of natural environments, in order to enlarge wild populations and stanch the hemorrhaging of biological wealth. There can be no purpose more enspiriting than to begin the age of restoration, reweaving the wondrous diversity of life that still surrounds us.
>
> E.O. Wilson (1992), *The Diversity of Life*

The United States has no national strategy to conserve biodiversity. Only recently have we even tried to forge such a strategy. The most earnest effort so far was the Keystone National Policy Dialogue on Biological Diversity, with which we introduced this book (see Preface). The Keystone Dialogue brought together a diversity of interests to formulate consensus recommendations for conserving biodiversity in the United States (Keystone Center 1991). Because of concerns about private property rights, the group agreed to limit its recommendations to federal lands. This group of 60 federal land managers, agency scientists, congressional staffers, academic biologists, environmentalists, timber and oil industry representatives, and other commodity interests was able to agree about basic goals and objectives for maintaining biodiversity on federal lands, even recognizing that biodiversity is valuable for its own sake: "It should be a national goal to conserve, protect, and restore biological diversity on federal lands: to sustain the health of ecological systems; to provide for human well being; and because of its intrinsic value" (Keystone Center 1991).

To implement this broad goal, the group made further recommendations, including suggestions to maintain viable populations and natural genetic variability, maintain representative examples of all ecosystems, integrate human activities with conservation, increase scientific understanding of biodiversity and conservation, achieve public awareness and understanding of biodiversity, and encourage the private sector to apply innovative approaches to conservation. These excellent recommendations were preceded, in a late revision at the insistence of commodity interests, by the phrase "to the greatest extent practicable" and were followed by vague discussions of how one might go about managing for biodiversity.

In our view, the Keystone Dialogue produced no recommendations specific enough to offer useful guidance to land managers. Instead of an urgent call for specific legislation, executive orders, or other meaningful actions, the final report called only for a "Federal Biological Diversity Policy and Coordination Committee." The charge of this group was simply to prepare a report on the state of biodiversity on federal lands. Such a report was later produced by the Departments of Interior and Agriculture (USDI–USDA 1992). In the introduction to this text, the secretaries of Agriculture and Interior claimed that "Americans can use and enjoy Nature's bounty within a legal and institutional system that respects private property rights, state–Federal relationships, and stimulates technological progress and sustainable economic growth." Throughout the report, biodiversity is treated as equivalent to natural resources, it is said to be in great shape and continually improving (thanks to government programs), and "the only way to simultaneously meet the needs of expanding populations and halt or reduce land conversion (to conserve biodiversity) is to increase the productivity of agriculture, forestry, and other human activities that use land as a resource base" (USDI–USDA 1992). The alternative approach of stabilizing population, reducing consumption, and conserving resources was apparently not considered.

We believe that it will take more than self-congratulatory reports by bureaucracies to protect and restore biodiversity in the United States. We need an aggressive program with ambitious long-range goals, full cooperation and integration among agencies and programs, specific indicators to measure the success of programs, and explicit listing of activities compatible and incompatible with biodiversity.

Because biodiversity conservation is largely a matter of how we treat the land, a national strategy for biodiversity must be land based. It must recognize real threats—logging, grazing, road building, mining, dam building, agriculture, housing development, off-road vehicles—and it must outline measures to reduce the extent of these activities, keep them out of natural areas, mitigate any damage done, and restore areas degraded by past abuses. In general, we need to interact with nature in a gentler and more sustainable way. And we need to compensate for past destruction.

A national strategy must recognize the connections between proximal threats to biodiversity and the ultimate causes: human population growth, poverty, misperception, anthropocentrism, cultural transitions, economics, and failures in policy implementation (Soulé 1991). It must address ultimate factors directly by providing incentives for controlling population and resource consumption, and by strong disincentives for not doing so. The carrot and the stick both have a place in conservation policy. A national strategy must also suggest alternative economies and lifestyles that are sustainable in the deepest sense of that term. Although we emphasize ecological aspects of conservation in this book, sociological, economic, and political approaches that dovetail with land conservation are sorely needed.

A national biodiversity strategy must also be hierarchical, not in the sense of authoritarianism, but rather planned and implemented at several spatial, temporal, and organizational levels, from a local watershed to the North American continent, from next week to 10,000 years from now. Management decisions at all scales must be informed by and consistent with the highest objective: maintaining global biodiversity in perpetuity. With such a strategy, we might truly reverse the trends of biotic impoverishment in America and across the world. Without it, these trends can only escalate as they have over the last few decades and, in fact, centuries.

Land Management Goals

Although some land managers consider virtually any manipulation of habitat to be an improvement on nature and certain environmentalists regard all management as arrogant and destructive, the true story is more complex. Land management can have positive, negative, or neutral effects on biodiversity. Management is positive if it serves to protect biodiversity from harm or helps restore an ecosystem previously damaged. It is neutral if it essentially mimics or substitutes for natural disturbance–recovery processes (a theoretical possibility, though not yet convincingly demonstrated anywhere). But management is negative if it contributes directly or indirectly to biotic impoverishment. A proper philosophy for management of public lands is that all actions must have ultimately positive or at least neutral effects on global, national, and regional biodiversity. Globally negative management should no longer be tolerated anywhere.

Conservation biology, as we noted earlier, is not value-free science. Rather, it is mission-oriented. Similarly, land management needs a mission. The mission we suggest is protecting biodiversity and letting natural processes operate while permitting compatible human uses in suitable areas. This mission is an ultimate goal. Goal-setting must be the first step in the conservation process, preceding scientific, technical, and political questions of how to conserve species and ecosystems in particular places. Primary goals for biodiversity conservation should be comprehensive and idealistic so that conservation programs have a vision toward which to strive over centuries (Noss 1987a, 1992b). A series of increasingly specific objectives and action plans should follow these goals and be reviewed regularly to assure consistency with primary goals and effective application.

Four fundamental objectives are consistent with the overarching goal of maintaining the native biodiversity of a region in perpetuity. These are to:

1. Represent, in a system of protected areas, all native ecosystem types and seral stages across their natural range of variation.

2. Maintain viable populations of all native species in natural patterns of abundance and distribution.

3. Maintain ecological and evolutionary processes, such as disturbance regimes, hydrological processes, nutrient cycles, and biotic interactions.

4. Manage landscapes and communities to be responsive to short-term and long-term environmental change and to maintain the evolutionary potential of the biota.

These goals were discussed by Noss (1992b) as the basis for establishing a network of connected wildlands throughout North America, an effort called The Wildlands Project, and were applied to a case study in the Oregon Coast Range (Noss 1992c, 1993a). We believe they provide an adequate foundation for land management planning in general. Alternative land uses can be considered vis-à-vis these objectives, analyzed as to their impacts, and pursued only when deemed compatible. Biology is a better "bottom line" for land use decisions than the traditional social and economic criteria for a very important reason—human cultural systems can adapt much more rapidly to new conditions than can species or ecological systems. We are the ultimate generalists and can change our niche readily to new circumstances, within broad biological limits. But ultimately, human cultures cannot persist without healthy ecological systems.

Conservation planning is conveniently pursued at a regional scale, using ecologically meaningful regional classifications (e.g., the ecoregions of Bailey or Omernick). Thus, conservationists (private citizens as well as agency professionals) in each region might develop strategies to fulfill the goals reviewed here. Coordination and physical linkage of habitats between regions can be accomplished by planning at still broader spatial scales, including national, continental, and global (Foreman *et al.* 1992, Noss 1992b). National planning should be mindful of global priorities, but should never declare any region a sacrifice zone, even if it is naturally low in species richness or has been long degraded by human activities. Global biodiversity is an aggregate of the unique floras, faunas, and processes of every region. Each region and its biota are an irreplaceable part of the biosphere.

Representation. A central goal of conservation is representing a broad spectrum of natural communities in a network of protected areas. Representation is an example of an ecosystem approach to conservation, because the focus is usually on habitats and species assemblages rather than on single species. Although widely touted and in many ways more efficient than single-species conservation, ecosystem conservation is problematic. One major difficulty is that ecosystems are more difficult to classify than are species. A related problem is that the public may not respond to ecosystems the way they do to individual species. After all, ecosystems are not particularly tangible and are seldom cute and cuddly (Noss 1991a). An appeal to charismatic species will still be needed in many cases to gain public support for protecting ecosystems.

Despite problems, we believe that the goal of representing all ecosystems, when expanded to multiple levels of biological organization and physical habitat gradients, is the most useful overriding goal of land conservation. Representation strategies and analyses will be discussed in some detail in Chapter 4.

Viable populations. Representing species and ecosystems in protected areas is one step toward saving them. But occurrence in a reserve is not the same as persistence. The representation objective must be complemented by the goal of maintaining viable populations of every species. A viable population is one that has a very good chance of persisting for a long time (see Chapter 2). In considering viability, we need not worry about each of the thousands of species that inhabit a region—this would be an overwhelming task. Most species can get along fine without human attention, and many thrive in human-disturbed landscapes. Rather, we must direct our attention toward those species most threatened by human activities (Diamond 1976). For species recognized as vulnerable, we should consider life history traits, habitat and area requirements, and other sensitivities that may predispose them to extinction. Some kinds of species should be analyzed individually: narrow endemics that are confined to only one or a few sites, carnivores that require huge areas of land, and any other species particularly vulnerable to human disturbance. Criteria for identifying species that warrant individual attention are reviewed in Chapter 4.

Ecological and evolutionary processes. The conventional emphasis on species, natural communities, and other patterns of biodiversity reflects a human tendency to categorize, to recognize discrete entities and boundaries, and to concentrate on the present: what we see today. But underneath all the patterns of biodiversity are fundamental ecological and evolutionary processes that operate on time scales of years to millennia.

Considering process is fundamental to biodiversity conservation because process determines pattern. Processes of disturbance, hydrology, nutrient cycling, dispersal, species interactions, and evolution determine the species composition, habitat structure, and ecological health of every site and landscape. These patterns, in turn, influence processes reciprocally. If natural fires are reduced in frequency due to a change in climate to cooler, moister conditions, the new community that develops may be resistant to fire, maintaining itself even when a warmer, drier climate returns. Invasion of a community by an aggressive exotic may strongly alter the composition of the community and ecological processes. A conservation plan that recognizes the interdependency of pattern and process and that strives to maintain fundamental ecological and evolutionary processes will more successfully conserve biodiversity than a strategy that focuses on preserving a species or community as a static entity.

Adapting to change. Process and change are intimately connected. Although equilibrium theories once dominated ecology, they have been largely replaced by dynamic paradigms. No natural community is static. Rather, communities change in many ways and at many different rates simultaneously, from daily fluctuations in weather and births and deaths of organisms to long-term trends in global climate, the position of continents, and the origin and extinction of species, genera, families, and higher taxa. Conservationists need to recognize the universality of change rather than drawing lines around areas and trying in vain to hold them forever in the condition in which they were found. Certainly changes caused by human activity that result in a net reduction in biodiversity must be fought aggressively. We might not always be successful, but it is our obligation to oppose destructive trends.

Other kinds of changes are essential to ecological health and should be considered part of the biodiversity that conservationists legitimately try to conserve. As stated by Botkin (1990):

> . . . to accept certain kinds of change is not to accept all kinds of change. Moreover, we must focus our attention on the rates at which changes occur, understanding that certain rates of change are natural, desirable, and acceptable, while others are not. As long as we refuse to admit that any change is natural, we cannot make this distinction and deal with its implications.

For example, species composition changes naturally with invasions and extinctions of species over hundreds and thousands of years. Apparently this change has been going on for as long as life. But the rapid changes in communities and ecological processes caused by human introductions of alien organisms are destabilizing and destructive to biodiversity (Elton 1958, Mooney and Drake 1989). Even if species diversity increases locally, such changes often result in reduced diversity globally, a process of global homogenization (Mooney 1988, Noss 1991c). The spatial scale of change must also be considered. Many changes that appear destructive at a local scale, such as a fire that destroys an old stand of trees, are part of a process that rejuvenates and diversifies the larger landscape. If only a few stands of old growth remain in a region, because the rest have been logged, then it may make sense to preserve those stands even against natural disturbances. But such defensive actions should be temporary. A long-term goal should be to restore large, diverse, connected landscapes where natural processes can operate unimpeded.

The Land Conservation Process: "Thinking Big"

The cornerstone of a national biodiversity strategy is land conservation: the prudent protection, management, and restoration of soil, water, air, and biota (see Leopold 1949). Land conservation is implementing goals of the kind discussed above: representing all ecosystems, maintaining viable populations, sustaining ecological and evolutionary processes, and being adaptable to change. Management is central to land conservation because all human actions affecting land represent management decisions.

An intelligent management decision depends on information. Thus, we envision a cyclic process with continual feedback from inventory, research, and monitoring (Fig. 3.1). Too often management has operated in isolation

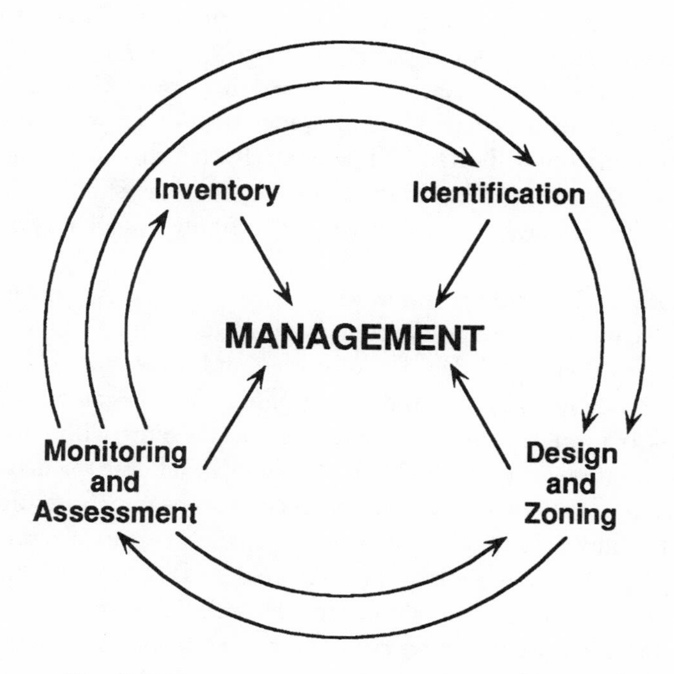

FIGURE 3.1 The land conservation process. Management must be informed and adaptable to new information. Biological inventory, identification of sites that require protection, design and zoning of landscapes, and monitoring and assessment of biodiversity and effects of management practices (which includes research) are all necessary parts of the process.

from potential sources of critical information. In our experience, no federal land management agency in the United States has adequate linkages among research, monitoring, and management programs; these programs are usually isolated administratively with little interaction between staffs. Current management has effects on biodiversity that vary from beneficial to disastrous. We need to eliminate the disasters and start moving all programs toward the beneficial. For this to happen, management must be ethical, ecologically informed, goal-oriented, and adaptable to new information. We are not proposing anything entirely new with this suggestion. Rudiments of the process we advocate go back decades. Many of the concepts and actions we suggest are already officially part of agency operations. Yet they remain largely uncoordinated and ineffective.

As a basis for a national strategy for maintaining biodiversity, the land conservation process must be implemented at many spatial scales: continental (in cooperation with our international neighbors), national, regional, and local. At all scales the key to making land conservation effective is to expand our thinking in space, time, and ambition—that is, to think big. Management decisions for a particular parcel of land need to be made within a context that is ultimately global in scope. Constraining our thinking within management units or agency boundaries will inhibit intelligent decisions and will often lead us astray. Rather than analyzing a property or management unit in isolation, the land manager should ask, "What can I do with this piece of land that will contribute to global conservation goals?"

Similarly, managers must think well beyond the next few fiscal years or other conventional planning cycles and consider effects of present management actions over years, decades, centuries, and millennia. These are the scales of fundamental ecological and evolutionary phenomena. We must also think big in terms of ambition and try to do more with conservation than we ever thought possible. Managers can help reframe the debate about proper land use and what steps are necessary for meaningful ecological recovery. To think only in terms of what is politically reasonable, practical, or financially profitable is shortsighted. At worst, a lack of ambition and acceptance of the status quo is an invitation to mass extinction.

The stages in the land conservation process we recommend are described below. This process is cyclic, rather than linear, and all stages can take place simultaneously.

Management. For the purposes of this book, management covers the spectrum from intense commodity production to hands-off preservation, each in its proper place and time. The only general requirement is that management of each individual area be considered within the larger context of biodiversity conservation. Considering a broader context may mean asking questions such as "What is unique about this piece of land? What is the regional and global status of species and communities in this parcel? How does this parcel interact ecologically with surrounding lands? Is this area an important migratory corridor for animals?"

Contextual management represents a major departure from traditional practices of managing land on a piecemeal or laissez faire basis. If we wish to maintain biodiversity, we must recognize that ecological processes transcend all human-designated boundaries. To manage an area in isolation from its landscape context is to invite biotic impoverishment (Noss and Harris 1986). Specific recommendations for managing forests, rangelands, and aquatic ecosystems will be presented later.

Inventory. An essential task in land conservation is to determine what you have and where it fits in a broader context. This task requires a biological inventory of the particular management unit and of as much as possible of the surrounding region. At a broader scale, a biological inventory is urgently needed for the entire nation.

For specific sites or regions, biological inventories can be constructed from information already published in the scientific literature, from unpublished lists and reports in agency files, or from interviews with field biologists and others who know the area well. In most cases, more field surveys will be needed to verify old records and search for organisms never before surveyed in the area. For example, a nature reserve might have been surveyed for birds, mammals, vascular plants, and butterflies, but have no complete lists of reptiles, amphibians, invertebrates besides butterflies, or bryophytes. Examining aerial photographs and satellite imagery is also part of the inventory process and is useful for mapping vegetation cover types and predicting species distributions from the availability of suitable habitat at landscape to regional scales.

Inventory must be a continuing process. We never will have complete information on the distribution and abundance of all biota in a region, much less across a nation or continent. Furthermore, the situation is always changing with modification of habitat, introduction of exotics, natural changes in climate, and other factors. The other steps in the land conservation process—identifying areas of high conservation value, designing and zoning the landscape, monitoring and assessment, and management—cannot wait for a complete inventory before proceeding. All public and private lands in the United States are being managed now, regardless of whether biological inventories have ever been conducted. Inventories are needed, however, to make informed decisions about future management. Otherwise, much biodiversity could be lost out of ignorance.

Identification. Identification refers to the process of evaluating sites by their conservation value and distinguishing areas that warrant special protection. Again, this process should be carried out at many spatial scales, from global to local. As a local example, if we have five private land parcels being considered for acquisition as a nature reserve and only enough money for one, how do we decide which to buy? Analogous scenarios for public lands are common. For instance, if political constraints limit us to designating only 20,000 acres as Research Natural Areas this year, which of many alternative sites should be protected? Conservation evaluation, the process of choosing among alternative sites for protection (Usher 1986), is a complex science in itself. Gap Analysis (Scott *et al.* 1991a, 1991b, 1993) and The Nature Conservancy's system of locating sites containing "the last of the least and the best of the rest" (Jenkins 1985) are examples of conservation evaluations. We will review the process of identifying areas and biological elements (species and communities) that require extra protection in Chapter 4.

Design and zoning. After areas most in need of protection are identified, management and land-use plans must be revised to reflect this new knowledge. Reserves that protect sensitive sites and species should be designed and the overall landscape zoned to optimize conservation potential. Design involves considering the size of reserve needed to maintain species and processes, ensuring that reserve boundaries conform to physiographic or ecological boundaries, and developing buffer zones to insulate sensitive areas from intensive land use and other human activities. As will be discussed in Chapter 5 (see also Noss 1992b), a model regional reserve network consists of core reserves, buffer zones, and corridors. Reserves form the backbone of the system, but by themselves are insufficient. Modern concepts of ecological planning emphasize integration of protection and production objectives across entire landscapes. Reserves are too few, too small, and too far apart to be viable unless enveloped in a landscape generally compatible with the needs of the native biota. Land-use planning and zoning for biodiversity must be applied to all lands at local, regional, national, and international scales (Brown 1988, Noss 1991a).

Monitoring and assessment. Monitoring and assessment go hand in hand with land management. The basic idea is that any land management practice or land-use regime is an experiment. We do not yet (and may never) know what we are doing. Hypotheses about the effects of practices on biodiversity and ecological sustainability must be formulated and tested in order to determine whether new management approaches work any better than those of the past. As Walters and Holling (1990) noted, "Every major change in harvesting rates and management policies is in fact a perturbation experiment with a highly uncertain outcome, no matter how skillful the management agency is in marshalling evidence and arguments in support of the change."

Monitoring and assessment, placed in a hypothesis-testing framework and using quantifiable indicators of biodiversity, are essential for measuring progress toward goals and comparing the long-term effects of alternative management practices (Noss 1990a, 1992d). But how do we apply what we learn from these experiments? Essentially, management experiments must function as an adaptive process (Holling 1978, Walters 1986). With adaptive management, we can learn from experience and modify management practices in accordance with information obtained from monitoring. Management policy must be based on the best current knowledge and be flexible enough to respond quickly to new information (see Chapter 9).

As shown in Figure 3.1, information derived from monitoring and assessment should feed back to decisions about inventory, identification, design, and zoning. We might, for instance, find that we need an inventory of soil biota and nutrients to help understand why tree growth rates are declining in our study area. Evidence of declining populations of forest interior birds in a reserve might call for reevaluating reserve design and perhaps selecting more parcels to add to the reserve or establishing buffer zones to minimize edge effects. Without monitoring and assessment, we are shooting in the dark.

Conclusion

Our intent in this chapter was not to deride all former approaches to conservation or to suggest that now we have all the answers. Rather, we have tried to show how failures of past approaches can be instructive for developing a new approach that, though untested, has a greater probability of maintaining biodiversity than anything tried so far. We, other conservation biologists, and society in general remain dangerously ignorant about natural ecosystems—all the more reason for a cautious and minimally intrusive kind of management now and in the future.

Although some failures of the past have reflected technical limitations or flawed scientific paradigms, we think it is now obvious that our greatest limitation has been philosophical. Past approaches were heavily anthropocentric, utilitarian, reductionist, shortsighted, and arrogantly faithful to technology. Although conservationists today—certainly including us—carry elements of this cultural baggage with them, they are trying to think bigger. Our preliminary national strategy is only an attempt to pursue the broad goals of biodiversity conservation in an ethically responsible and scientifically defensible way. The following chapters provide more detail on how we see this strategy being implemented.

CHAPTER FOUR

SELECTING RESERVES

> We need the tonic of wildness,—to wade sometimes in marshes where the bittern and the meadow-hen lurk, and hear the booming of the snipe; to smell the whispering sedge where only some wilder and more solitary fowl builds her nest, and the mink crawls with its belly close to the ground. At the same time that we are earnest to explore and learn all things, we require that all things be mysterious and unexplorable, that land and sea be infinitely wild, unsurveyed and unfathomed by us because unfathomable. We can never have enough of nature . . . We need to witness our own limits transgressed, and some life pasturing freely where we never wander.
>
> Henry David Thoreau (1854), *Walden*

In an ideal world, all human actions would be compatible with conservation of biodiversity. We would not have to think about protecting anything. We would live lightly on the land, limit our reproduction and use of resources, and develop only technologies that are nondestructive and sustainable. We would stay away from wading bird rookeries, eagle nests, bear dens, rare plant populations, and other sensitive sites out of respect. We would be humble, gentle, and wise in our dealings with other creatures.

We do not live in an ideal world. Rather, we live in a world where humans are breeding like rabbits, often ruthlessly destroying the natural habitats around them, and heedlessly dumping poisons into the atmosphere, oceans, and fresh waters. As noted earlier, direct habitat alteration by humans is the greatest peril for biodiversity. Habitat alteration is far more pervasive and often more subtle than the devastation that comes immediately to mind—slashing and burning of tropical forests for conversion to cattle pastures, construction of megadams, or the ugly urban sprawl of southern California. It occurs also on our farmlands, on our national forests and other public lands, and in our own backyards. Habitats are altered with every road built, with every clearcut, with every irrigation ditch, with every strip mine, and with too many cows. Biodiversity will only be conserved when we make a conscious decision to protect the land and keep much of it "pasturing freely where we never wander."

Protecting every acre in nature reserves is neither feasible nor necessary to conserve biodiversity. As will be discussed in Chapter 5, a sensible approach is land-use zoning, where the most sensitive sites, representative sites, and hot spots of biodiversity (usually defined as areas containing high native species richness or high concentrations of endemic species) are given the most protection. These sites then can be shielded from intensive land uses by one or more buffer zones (UNESCO 1974, Harris 1984, Noss and Harris 1986, Noss 1987a). Selecting sites that warrant the highest degree of protection is the most vital task in the entire land protection process, because these sites are the core areas around which a compatible land-use system can be built. Losing hot spots of biodiversity to development is unacceptable, for these areas are usually irreplaceable. But just as necessary as capturing hot spots in a reserve system is assuring that all species and ecosystems native to a region are adequately represented.

As we explore ways to identify biodiversity hot spots and other priority conservation areas in this chapter, keep in mind several themes: (1) Because hot spots occur at various spatial scales, conservation evaluations should also occur at several scales, from local to global. (2) Because biodiversity occurs at several levels of organization, hot spots should also be sought at several levels, for example, genetic, species, community, and landscape. (3) Protecting the most diverse sites is not enough. Rather, all native ecosystem types and species must be represented in a system of protected areas. (4) Selecting ideal sites for conservation does not tell us much about how to design reserves to protect these areas, how big reserves should be, how they should be connected, or how they should be managed. The design question, along with some general principles for management, will be explored in Chapter 5. Management guidelines for forests, rangelands, and aquatic systems will be reviewed in Chapters 6, 7, and 8, respectively.

Biogeography as a Guide to Protection

Biogeographical information is critical to conservation evaluation. Species, genes, and other elements of biodiversity are not distributed randomly across the earth, nor do they occur in a uniform pattern. Rather, they respond to environmental gradients—including climate, substrate, and topography—and reflect a long history of species colonizations and extinctions, plate tectonics, and other global processes. By paying attention to distributions of organisms, conservationists can identify sites of greatest importance for protection. Basing reserve locations on biodiversity qualities would represent something of a revolution in conservation because most reserves, as noted earlier, were selected on the basis of scenic, recreational, and economic criteria and are usually areas with few extractable resources and low biodiversity (Harris 1984, Foreman and Wolke 1989, Noss 1990b). Where are our million-acre parks of tallgrass prairie and old-growth forest?

Hot spots of biodiversity can be recognized at several spatial scales. Globally, the humid tropics stand out as biodiversity hot spots, with the greatest richness for many taxonomic groups in Central and South America (Erwin 1982, McNeely *et al.* 1990, Gentry 1992). But the tropics are not the only hot spots. Some taxa reach their greatest richness at higher latitudes. Diversity of stream-dwelling invertebrates seems to be generally higher in the temperate zone than in the tropics, although more surveys are needed to establish this pattern with certainty (MacArthur 1972, Allan and Flecker 1993). Soil invertebrate faunas also may be richer in temperate forests than in the tropics. One-foot-square samples of old-growth forest soil and litter often yield 200–250 species of invertebrates in the Pacific Northwest (Moldenke and Lattin 1990), more than found in tropical samples.

Within the temperate zone and the United States in particular, species richness peaks in different regions depending on the group (Table 4.1). General explanations for these geographic patterns of species richness remain controversial and not all taxonomic groups follow the same pattern. For example, little relationship exists between tree and vertebrate species richness (Currie 1991). Across the western United States, the number of vertebrate species increases as vegetation structure (for example, different layers of vegetation) becomes more complex, as environmental conditions become more varied, and as evolutionary history has promoted diversification (Vale *et al.* 1989). More generally, species richness is related to the amount of energy flowing through an ecosystem (Currie and Paquin 1987, Currie 1991).

Another important measure of biodiversity is endemism, the occurrence of species with very narrow ranges (say, less than 50,000 km^2; Terborgh and Winter 1983, Gentry 1986). Centers of endemism are areas where many narrow endemic species occur together; they range in size from single outcrops to entire physiographic regions. Endemism correlates generally but not completely with species richness (Gentry 1992). Extensive studies comparing areas of high species richness and centers of endemism in the United States have not been completed, but there appears to be a positive relationship. The greatest numbers of endemics are typically on true islands or island-like terrestrial habitats, including isolated mountain peaks, desert springs, serpentine substrates, and other unusual habitats. Thus, endemics are usually concentrated in habitats that are themselves localized (Stebbins

TABLE 4.1 Hot Spots of Species Richness in the Continental United States

Taxon	Areas of greatest species richness
Vascular plants	California, followed by Texas, Arizona, Oregon, and Florida (in that order) (Kartesz 1992)
Trees	Southeastern Coastal Plain and Piedmont (Currie and Paquin 1987), Northern Florida (Platt and Schwartz 1990)
Coniferous trees	Northwestern California–Southwestern Oregon (Wallace 1983)
Mollusks	Tennessee River system (Tennessee, Alabama) and Coosa River system (Alabama) (Stansbery 1970)
Tiger beetles	Southern Great Plains and West Gulf Coastal Plain (Texas) (Pearson and Cassola 1992)
Butterflies	Western Great Plains and Central Rocky Mountains (Colorado) (Pearson and Cassola 1992)
Fish	Cumberland Plateau in the Tennessee and Cumberland River Drainages, and secondarily, Ozark Plateau, Ouachita Mountains, and Atlantic Coastal Plain (McAllister *et al.* 1986)
Amphibians	Southern Appalachians and Piedmont, secondarily Southeastern Coastal Plain (Kiester 1971, Currie 1991)
Reptiles	Gulf Coastal Plain (Eastern Texas) and, secondarily, Southeastern Coastal Plain and Southern Arizona–New Mexico (Kiester 1971, Currie 1991)
Birds (Breeding)	Sierra Nevada, Southeastern Arizona–Southwestern New Mexico (Cook 1969)
Mammals	Sierra Nevada and, secondarily, Southern Cascades and Desert Southwest (Simpson 1964, Currie 1991)

1980), which is not surprising since geographic isolation of populations is known to foster speciation (Mayr 1970).

States with the highest numbers of endemic plants (Fig. 4.1; Gentry 1986) generally have the highest plant species richness (Table 4.1), except for Hawaii, which has the second highest number of endemics but the fourth *lowest* number of native plant species (Kartesz 1992). This is a predictable consequence of Hawaii's isolation. Few species were able to colonize the distant archipelago, but those that did evolved into a number of new species found nowhere else. Regions with high levels of endemism for terrestrial biota include the southern Appalachians, Apalachicola lowlands and Lake Wales Ridge of Florida, Ozark Mountains, Mediterranean-climate region of California, Klamath Mountains, and isolated ranges in the Intermountain West and Southwest (Gentry 1986). For freshwater fish, centers of

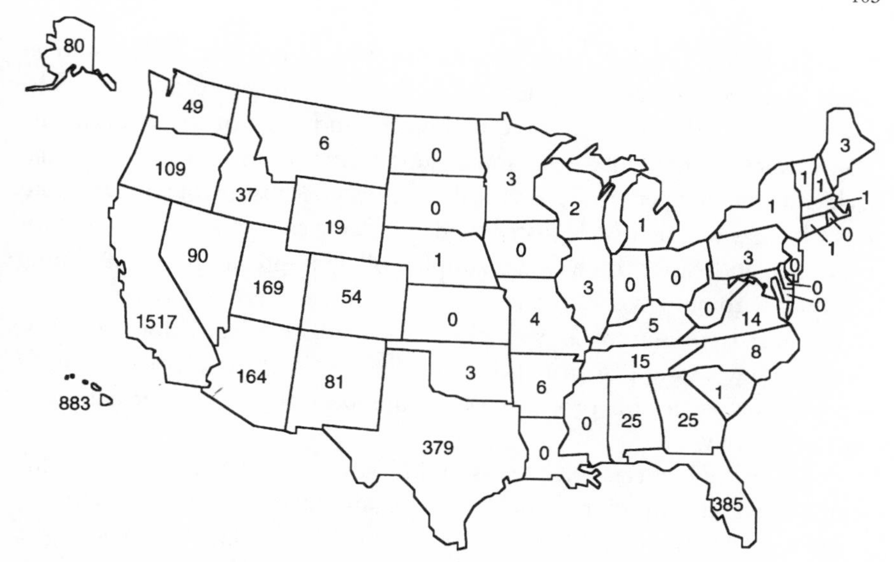

FIGURE 4.1 Number of plant taxa endemic to each state (from Gentry 1986). Note that some regions of relatively high endemism that overlap state boundaries (e.g., Southern Appalachians) are obscured when endemics are tallied by state. Used with permission of the author and Sinauer Associates.

endemism include northern California–southern Oregon, Tennessee and Cumberland rivers, Atlantic Coastal Plain, Ozark and Ouachita mountains, the Rio Grande–Rio Pecos junction, Bear Lake in Idaho and Utah, and Death Valley (McAllister *et al.* 1986). No other place in the world has as many endemic freshwater mussels as the Tennessee River system or as many endemic freshwater gastropods (snails) as the Coosa River drainage in Alabama (Palmer 1985). Each of these regions of high endemism has habitats with a history of isolation.

Because they are localized, endemics are often at high risk of extinction. A 1988 report from the Center for Plant Conservation, based on a survey of 130 botanists, identified 680 plant species in the United States and Puerto Rico that face a high probability of extinction by the year 2000 (cited in Holsinger and Gottlieb 1991). Not surprisingly, 75 percent of the species are from California, Florida, Hawaii, Texas, and Puerto Rico—all hot spots of endemism. Of the endemic snails in Alabama's Coosa River system, 62 species are apparently extinct (Palmer 1985).

Centers of species richness and endemism must be included in reserve networks. In the case of endemics, failure to protect the single site or small number of localities where a species occurs can lead to extinction. Such tragedies are frequent in the tropics. For example, an isolated ridge in

Ecuador called Centinela was found to contain at least 38 and possibly as many as 90 new species of plants, all but a few confined to that area. The forest on the ridge was destroyed by loggers, and most of these species are now extinct (Gentry 1986, Diamond 1990). Centers of species richness that lack endemics are perhaps less critical to protect because most species found in such areas will be widely distributed and many can get along fine in multiple-use landscapes. However, examples of widespread species declining abruptly with human disturbance or persecution are certainly known (Scott *et al.* 1991a). Representing all species in protected areas (see below) is a prudent strategy to ward against such losses, and areas of high species richness offer opportunities for protecting large numbers of species efficiently and cost-effectively.

Areas of high species richness should be recognized separately for different taxa and for each region. Otherwise, all our reserves would be in the tropics (or, for the United States, they might all be in California). Tundra is no less precious than tropical rainforest, although it contains far fewer species. Maintenance of biodiversity at global, continental, and national scales requires protection of the characteristic faunas, floras, habitats, and processes of each region. The tundra, for instance, may have low species richness, but most arctic plants are not found beyond the limit of tundra and northern boreal forest (Murray 1992) and would not be saved by temperate or tropical conservation efforts. Over time, faunas and floras have changed and will change again. We should let nature determine the rate and direction of these changes. The rate of change imposed by human civilization is faster than many species can tolerate.

Representation

We have introduced two major criteria for assessing the conservation value of sites: species richness (or diversity) and endemism. Other criteria commonly employed in conservation evaluations include naturalness, rarity, area (extent of habitat), threat of human interference, amenity value, educational value, scientific value, and representativeness (Margules and Usher 1981, Usher 1986). Each of these criteria is useful in various situations, but we will not discuss them here. Instead, we focus on representation as perhaps the most comprehensive of all conservation criteria. Representation is based on a simple but compelling idea: "a prerequisite for preserving maximum biological diversity in a given biological domain is to identify a reserve network which includes every possible species" (Margules *et al.* 1988). Representation is subtly different from the criterion of representativeness (Noss 1992b).

Under the latter concept, sites are sought that represent archetypal communities. Once represented as a specimen in the collection, a community type is considered protected. In contrast, representation, as we see it, means capturing the full spectrum of biological and environmental variation with the understanding that this variation is dynamic and not easily classified. Because biodiversity occurs at multiple levels of organization, conservation programs ideally should seek to represent all genotypes, species, ecosystems, and landscapes in protected areas. In practice, vegetation usually provides a good surrogate for the rest of biodiversity in the short term (Scott *et al.* 1993), so long as variation within classified types is recognized and sensitive and endemic species are given the extra attention they deserve.

Representation is one of the oldest goals of conservation. Early efforts to identify and preserve a broad spectrum of natural communities in North America were led by the Committee on the Preservation of Natural Conditions of the Ecological Society of America. In 1926, the Committee published the *Naturalist's Guide to the Americas* (Shelford 1926), a description of biomes and their remaining natural areas. This committee evolved into a separate organization, the Ecologists' Union, which became The Nature Conservancy in 1950. Another committee of the Ecological Society of America, the Committee for the Study of Plant and Animal Communities, assessed the adequacy of protection of ecosystems in three classes of nature sanctuaries. The highest class was fully protected areas, with virgin vegetation and being large enough to contain all the animal species in self-maintaining populations historically known to have occurred in the area (Shelford 1933, Kendeigh *et al.* 1950–51). By 1950, no first-class sanctuaries existed for deciduous forests, prairies, or many other biomes in North America. However, opportunities remained in some southern swamps, boreal forests, higher elevations in western mountains, desert, and tundra (Kendeigh *et al.* 1950–51). Most of those opportunities have since been erased by resource extraction. Today, if all ecosystems are to be represented in protected areas, the emphasis must be on restoration or "wilderness recovery" (Noss 1991d, 1992b).

Coarse and Fine Filters

Most strategies for representing ecosystems have had to be content with smaller natural areas that harbor less than complete biological communities (e.g., no large predators). Much of this effort has been led by The Nature Conservancy, by state natural areas programs, and by a professional society, The Natural Areas Association. The Nature Conservancy has called its community-level conservation strategy a "coarse filter" (Noss 1987b) and has estimated that 85–90 percent of species can be protected by conserving examples of natural communities, without having to inventory and manage each species individually. This interesting idea seems reasonable but has not been tested empirically. Testing it would be impossible without a thorough catalog of all species in the region of concern, including bacteria, soil invertebrates, and other cryptic forms. Such data are not rapidly forthcoming, as perhaps only one-third of species and their developmental stages in the United States have been described (Kosztarab 1986).

A potentially serious limitation of the coarse filter or any representation strategy based on vegetation is that natural communities are not stable; they change as species respond more or less independently to environmental gradients in space and time. For instance, when climate changes, species track shifting habitat conditions at different rates determined by their dispersal capacities and other aspects of autecology (Davis 1981, Graham 1986). The plant associations we describe and classify today are made up of species that may have been apart for a longer period of their evolutionary histories than they have been together. Perhaps the ideal way to represent biodiversity at the ecosystem level is to maintain the full array of physical habitats and environmental gradients in reserves, from the highest to the lowest elevations, the driest to the wettest sites, and across all types of substrates and topoclimates (Noss 1987b, 1992b; Hunter *et al.* 1988). Physical gradients must be unfragmented so that species can shift distributions in response to environmental change without encountering barriers to movement.

Despite problems, vegetation-based coarse filters are workable in the short term, pending dramatic climate change, because vegetation integrates underlying physical habitat variables. Practical advantages of a coarse filter include efficiency and cost-effectiveness (it is easier to deal with dozens or even hundreds of ecosystem types than thousands or millions of species) and the assumed ability to protect species we know nothing about and could not begin to inventory individually. Furthermore, a coarse filter can be applied at any level of classification hierarchy, including landscape types based on physical habitat gradients and patterns rather than on species composition (Noss 1987b).

Given our present state of knowledge, ecosystem approaches cannot be counted on to maintain all of biodiversity. Some species will always fall through the pores of a coarse filter. Hence, a fine filter of rare species inventory and protection planning is needed as a complement. The fine filter is also an application of the representation criterion and is usually focused on narrow endemics, wide-ranging animals, and other species not predictably associated with specific vegetation or habitat types. Representation of plant community types, if in isolation from one another, will do little to maintain viable populations of grizzly bears. The fine filter, focused on sensitive or localized species, can be considered a safety net and is often employed through the provisions of the Endangered Species Act or state equivalents.

No conservationist in the United States could help but notice the limitations of the coarse and fine filter approaches as currently employed. The fine filter has bogged down in sheer detail. There are too many rare and endangered species to deal with individually, at least given current and prospective agency budgets and private conservation dollars. On the other hand, the coarse filter has run into the age-old problem of classification. No accepted classification of communities or ecosystems exists for the United States. As noted earlier, ecosystems are more difficult to classify than species. Compared with species, ecosystems are short-lived, ambiguous, artificial, and perhaps intangible (Noss 1991a, Scott *et al.* 1991a). At which level of classification hierarchy should natural communities be recognized, inventoried, and protected? In California, for example, should we recognize 52 wildlife habitat types (Mayer and Laudenslayer 1988), 375 natural communities (Holland 1987), or something in between?

Ecosystems at any level of classification can become endangered (Noss *et al.* 1994). Therefore, we recommend construction of hierarchical (multi-level) community classifications with cross-walks to other classifications. The classification used in a particular case will depend on which level of hierarchy works best for the spatial scale or level of detail desired. In some situations, we will be interested in the status of biomes (tundra, grassland, deciduous forest), whereas in other cases we need detailed accounts of plant associations such as the *Artemisia tridentata–Atriplex canescens–Sarcobatus vermiculatus/(Oryzopsis hymenoides)* community on Great Basin dunes in Oregon. The Nature Conservancy is developing a hierarchical classification of vegetation consistent with a global UNESCO framework and has completed a preliminary series-level (i.e., dominant plant species) classification for the western United States (Bourgeron and Engelking 1992).

Assessing Representation

Representation of species and ecosystem types in reserves can be assessed systematically, removing some of the biases of traditional subjective approaches. The Australians have been the intellectual leaders in this area (for example, see Kirkpatrick 1983, Margules *et al.* 1988, Margules 1989, Pressey and Nicholls 1989, and Bedward *et al.* 1992). Iterative computer algorithms (programs that repeat a sequence of analytic steps many times) have been developed that define the smallest number of sites needed to represent each species once, twice, or however many times desired. The same algorithms can be applied to classified community types.

The more biodiversity one wants to protect, the more land area is needed. In Idaho, the Gap Analysis project (discussed later in this chapter) determined that 4.6 percent of the state's area (represented by a hexagonal grid) was necessary to represent all vertebrate species at least once; 7.7 percent was needed to represent all endangered, threatened, and candidate species of vertebrates and plants (Scott *et al.* in press). This proportion of the landscape is typical of what many regions already have set aside in reserves, though existing areas usually fail to meet representation objectives because their locations were not biologically determined. In a study of wetlands in the Macleay Valley floodplain of Australia (Margules *et al.* 1988), 4.6 percent of the total number of wetlands was required to represent each plant species at least once, but these sites constituted 44.9 percent of the total wetland area.

Studies that have considered a greater number of taxa have found that much more area is needed to represent all species in reserves. To represent all bird, mammal, reptile, and plant species on 18 islands in the Gulf of California, 99.7 percent of the total area must be protected. For chaparral canyons around San Diego, 62.5 percent of the available area is needed to represent each species of bird, mammal, and plant at least once (Ryti 1992). These figures are high in part because species distributions in these particular regions are mostly narrow and localized. Furthermore, representing a species in just one site of undetermined size is not enough. Especially for large vertebrates, large or interconnected blocks of land are needed to assure population viability (see Chapter 5).

Representing community types rather than just species would do more to capture taxa not currently well inventoried, such as invertebrates. This is a compelling argument in favor of a coarse filter. Vegetation may be a good coarse filter for little-known taxa. In Idaho, only 8 percent of the land area is needed to represent each of 119 vegetation types (Scott *et al.* in press). In the Australian study cited above (Margules *et al.* 1988), 75.3 percent of the total area was required to represent all wetland types *and* all plant species at least once. And again, redundancy (representation more than once) should be built in to ward against catastrophe and to enhance the probability of metapopulation persistence (Gilpin and Hanski 1991).

There seems to be no getting around the fact that a significant amount of habitat must be protected in order to represent biodiversity adequately. As concluded by Margules *et al.* (1988), "[T]he belief that biological diversity is 'reasonably secure' or 'as well taken care of as possible' with the dedication of one or a few well chosen reserves in an ecological domain is unfounded. The reality is that a very large number of reserves seems to be necessary to secure biological diversity." Reserve size, multiplicity, and other considerations for determining total network area will be discussed in more detail in Chapter 5.

When assessing the relative representation of species or ecosystem types in reserves, scale considerations are paramount. State or other political boundaries are inadequate for evaluating representation because biogeography does not conform to artificial boundaries. We recommend that representation of species be assessed within the context of the entire historical and present distributional ranges of each taxon. However, representation need not be assessed for every known species individually. Rather, the fine filter should be concentrated on rare and endemic species, those known to have declined due to human activity, or those that are at risk of imminent decline. Species in these categories that are more distinct taxonomically (for example, the only member of a genus, family, or order) might be weighted higher than species with many close relatives (Vane-Wright *et al.* 1991). Giving distinct taxa extra attention will help maintain total genetic diversity, as such species can be assumed to have more distinct genomes.

Representation is probably best handled by focusing on vegetation and habitat types, environmental gradients, hot spots of species richness, and centers of endemism. Centers of taxonomic distinctness should also be recognized where possible (we know much more about the systematics of some groups than others). All of these entities are best assessed at regional and global scales. A nested hierarchy of regions (i.e., regions within regions) may be optimal for planning, proceeding top-down in conservation evaluations from biosphere–continent–ecoregion–landscape–site. That is, representation of ecosystems is evaluated at each of these scales, and hot spots of richness and endemism are also identified at each scale. The richest sites globally, nationally, and regionally should be recognized as most critical for protection, but no ecosystem types should be left unrepresented.

The United States has no comprehensive program for representing ecosystems in protected areas. Recent estimates of the fraction of major terrestrial ecosystem types not protected in the United States range from 21 to 53 percent (Shen 1987). This level of representation is surely inadequate to meet biological conservation goals. Yet the question of what is adequate representation is a quagmire. Science cannot tell us precisely how many times or in what sized reserves each species or ecosystem type must be represented to be viable. Detailed, region-specific studies may come close to answering these questions, but conclusions may not transfer well between regions. Perhaps the optimal approach is to set priorities on the basis of irreplaceability (sites that, if lost, cannot be compensated for elsewhere), hot spots of richness and endemism, poorest current representation, and urgency of threat. As top-priority sites are protected, work down the list until as much is represented and as many times as possible—the more, the better. Meanwhile, biodiversity monitoring (tracking the status of species and community types) can help inform us when we have protected enough land and are managing it properly.

Tools for Inventory and Evaluation

We have emphasized a top-down hierarchical approach to conservation evaluation. There is no single spatial scale at which all conservation decisions should be made, although the biosphere is the broadest context within which we decide what to protect. The status of species, ecosystems, and other elements of biodiversity at a global scale must take precedence over their status at local scales. Otherwise, species that are rare locally but common globally may get more attention than they require for persistence; or more seriously, species common locally but rare globally may not get enough attention to assure their survival. We should conduct conservation inventories and evaluations at several scales, recognizing that the broader the scale (the larger the area considered), the lower the resolution (the less detail that can be observed). Global inventories can tell us about "megadiversity countries" (Mittermeier 1988, McNeely *et al.* 1990) and other big hot spots, and identify places where deforestation and other threats to biodiversity are most rampant. Continental and national inventories provide more detail on the status of elements of biodiversity, accompanied by higher resolution maps. Regional evaluations provide more detail still. Biogeographically defined regions are a convenient scale for many conservation planning exercises (Noss 1983, 1987a, 1992b; see Chapter 5). Finally, the landscape and local scales (hundreds to hundreds of thousands of acres) are where specific habitat requirements and population viability can be assessed (except for large carnivores and other wide-ranging animals, which must be evaluated over tens of millions of acres), and protective measures and other management actions can be implemented.

Here we discuss inventory and evaluation at national, regional, and landscape scales. Two major sources of information are emphasized below: the natural heritage programs (conservation data centers) established in each of the 50 states, Canada, and many Latin American countries by The Nature Conservancy, and the Gap Analysis project of the United States Fish and Wildlife Service. The heritage programs are ongoing inventories with continually improving coverage of distributions of rare species and natural communities. The Gap Analysis project has been initiated in over half the states and is scheduled for completion by the start of the next century.

Natural Heritage Programs

A breakthrough in biodiversity inventory occurred with the establishing of state natural heritage programs by The Nature Conservancy, beginning in 1974. These programs, now mostly incorporated within state agencies, usually represent the best single source of information on biodiversity in each state, particularly on rare species and communities (Jenkins 1985, 1988; Noss 1987b). Previous efforts to determine conservation priorities on the basis of biological data were hampered by inefficiency. Sites could not be compared objectively due to an "apples and oranges" problem: biological data were mixed with information on recreational potential, scenery, and other values. Because data were not collected systematically, coverage was uneven and incomplete. Sites known to contain rare species were difficult to rank because the relative rarity of species, something that could only be determined by a comprehensive view, was usually unknown. If one site contains five rare species and another only two, is the first site more important? Not necessarily, if the species at the second site are more imperiled globally or are present in populations more likely to be viable or defensible in the long term.

The Nature Conservancy, led by its scientific director, Bob Jenkins, established the heritage programs to correct these deficiencies. For the first time, sites could be compared on the basis of objective and standardized information about the relative rarity of species at state and global scales, the quality of their occurrences at sites, and other parameters. In the heritage system, sites are not the functional units. Rather, the system is based on "elements of diversity," mostly rare species and natural community types, but also wading bird rookeries, geological features, and other attributes of conservation interest. These elements are ranked, mapped, and tracked throughout their ranges. Sites can then be compared on the basis of their attributes and their contribution to biodiversity conservation at state, national, and global scales. As described by Jenkins (1985):

> By listing, classifying, and characterizing the *elements* rather than the natural areas where they occur, the inventories can determine relative endangerment, track down the finest occurrences on the landscape, and identify conservation priorities in the state. The system makes it possible for heritage scientists to compile a list of places that support "the last of the least, and the best of the rest"—those elements that are rare, endangered, or otherwise sensitive to destruction.

The significance of a species or community type in the heritage system is denoted by its priority rank, with each element assigned a rank at both global and state scales. Ranks are frequently updated on the basis of new information. States differ somewhat in their interpretation of global ranking criteria; when applied quantitatively (e.g., California Natural Diversity Data Base, unpublished), the criteria are:

G1 = Less than 6 viable element occurrences (EOs) or less than 1000 individuals or less than 2000 acres.

G2 = 6–20 EOs or 1000–3000 individuals or 2000–10,000 acres.

G3 = 21–100 EOs or 3000–10,000 individuals or 10,000–50,000 acres.

G4 = Apparently secure; this rank is clearly lower than G3 but factors exist to cause some concern; that is, there is some threat, or a somewhat narrow habitat.

G5 = Population demonstrably secure to ineradicable due to being commonly found in the world.

Other ranks may be assigned for historical occurrence (GH), presumed extinct (GX), questionable taxonomic status (G#Q), and other situations (Master 1991). Thus, elements (species or community types) are ranked primarily according to population size, number of occurrences (populations or locations), or area covered. Furthermore, The Nature Conservancy's "Element Global Ranking Record" often contains information on trends, threat, and fragility (Master 1991) that may influence rankings. Elements ranked G3 or higher (or T3 and higher, the "T" indicating a subspecies or other taxonomic category) are of greatest conservation significance. The list of G3 and above elements typically reflects current biological opinion better than do the lists of endangered and threatened species kept by state or federal agencies. State ranks (e.g., S1, S2) are usually based on identical criteria except that the status of each element is considered separately within each state. Global ranks are more significant biologically because they correspond to the total range of a species, but state agencies often concentrate on elements ranked high within their state's boundaries. Elements ranked high at a state level but lower globally are often peripheral or disjunct populations that may be of evolutionary significance. We can expect such populations to diverge genetically over time. However, concentration on peripherals at the expense of the characteristic regional biota can lead to unexpected losses of diversity (Noss and Harris 1986, Noss 1992a).

The heritage programs are convenient places to shop for biodiversity information. Heritage program databases consist of computer files, manual files (including relevant technical literature on elements and sites), maps (with all 7.5-minute USGS quandrangle maps for a state with coded element occurrences), and often geographic information systems (GIS). For a nominal fee, agencies or individuals can receive a printout with requested information for any element or geographical area (e.g., site, watershed, physiographic region, ecoregion) of concern. If the heritage program has a GIS or other computerized geographic database, maps of any area with element occurrences can also be requested.

Gap Analysis

The most significant national effort to supplement heritage-style inventories by proactive, ecosystem-level approaches is the Gap Analysis project of the U.S. Fish and Wildlife Service. Gap Analyses are being conducted state by state, carried out through the Cooperative Research Units and cooperating state and federal agencies and universities (Scott *et al.* 1991a, 1991b, 1993). Gap Analysis is basically an assessment of representation of vegetation types and species in protected areas, using satellite imagery, ancillary data on vegetation, wildlife–habitat association models, and GIS mapping. Gaps in the representation of species, ecosystems, and hot spots of species richness are selected as priorities for protection. Thus, Gap Analysis is a coarse-filter approach to conservation. Analyses to determine priority sites for efficient representation are using iterative computer algorithms (Scott *et al.* in press), such as those developed by Australians (e.g., Margules *et al.* 1988).

Conducted in Hawaii, the first Gap Analysis project in the United States focused on endangered forest birds (Scott *et al.* 1986). Range maps were overlaid manually to produce maps of endangered species richness. The visually striking result (Fig. 4.2) showed that existing preserves almost completely missed the hot spots of endangered species richness. Several of the most critical areas found in this study have since been protected by The Nature Conservancy and state and federal agencies. This project demonstrated a key principle of pragmatic conservation: a visually compelling map, backed up by data, can do more to stimulate protective action than any number of words. Simply knowing where hot spots are is not enough to convince decision makers to designate reserves. You have to demonstrate your case convincingly, which usually means visually.

The second Gap Analysis, for Idaho, was the first project based on vegetation maps produced from satellite imagery and using GIS analysis. The Idaho methodology has since been refined into a relatively standardized format for projects being carried out throughout the United States (Scott *et al.* 1993). The major elements of this process are summarized below.

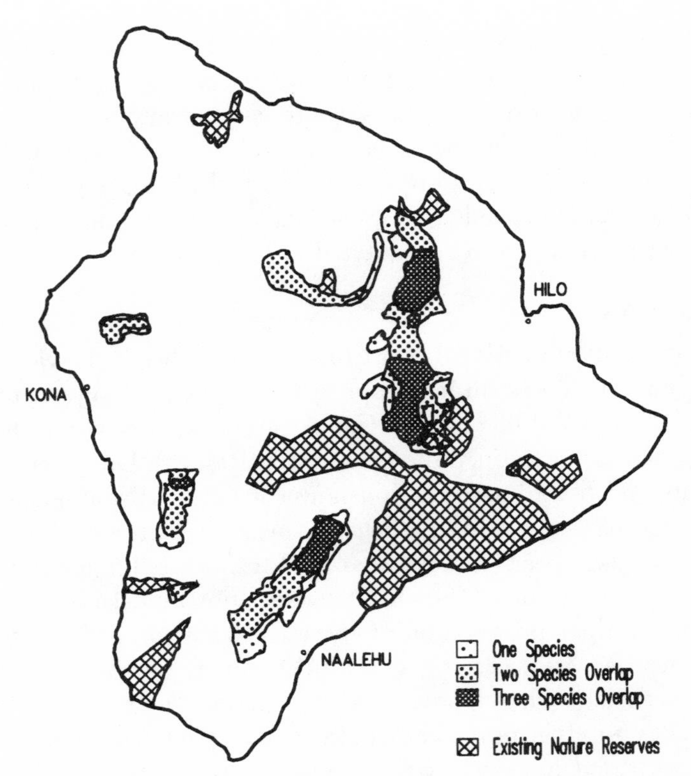

FIGURE 4.2 Distribution of endangered Hawaiian finches in relation to existing nature reserves on the island of Hawaii in 1982 (from Scott *et al.* 1993, as adapted from Kepler and Scott 1985). Note that "hot spots" of high endangered species richness were missed by existing reserves. Since these data became available some of the hot spots have been protected in new reserves. Used with permission.

Vegetation mapping. The vegetation map is the fundamental data layer in Gap Analysis. All inferences about species richness and representation follow from analysis of the vegetation map. Unlike previous efforts to analyze representation in the United States (e.g., Crumpacker *et al.* 1988), Gap Analysis is based not on potential natural vegetation but on *actual* vegetation, including human-created habitats. Current satellite imagery, chiefly LANDSAT Thematic Mapper (TM) scenes, is the basic data source. Vegetation polygons [areas of relatively homogeneous cover of at least 100 ha (247 acres)] are delineated and labeled through a combination of visual photointerpretation of satellite photographic images, digital classification of satellite data, and reference to aerial photographs and existing vegetation maps (Scott *et al.* 1993). Ground-level field verification of sample sites is used to proof the maps. The vegetation map corresponds to a classification of plant communities to the series (dominant plant species) level, and is being standardized through cooperation with The Nature Conservancy (Bourgeron and Engelking 1992). The vegetation map and all subsequent Gap Analysis maps are produced at scales from 1:100,000 to 1:500,000; thus, small landscape features are not portrayed.

Species range maps. We are all familiar with range maps in field guides. We know from using these maps that much of the area within depicted range boundaries is not inhabited by the species in question because the habitat is not suitable. Even for a widespread species, distribution is usually patchy within the range. In addition, much of what was formerly suitable has been converted to farmland, conifer plantations, urban areas, and other unacceptable habitats. Gap Analysis refines range maps for species by identifying polygons of suitable vegetation within overall range boundaries. Species are associated with particular vegetation types on the basis of simple habitat relationship models. A GIS overlay of the species' known distribution (from range maps, museum specimens, etc.) and the vegetation-derived map of suitable habitat provide a map of predicted current distribution. Particularly for poorly surveyed regions, the GIS-based predicted distribution may be more accurate than empirical data. In Idaho, for example, the GIS map predicted sharp-tailed grouse to occur in parts of the state not known to be occupied by this species. Later field surveys have confirmed the GIS prediction (Scott *et al.* 1993). In Gap Analysis, species distribution mapping is confined to taxa for which distributions are fairly well known; in most states, this means vertebrates, and sometimes butterflies and trees.

Species richness maps. The power of a GIS is especially useful when one wishes to overlay species distribution maps to produce maps of species richness. Such maps can be constructed to show the number of species expected for each vegetation polygon. Or a grid can be overlaid across polygons to show expected species richness across environmental gradients. We believe that the latter approach is more realistic (Fig. 4.3). Hot spots of high species richness usually show up clearly in such analyses. Species richness maps can be produced for any group of interest: reptiles, mammals, arid-land rodents, diurnal raptors, game species, or any other combination of species for which individual range maps have been prepared. Typically, hot spots are areas of high habitat diversity, such as steep elevational gradients or soil mosaics. Species-rich areas are important because they represent opportunities to protect large numbers of species

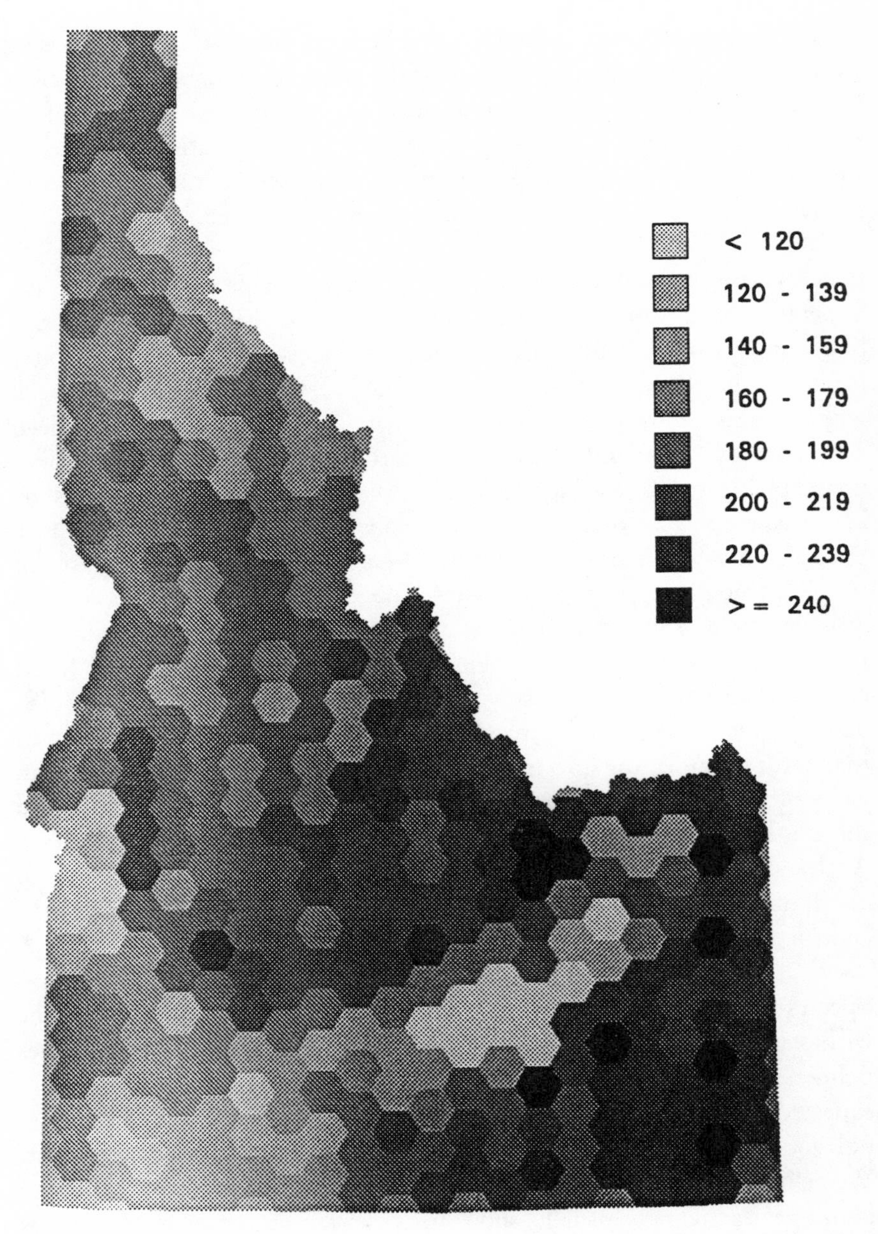

FIGURE 4.3 Number of vertebrate species per 635-km² hexagon in Idaho (from Scott *et al.* in press). Landscapes with high species richness in each region of the state make efficient sites for location of reserves. Used with permission.

efficiently. Introduced and weedy species are usually screened out before hot spots are identified. Otherwise, decisions would be biased by disturbed habitats brimming with weedy species.

Aquatic, wetland, and rare species. Some kinds of habitats and species do not lend themselves to the kind of mapping just described. Because the minimum mapping unit in Gap Analysis projects is usually 100 ha (the smallest area discernible for the scale of map produced), small, patchy habitats do not show up. Some patchy habitats, for example vernal pools in a matrix of coastal sage scrub in California, can be included as attributes in polygons of the vegetation type in which they commonly occur. The label for these vegetation types would simply note that vernal pools are a common inclusion. For streams, riparian areas, lakes, and associated species, using digital line graphs for mapping worked well in Idaho (Scott *et al.* 1993). Many states have National Wetland Inventory maps, which can be used to map wetlands and wetland-associated species. Rare species, whether terrestrial or aquatic, are generally localized and cannot be mapped on the basis of habitat relationships; they require a fine filter. Data from natural heritage programs on occurrences of rare plants and animals (say, those ranked G3 and higher; see above) can be entered into the GIS as point locations. Clusters or concentrations of points show hot spots of rare species richness. Centers of endemism can be delineated by mapping point localities of rare species that meet the definition of narrow endemics (e.g., total range less than 50,000 km^2; Terborgh and Winter 1983, Gentry 1986).

Land ownership and management status. A major concern is how habitats of high conservation value are being managed. If biodiversity hot spots are already located within national parks or other reserves, we may have some confidence that they are being protected (albeit most reserves were not established to maintain biodiversity and many are managed poorly). If, on the other hand, our biodiversity hot spots are on multiple-use public lands or private lands, we have greater cause for concern. Each private or public land parcel is managed somewhat differently, depending on the individual land manager, access, budget, political or economic pressure for exploitation, and other factors. However, for broad-scale inventories such as Gap Analysis, it is helpful to recognize classes of protection or management status. Currently, four classes are recognized (see Scott *et al.* 1993): (1) Management Status 1: most national parks, Nature Conservancy and Audubon Society preserves, some wilderness areas and national wildlife refuges, research natural areas, and other areas managed for their natural values; (2) Management Status 2: most wilderness areas, national wildlife refuges,

BLM areas of critical environmental concern, and other areas generally managed for natural values, but receiving some uses that degrade natural qualities; (3) Management Status 3: most undesignated public lands, such as national forests, BLM lands, state parks, and other lands with some legal mandates for conservation but many potentially damaging uses; (4) Management Status 4: private or public lands without legal mandates to protect natural qualities and managed primarily for intensive human uses. A management status overlay for each state Gap Analysis places each land parcel in one of these classes.

Finding the gaps. The next step is the Gap Analysis per se. A comparison of maps showing biodiversity hot spots and other priority sites with the management status overlay shows gaps in protection (e.g., Fig. 4.2). Unprotected and underrepresented vegetation types and hot spots are areas that warrant immediate conservation action.

Limitations of Heritage Programs and Gap Analysis

We have described two major sources of biodiversity information for the United States. These programs can provide agencies, organizations, corporations, and private citizens with a tremendous amount of high-quality information on which to base conservation or development decisions. However, both the heritage programs and Gap Analysis have a number of limitations that must be recognized by users of the data. Otherwise, misinterpretations may occur.

Heritage programs. Perhaps the most serious limitation of the heritage approach is the problem of false negatives or "white holes" in the database. Often, absence of element occurrences for an area does not mean that nothing significant occurs there. It may only mean that no one has looked. Thus, the database has a strong bias toward well-surveyed sites, such as those near a major university, field station, or road. Sites that have been surveyed intensively without finding anything do not show up in the database. The heritage programs are designed to work through successive approximation (Jenkins 1985), that is, they are continually adding and refining records. Eventually, biologists will have surveyed a state or other region thoroughly enough so that all white holes should disappear (though even then, new populations of rare species may become established naturally and all populations will fluctuate). Until that time, and especially for large western states with limited field surveys, the unevenness of coverage must be taken into account when using heritage databases to make conservation decisions.

Another drawback of the heritage system is that it is based on mappable point locations of rare species and community types. Thus, it emphasizes plant populations, small animals, and narrowly defined natural communities that are easy to map. Animal species with a low population density or large individual home ranges are not tracked well by the system. Some of our most imperiled taxa—grizzly bears, wolves, the Florida panther—would be missed by an evaluation based solely on heritage data. Because large carnivores may play keystone roles in natural ecosystems (Terborgh 1988) and are a critical test of society's commitment to conservation, their absence is not trivial. Furthermore, much of biodiversity occurs as patterns and processes at scales beyond the individual plant community, as described earlier. Landscape gradients represent a higher order of biodiversity that is fundamental to the survival of animals that require heterogeneous environments (Noss 1987b). An *Ambystoma* salamander, for example, needs not only forest with large logs to hide under, but also ponds for breeding and larval development, and uninterrupted movement routes in between. Heritage data must be augmented by spatial databases to capture these broader expressions of biodiversity.

A final problem with heritage data is the heavy emphasis on rarity. Although rare species are often most vulnerable to extinction, two species of equivalent rarity are not necessarily equally endangered. Many life history factors must be considered, among them vulnerability to different kinds of human disturbances. Extent of past decline is often a good indicator of vulnerability and ecosystem-level effects (Noss *et al.* 1994). Major declines of keystone species, such as beaver, bison, prairie dogs, gopher tortoises, wolves, or perennial bunchgrasses, are more significant ecologically than the rarity of an endemic plant, which perhaps has not changed much in distribution since the Pleistocene. Some rare species are not particularly vulnerable to further decline, but others are.

The emphasis in the heritage system on point localities and rarity has led to protection of small natural areas that may not be viable in the long term (Noss and Harris 1986). Therefore, many conservation biologists are calling for landscape-level approaches that strive to protect ecosystems and whole assemblages of species before they become critically rare (Noss 1983, 1987a; Harris 1984; Scott *et al.* 1987, 1991a, b, 1993).

Gap Analysis. Gap Analysis also has limitations as a source of biodiversity information. The major limitation is simply its coarse level of resolution. To provide a national or regional picture of the status of biodiversity cost-effectively, maps cannot contain too much detail. Thus, habitats smaller than the minimum mapping unit (100–200 ha) are not shown. This is problematic when fine-grained habitat patchiness is an important element of an area's biodiversity; but as noted above, such patchiness can at least be acknowledged by attaching attributes to vegetation polygon descriptions.

Other limitations of Gap Analysis include (1) seral stages of forests are not discriminated (except early successional stages versus closed-canopy stands, thus equating tree plantations with old growth); (2) ecotones between vegetation types are missed; (3) species distribution maps are predictions only (not rigorously field-verified); (4) habitat quality is not indicated (Scott *et al.* 1993). These limitations are not trivial.

Probably the best way to view Gap Analysis is as a first-cut, coarse-filter assessment of biodiversity representation nationwide. Areas that stand out as hot spots on the basis of Gap Analysis should then be subjected to more detailed study. Conservation planners should not interpret Gap Analysis results as suggesting that conservation will be easy and will require protecting relatively little land. For example, an analysis of the Idaho Gap data shows that 96 percent of the vertebrate species native to the state can be "captured" in just seven (1.8 percent) of 389 equal-area hexagons (each 635 km^2 in size) that cover the state. Does this mean that populations of all vertebrates in Idaho can be maintained by protecting only 4445 square kilometers (1.1 million acres)? Of course not. As will be discussed in Chapter 5, maintaining viable populations of vertebrates (as measured by the needs of the most demanding, in this case, the grizzly bear) requires protection of tens of millions of acres, well connected across the Northern Rocky Mountain region. In any case, a comprehensive biodiversity analysis of any region will require a plurality of databases and evaluation criteria. There are no easy solutions to a problem as complex as conserving biodiversity.

A Case Study: The Oregon Coast Range

In real-world conservation, we never have enough data to make fully informed decisions about conservation priorities. One must work with the best available information and with whatever technology is affordable. No single database will answer all important questions about protection needs. Therefore, a pluralistic approach using a variety of databases, maps, and evaluation procedures is essential. One of us recently completed a preliminary conservation plan for the Oregon Coast Range (Noss 1992c, 1993a). The project was funded by grants from private foundations to a regional conservation group, the Coast Range Association. The project budget (about $20,000) was a fraction of what an agency might spend planning a single timber sale. Yet, this study provides an example of applying diverse databases to a complex planning problem at a regional scale.

Goals

The Coast Range project is a case study for The Wildlands Project, a cooperative effort of conservation biologists and environmental activists to restore biodiversity and wilderness in North America (Foreman *et al.* 1992, Noss 1992b, Mann and Plummer 1993). The Wildlands Project is long range (planning over decades and centuries) and highly ambitious. The Oregon Coast Range plan was concerned with implementing the goals of The Wildlands Project at a regional scale and serving as a model for development of similar plans elsewhere. The ecological goals of the project are those introduced in the preceding chapter as being requisite for a national strategy for biodiversity: to represent all ecosystems, maintain viable populations, maintain ecological and evolutionary processes, and be adaptable to change. These goals may take decades to meet and will require reducing human population and resource consumption regionally and globally. The process for fulfilling these goals is largely one of establishing a series of regional reserve networks that include zones with varying types and intensities of compatible human use.

Study Area

The Oregon Coast Range Bioregion (Physiographic Province) encompasses 2.2 million hectares (5.3 million acres) and is topographically diverse, with elevations ranging from sea level to about 1225 m (4020 ft). Major habitats include sea stacks, sandy beaches, rocky coastal cliffs, coastal headlands, tide pools, mud flats, salt marshes and estuaries, streams and rivers of various sizes, grass balds, and many types of forest. A little over half of the Coast Range is in federal or state ownership. The outstanding vegetational feature of the Coast Range is its forests, which are among the most productive in the world and are distinguished by conifer dominance and large trees and amounts of woody debris.

The Coast Range has been heavily altered by human activities, including logging, road building, agriculture, and development. Some ecosystem types are virtually gone. For example, 96 percent of the coastal temperate rainforests in Oregon have been logged (Kellogg 1992). Currently, 30 animal taxa and 34 plant taxa in the Coast Range are listed by the Oregon Natural Heritage Data Base (1989) as endangered throughout their range, threatened throughout their range, endangered or threatened in Oregon but more common or stable elsewhere, of concern in Oregon, or limited in abundance throughout their range. A few of these rare species are federally listed as threatened or endangered; many more are federal candidates for listing or are state-listed and face a moderate to high probability of extinction if landscape degradation continues. The California condor, grizzly bear, gray wolf, southern sea otter, hoary elfin butterfly, North Pacific plantain, and possibly the Pacific fisher and wolverine (last reported in the early 1980s) are regionally extinct (Oregon Natural Heritage Data Base 1989, Noss 1992c).

Data and Mapping

The basic task in this study was to distinguish areas of critical ecological value throughout the Coast Range by mapping concentrations of rare and vulnerable species, significant areas of old growth and other late-successional forests, outstanding examples of natural communities of all types, important watersheds for native salmonids and other aquatic life, large relatively unfragmented landscapes, and linkages between biologically significant areas. These features were mapped at a scale of 1:126,720 (1/2 inch = 1 mile) on nine base maps. Hot spots of biodiversity located by this mapping process were recognized as potential reserves. All mapping in this study was done manually, as the small budget did not allow for GIS analysis. Seven major data sets were overlaid on base maps using mylar sheets, or visually compared, to create maps showing biological significance of landscapes.

Heritage program data. An important advantage in the Coast Range study was the availability of relatively complete heritage data. Compared with most other regions, the Oregon Coast Range is well known biologically. Data on element occurrences (EOs) for the region were obtained in the form of a computer printout from the Oregon Natural Heritage Program. The printout contained 656 records of species and natural communities ranked G3/T3 or higher at a global scale or S2 or higher at a state scale, with information on location, mapping precision, survey site name, date last observed, managed area name (if applicable), and other data for each EO. Occurrences not confirmed since 1970 were not mapped, nor were those of dubious biological value (for instance, sightings of single peregrine falcons).

Spotted owl locations. The heritage program database did not include records for the northern spotted owl, a species federally listed as threatened. The Oregon Department of Fish and Wildlife (ODFW) provided maps of all recent records of northern spotted owls in the Coast Range. These data included 1990–92 records from Bureau of Land Management (BLM), Siuslaw National Forest, and state lands. Some occurrences were on private lands next to state or federal lands. Spotted owl locations were transcribed onto overlays directly from ODFW maps.

Marbled murrelet locations. The marbled murrelet is another bird recently listed as federally threatened. The heritage program printout contained marbled murrelet data through 1989, most of which was collected on the Siuslaw National Forest. Since 1989, the most extensive murrelet surveys have been conducted by ODFW on state lands in 1992, not yet in the heritage program database. Printouts were obtained from ODFW of all records from this database, which included township–range–section and information on type of detection. Records were mapped only when they contained information suggesting that the site was occupied by a breeding pair.

Gang of Four data. The Coast Range project did not have access to Gap Analysis data, as the Oregon Gap Analysis was still in its early stages, so ecosystem representation could not be fully assessed. Gap Analysis data will later be used to refine proposed reserve locations and boundaries. Map data did exist for old-growth forests. Old growth is a major conservation issue in the Coast Range, as elsewhere in the Northwest. The most complete maps of old-growth and late-successional forests in the Coast Range are those produced by the congressionally appointed Scientific Panel on Late-Successional Forest Ecosystems, or the "Gang of Four" (Johnson *et al.* 1991). The Gang of Four restricted their analysis to federal lands in the Pacific Northwest, where most of the remaining old growth occurs. For the Coast Range, those lands comprise the Siuslaw National Forest and the four BLM districts. Data consisted of 1:126,720-scale mylar overlays showing late-successional/old-growth forests in three categories defined by Johnson *et al.* (1991): (1) most ecologically significant; (2) ecologically significant; and (3) owl additions, areas added to meet the requirements for spotted owls. Each Gang of Four map was overlaid for evaluation.

Watershed reserves. The Oregon Chapter of the American Fisheries Society (AFS) provided maps and descriptions of candidate watershed reserves and priority restoration areas. The AFS used several criteria for selecting watersheds: (1) relatively unaltered and characteristic reference areas; (2) genetic refuges, such as areas with little history of hatchery stocking; (3) aquatic species richness; (4) ecological function, for example streams or springs important for maintaining water quality or low temperatures throughout a basin; (5) connecting corridors, such as streams or watersheds that link protected areas or habitats necessary to support different life history stages; (6) areas highly sensitive to disturbance or human uses, such as geologically unstable watersheds; and (7) scientific value. Boundaries of draft watershed reserves were transcribed directly onto overlays.

Wildlife corridors. A local conservationist, Catharine Koehn, mapped probable wildlife movement corridors through the south-central Coast Range on the basis of road mortality data collected by the Oregon Department of Transportation and county highway employees. Roadkills were often distributed linearly across a landscape, so were used to infer wildlife movement routes. Most mortality data were for black-tailed deer but also included elk, black bear, bobcat, and other species. The corridors delineated in this preliminary mapping follow ridgelines for most of their length.

Undeveloped landscapes. Roadless areas, areas of low road density, and other relatively undeveloped landscapes were evident on base maps. When not located within hot spots defined by biological and ecological criteria, these areas were assessed for possible inclusion within boundaries of proposed core reserves or buffer zones.

Many wild areas will not show up in conservation evaluations of the type described in this chapter, often because of difficult access and a resulting lack of biological surveys. However, roadless areas and other undeveloped landscapes can be expected to contain reclusive species sensitive to human activity and are thus of high value for conservation (Noss 1991b, 1991d). Also, many such areas are inherently sensitive to human disturbance (for example, with steep slopes or unstable soils) and need to be protected from development to maintain healthy watersheds and fisheries. Wild areas are also inherently or spiritually important. They may help inspire a sense of humility and respect for nature (Noss 1991b, 1991d), something desperately needed in our society.

Reserve Mapping

A top overlay on each base map was prepared to show potential reserve and buffer zone boundaries. Three categories of conservation areas were considered: (1) Class I Reserves, areas of highest priority for protection and restoration; (2) Class II Reserves, which might accommodate a greater variety of human uses; and (3) Multiple-Use Buffer Zones, areas permitting a still broader range of uses. Potential reserve locations were delineated mainly on the basis of the following features assessed from biological/ecological data layers: (1) clusters or concentrations of EOs, spotted owl locations, and/or marbled murrelet locations; (2) presence of significant old-growth or other late-successional forest; and (3) presence of significant watersheds for aquatic biodiversity. Sites found to have high spatial overlap or concentration of these criteria were considered potential Class I (high priority) Reserves. Class II Reserves usually enclosed Class I Reserves. Their locations were based on the same criteria, but they usually contain a lower density of rare species occurrences, old growth, or other conservation indicators. Potential Multiple-Use Buffer Zones often contain EOs, spotted owl locations, and other features of importance, but in a less-concentrated, more-dispersed pattern. Many buffer zones were designed to provide connectivity between high-priority sites within the region, or to link the Coast Range to other regions. Such linkages are especially important for promoting reintroduction and population viability of carnivores.

Site Ranking

The mapping process located sites throughout the Coast Range that can be considered hot spots of biodiversity. A more detailed site ranking process was conducted for potential Class I Reserves. Although all these sites are considered of high priority, some are richer in biodiversity than others, more threatened, or otherwise demand more urgent protection. Ranking on a number of discrete criteria allows for comparisons among sites and for weighting of certain criteria where desired.

Sites were ranked for each of five criteria on a scale of 0 to 5, from least to most significant. Numerical scores were assigned on the basis of judgments about the relative value of sites for each criterion. Values were summed to yield a total score for each site. Site ranking criteria are described below (see Noss 1992c and 1993a for details).

Element occurrences (EOs). This criterion was applied in two forms: (1) number of species and communities ranked G3 and above, T3 or above, or S2 and above within site boundaries; and (2) total number of occurrences (EOs) of these elements within site boundaries. Number of elements had first priority in ranking, but sites with more total EOs ranked higher than other sites with the same number of elements but fewer occurrences. Murrelet records mapped from the ODFW database were added to element occurrences for each site for this criterion. Spotted owls records, however, were considered under a second criterion.

Spotted owls. This old-growth associated species was considered separately. More intensive surveys have been conducted for spotted owls than for other rare species in the region, and it is an excellent old-growth indicator species (although it also nests in younger natural forests that retain large structures that have persisted through disturbance). Scores for each site were based on the total number of owl locations and, secondarily, on density [number of locations per 1000 acres (404.7 ha)].

Old growth. Sites were ranked according to the amount and density of ecologically significant late-successional forests found within their boundaries, according to maps produced by Johnson *et al.* (1991). This criterion overlaps partially with the preceding criterion because spotted owls are largely old-growth dependent species. However, the two criteria are not identical. Site tenacity may cause owls (plus murrelets and other old-growth species) to return to a former breeding site even after most of the appropriate habitat has been destroyed. Also, birds may be missing from suitable habitat by chance or due to factors unrelated to habitat suitability, such as predation, competition, or persecution by humans. Finally, although spotted owls have been well surveyed, some areas of extensive late-successional forest mapped by Johnson *et al.* (1991) have few if any spotted owl records; these tended to be less accessible sites, suggesting that owl surveys are incomplete.

Connectivity. This criterion assessed the importance of sites in contributing to regional connectivity. Sites were ranked high if they fell within a regional wildlife corridor system (as determined by road kills and other records), encompassed a significant part of a critical riparian corridor mapped by the American Fisheries Society, or otherwise formed natural linkages in a network of significant sites.

Watershed. The value of a site to fisheries and aquatic biodiversity was assessed by the extent to which it included or was included within a draft AFS watershed reserve (as defined by the criteria explained above), or by its position in relation to significant water bodies or watersheds.

Total score. The total score for each site is the sum of all criterion scores. This scoring assumes that each criterion is of equal value, although old forest features are emphasized because two criteria (spotted owls and old growth) apply. Criteria could be weighted to produce different rankings if other criteria are deemed more important (see Duever and Noss 1990). The highest possible value is 25.

Results

This conservation evaluation produced 31 proposed Class I Reserves embedded in a more or less continuous network of proposed Class II Reserves and Multiple-Use Buffer Zones throughout the Coast Range (Fig. 4.4).

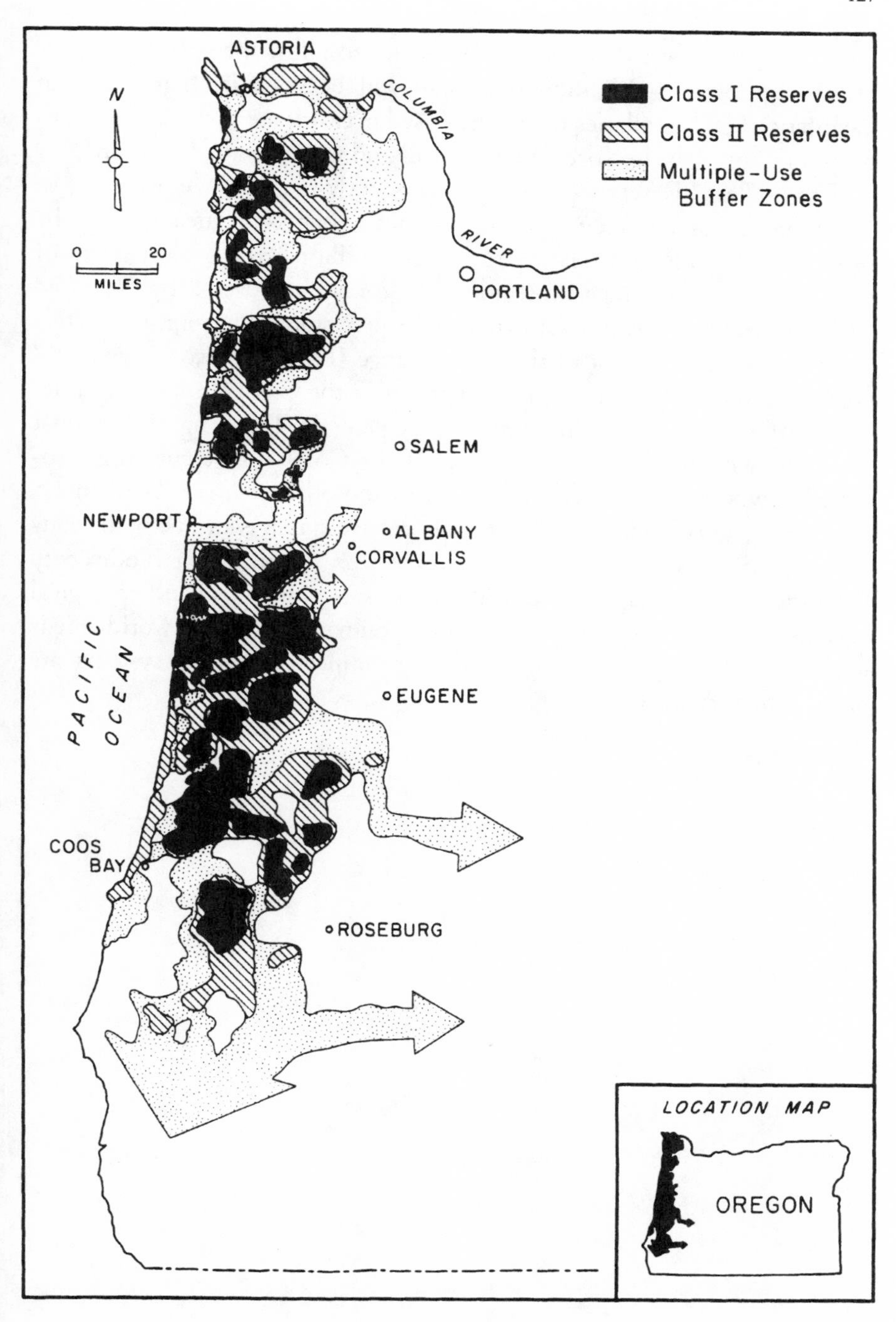

FIGURE 4.4 A reserve network proposed to capture hot spots of biodiversity, large areas of intact native forest, important watersheds, and to meet other conservation goals in the Oregon Coast Range (from Noss 1992c and 1993a).

Each Class I site was described by Noss (1992c). The theoretical and empirical support for this kind of design will be considered in detail in Chapter 5. The boundaries of the proposed network are vague and need to be refined through site-specific inventories and more detailed mapping.

The 31 Class I Reserves proposed in this study range in size from 1554 to 50,635 ha, averaging 16,338 ha (40,371 acres). The total area covered by proposed Class I Reserves is 506,483 ha (1.25 million acres), or 23 percent of the Oregon Coast Range Bioregion. The total area covered by proposed Class II Reserves was not estimated with precision, but is approximately 567,000 ha, or 26 percent of the Coast Range. Thus, the two reserve categories together encompass about 50 percent of the Coast Range. Multiple-Use Buffer Zones cover an additional 25 percent. These figures contrast strikingly with the less than 3 percent of the Coast Range currently protected in wilderness, research natural areas, and other reserves (Noss 1992c, 1993a). Yet, 50 percent of a region in protective status is in line with many recent estimates of what it takes to protect and restore native biodiversity (Noss 1992b). In Chapter 5, we will explore some biological and ecological reasons why the status quo level of protection nationwide and worldwide is unacceptable, as well as why large, interconnected reserve systems are needed in every region.

CHAPTER FIVE

DESIGNING RESERVE NETWORKS

> Wilderness complements and completes civilization. I might say that the existence of wilderness is also a compliment to civilization. Any society that feels itself too poor to afford the preservation of wilderness is not worthy of the name of civilization.
>
> Edward Abbey (1982), *Down the River*

In Chapter 4, we explored some criteria, databases, and techniques for recognizing areas of high conservation value. We emphasized selection of biodiversity hot spots (centers of endemism, rarity, or high species richness) and sites needed to assure adequate representation of all ecosystems and species in reserves. We also noted that it is a good idea to protect roadless areas and other wildlands, even or perhaps especially when we know little about their biological value.

Identifying hot spots, other important sites, and gaps in their representation in reserves is only the first step toward protecting these areas. Now we must design a system of reserves or a land-use regime that will maintain the dynamic biodiversity of these areas and the entire region in which they exist in perpetuity. No one knows precisely how to do this. But we can assess what has been learned from 100 years of trial and error in land conservation and piece together a strategy that has a high probability of success.

Sometimes the process of identifying hot spots leads directly to some kind of reserve proposal, without much more effort in design. For example, a serpentine or limestone ridge full of endemic species is an instant candidate for a reserve, and its boundaries are naturally defined. As another case, if we are interested in protecting old-growth forests, a map showing the location of remaining stands can be used to determine the location of large, relatively unfragmented stands that are of highest value for conservation, a process followed in the Oregon Coast Range project described in Chapter 4.

Although the general locations of reserves show up immediately in many conservation evaluations, defining reserve boundaries usually requires additional work. For instance, we may need to consider buffer zones to shield sensitive sites from external influences. We should estimate the sizes of populations that might inhabit a site and the ability of animals or plant propagules to travel between sites. We should assess the area of habitat needed to support viable populations or metapopulations. If we are using Gap Analysis data, we must examine in greater detail the broad areas recognized as important on a statewide scale to see how species and vegetation types are distributed on the landscape and where feasible reserve boundaries and linkages might be located.

Usually, the process of finding important sites (Chapter 4) addresses only the first of our four major conservation goals: representing all native ecosystems across their natural range of variation. To maintain viable populations and ecological and evolutionary processes and implement a land-use regime that will allow organisms to adapt to changing environments, we must consider issues of reserve size, proximity, connectivity, other aspects of pattern, and perhaps above all, how to manage the overall landscape. These issues are often more difficult and politically contentious than simply distinguishing areas of conservation value.

Nature Reserves or Multiple Use?

A controversy has been fermenting among managers and biologists about how best to conserve biodiversity. The debate is most intense concerning federal lands (Grumbine 1990a, Brussard 1991). Two major approaches have been proposed: (1) establish more or bigger parks, wilderness areas, and other reserves; or (2) manage better the semi-natural matrix (multiple-use public and private lands) that covers most of our country (Brown 1988).

These two options are not mutually exclusive. Most conservation biologists will probably agree that in most regions we should pursue both options in tandem. We need more and bigger reserves *and* more ecologically sensitive management of other lands. Biologists differ greatly in the emphasis they give to each option but not over the need for both. Nature reserves—defined here as areas managed primarily for their natural values—are central to land-use planning because they are the places that have the most to lose if not managed properly. They are also benchmark areas with which lands exploited by humans can be compared. If selected, as they should be, to represent hot spots of biodiversity and other critical sites, nature reserves are by definition irreplaceable or very nearly so.

Potentially destructive development or management practices should not be allowed in nature reserves. Sites outside reserves and of lesser conservation value can afford greater management experimentation, such as with innovative forestry techniques designed to provide commodities for people as well as to maintain most elements of biodiversity. Because land-use experiments may succeed or fail, reserves must be able to sustain species unlikely to persist in multiple-use areas. Uncertainty about the long-term impacts of management practices on biodiversity is reason enough to be conservative and place as much area as possible in wilderness or other strictly protected reserves. These reserves then can function as benchmarks for management experiments. As noted by Aldo Leopold (1941), wilderness provides a "base-datum of normality" for a "science of land health." It is an imperfect baseline because human impacts often cross boundaries and all natural communities change over time, but it is the best we have. Scientists shudder to think of experiments without controls, but this is what happens with much of our land management (Noss 1991b).

We do not believe that resource management is inherently destructive, but so far the record of its effects on biodiversity is rather bleak. Some new techniques are promising, but it is too early to say that they are safe. We can predict that many native species will persist on multiple-use lands with current practices, at least in the short term. But protecting many species is not good enough. We must strive to maintain *all* native species. Unless we slow the extinction rate to natural levels (sometimes estimated as about one species lost globally each year, balanced by slightly more than one new species created per year) we remain "Man the exterminator" (Diamond 1982).

Reserve Design in the United States: A Brief History

Soulé (1987) felt that most of conservation history has been concerned with the "protection of whole systems." In contrast, Simberloff (1991) stated that "[u]ntil the advent of the 'new conservation biology' . . . , refuges were usually chosen on the basis of habitat. The key was extensive knowledge about the autecology of species of interest, especially their habitat requirements, followed by a search for the most substantial amount of that habitat one could find." We disagree with both of these assessments. Until recently, science—whether "whole system" science or autecology—rarely entered into the conservation process.

A glance at our present reserve system in the United States shows little influence of science in selection or design. Many national parks are essentially square. Few conform to watersheds, mountain ranges, or other physiographic or biogeographic features that define natural regions. Most parks are too small to maintain viable populations of the largest animals that inhabit them (Newmark 1985, 1987). The lack of attention to science is reflected in the enabling legislation for our major reserves. As noted in Chapter 1, the national parks were created to conserve scenery and other natural objects, and secondarily to provide for public enjoyment of these things. Wilderness areas were established to preserve areas "where the earth and its community of life are untrammeled by man" (in Zaslowsky 1986), with little attention to ecological or scientific criteria (Nash 1984). Although the Wilderness Act of 1964 includes scientific value among many potential reasons for designation, scientific value is not mandatory or preeminent (Davis 1988). Virtually all agency assessments of wilderness proposals in recent years have focused on expected recreational visitor days (Noss 1990b).

National wildlife refuges were established on grounds that are ostensibly more scientific. The refuges were set aside to "provide, preserve, restore, and manage a national network of lands and waters sufficient in size, diversity and location to meet society's needs for areas where the widest spectrum of benefits associated with wildlife and wildlands is enhanced and made available" (National Wildlife Refuge System Act of 1966, cited in Zaslowsky 1986). The four main purposes of the refuges were to protect habitat of endangered species, to perpetuate migratory bird populations, to preserve natural diversity of all animals and migratory birds, and to engender an understanding and appreciation of wildlife (Zaslowsky 1986). However, the design and management of refuges have fallen far short of these lofty goals (Defenders of Wildlife 1992, Norris 1992, Curtin 1993). The independent commission appointed by Defenders of Wildlife concluded that refuge system planning is weak, lacks a scientific basis, and rarely coordinates with threats and opportunities arising from management of adjacent lands (Norris 1992).

Other kinds of protected areas suffer from similar problems. Research natural areas, established on federal lands primarily to represent natural communities and to serve as baselines for comparison with manipulated sites, are far too small to serve this purpose adequately (Noss 1990b and see below). They are also affected by clearcutting and other incompatible activities on adjacent lands.

Incongruous Boundaries

The incongruity between natural boundaries and reserve boundaries has led to major problems for reserve managers. Everglades National Park was one of the first and only national parks set aside to protect an ecosystem rather than mainly to protect scenery. Yet this 1.4 million acre park, bounded by the 586,000 acre Big Cypress National Preserve and 862,000 more acres in three water conservation areas (Fig. 5.1), is failing in its mission. The symptoms of degradation are many. In the 1870s wading birds, including 16 species of herons, bitterns, ibises, spoonbills, and the endangered wood stork, had combined populations as great as 2.5 million birds (Robertson 1965). Today, populations have been reduced by perhaps 90 percent and the nesting population in the national park is an ever smaller fraction of the regional total. Breeding pairs of wood storks have declined steadily in south Florida since the 1960s (Ogden *et al.* 1987). Exotic trees like Brazilian pepper and melaleuca now dominate much of the Everglades and have led to changes in hydrology, fire regimes, and biota.

What went wrong in the Everglades? The simple answer is that the park and adjacent conservation lands are not big enough to encompass the functional ecosystem that sustains the Everglades. The park was treated as a self-contained unit and expected to take care of itself with little effort from

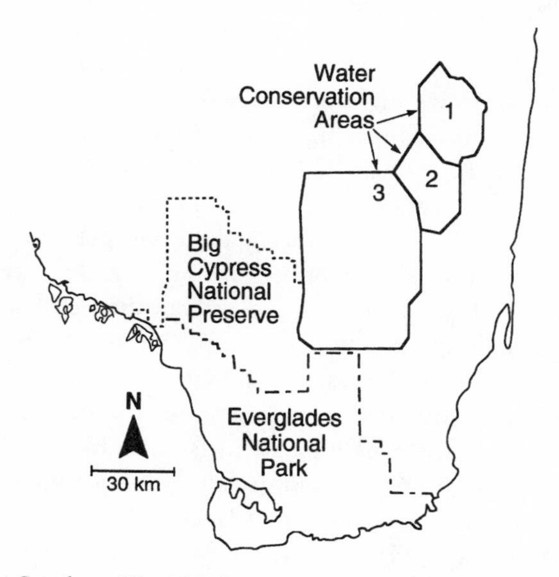

FIGURE 5.1 Southern Florida, showing major contiguous reserves in the Everglades ecosystem.

managers and little attention to what happened beyond its boundaries. Yet, the Everglades ecosystem actually encompasses most of south Florida. The hydrological regime of the Everglades begins naturally in the chain of lakes just south of Orlando, some 200 miles north of the Park. Canals, levees, and withdrawals and deliveries of water for agriculture and cities (the timing of which does not coincide with natural seasonal cycles) have altered the Everglades ecosystem beyond the ability of many native species to adapt.

The underlying problem in the Everglades, as with most other reserves throughout the United States, is the failure to take a regional perspective and coordinate efforts among a multitude of agencies and landowners (Kushlan 1979, 1983; Harris 1990). At least ten state and federal agencies have management jurisdiction over the Everglades ecosystem (Harris 1990). To perform naturally, the Everglades cannot be partitioned by artificial boundaries and multiple mandates. Rather, millions of acres—from Orlando to Florida Bay and from the Gulf of Mexico to the Atlantic Ocean—must be managed as a functional ecosystem. Although creating a park that encompasses most of South Florida is not politically feasible today, ecological management must indeed span this vast area. To this end, Harris (1990) has suggested the creation of a regional biosphere reserve. Although the park is already designated a biosphere reserve, the functional ecosystem was not included in the boundaries.

A second example of incongruous boundaries is provided by Yellowstone National Park, the oldest, largest, and most famous of all our parks. At 2.2 million acres, Yellowstone might be expected to be fairly secure. Yet, it is a grossly incomplete sample of the 14 to 19 million acres that constitute the Greater Yellowstone Ecosystem (GYE; Fig. 5.2). The GYE is rightly considered a showcase of American wilderness. Its values are summarized nicely by Glick *et al.* (1991): "What is significant about Greater Yellowstone's biological diversity is not the sheer numbers of species, and not even the abundance of many of the species that are present, but the fact that Greater Yellowstone's natural diversity of species is still essentially intact." Yet the famous biodiversity and wildness of Yellowstone are at risk. The park is much too small to maintain viable populations of some key species, most notably the grizzly bear (Shaffer 1992). In fact, an area much larger than even the GYE is needed to maintain grizzly bears in the long term. The Park is also not large enough to exist in balance with its disturbance regime, as natural fires in the GYE are characteristically large and catastrophic (Romme and Knight 1982).

The size of the park would not be such an issue if the lands surrounding it were managed in an ecologically responsible manner. Some 28 state and federal agencies and committees share management responsibility for the

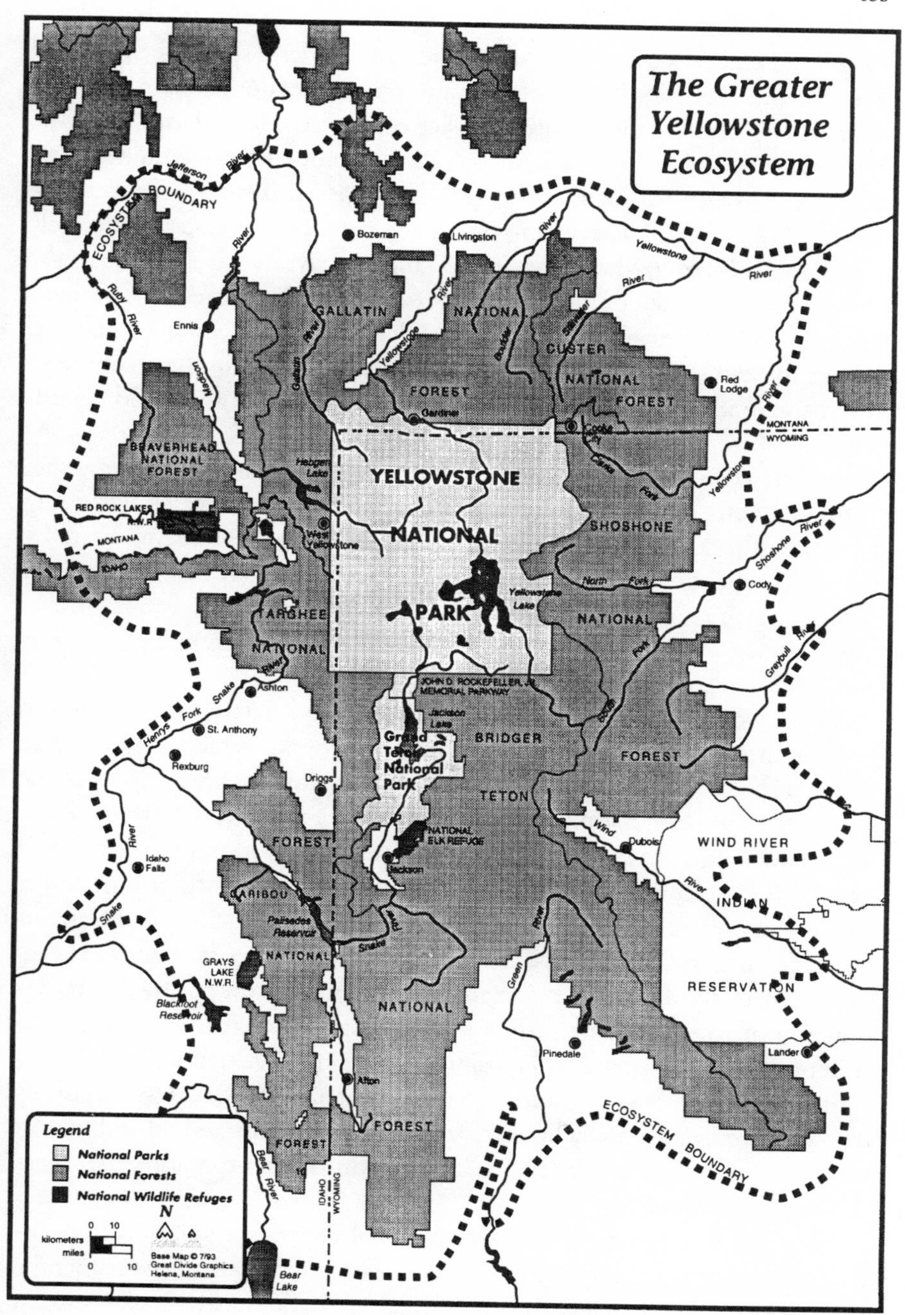

FIGURE 5.2 The Greater Yellowstone Ecosystem, variously defined as 14 to 19 million acres in size. From Great Divide Graphics, Helena, MT.

GYE, and each has a different idea about how the land should be managed (Goldstein 1992). Biodiversity conservation is not a major concern for many of these agencies. The U.S. Forest Service controls the largest area, about 60 percent of the GYE, but its lands are divided among three administrative regions and seven national forests. The most visible insult to the integrity of the park is heavy clearcutting on adjacent national forests, particularly the Targhee National Forest in Idaho (Fig. 5.3). More troubling than the visual impacts created by clearcutting next to park boundaries is what many consider to be the managerial hubris involved. Ironically, all seven national forests in the GYE run deficit timber programs, meaning it costs more to log than comes back to the government in revenue from the wood (Goldstein 1992).

Associated with clearcuts on national forests around Yellowstone are roads that provide access to other disruptive human activities, including poaching. Illegal shooting as well as killing of alleged "problem" bears by managers are the major sources of mortality for the grizzly bear in the GYE (Mattson 1990). Removal of roads and sheep allotments, where grizzly mortality is concentrated, is probably essential to grizzly recovery (Mattson and Reid 1991).

Livestock production, expecially of sheep and cattle, is perhaps the greatest threat to biodiversity in the GYE. Roughly 50 percent of the public lands in the GYE, including all three national wildlife refuges, BLM lands, portions of all seven national forests, and even Grand Teton National Park, are leased for livestock. The impacts of livestock production extend well beyond the direct effects of grazing on vegetation. Livestock production is largely responsible for the extirpation of the wolf and black-footed ferret in the GYE, the endangerment of the grizzly bear, declines in bighorn sheep, management conflicts with bison and elk, decline of native fishes due to dewatering of streams for irrigation and degradation of riparian zones, invasion of weedy plants, and soil erosion (Noss 1991e, Wuerthner 1992).

In addition to these problems, the GYE is threatened by subdivisions and development, virtually unrestricted tourism, mining, and oil and gas development (Glick *et al.* 1991). Each of these threats has the potential to escalate greatly in coming years.

How can we protect our premier national park? The solution is essentially the same as for the Everglades or any other park we might name: the entire regional landscape must be managed in a coordinated fashion toward the objective of biodiversity conservation. Agencies and citizens must work together, perhaps by establishing a lead agency with strong conservation credentials and overall management authority, to assure that activities incompatible with biodiversity conservation cease throughout the GYE. No

FIGURE 5.3 The boundary between Yellowstone National Park (left) and Targhee National Forest (right), illustrating lack of cooperation by the USDA Forest Service for managing its lands as a buffer zone for this biosphere reserve. Photo by George Wuerthner.

matter how difficult this objective may seem politically, the alternative—biotic impoverishment—is unacceptable.

Like the Everglades, Yellowstone has already been designated a biosphere reserve. But as with the Everglades, the boundaries of the biosphere reserve stop with the park and fail to encompass the greater ecosystem. Furthermore, as Figure 5.3 vividly attests, the park lacks a fundamental feature of a model biosphere reserve—buffer zones. The absence of buffer zones is due to lack of cooperation by the Forest Service and ultimately to lobbying from commodity interests (Keiter 1989, Goldstein 1992). An interagency effort to improve cooperation was quashed by political pressure (Milstein 1991). The Greater Yellowstone Ecosystem remains uncoordinated and unprotected.

These two examples of incongruous boundaries, simplified for brevity, show that biodiversity conservation will not succeed if constrained by political boundaries and the disparate mandates of multiple agencies (Grumbine 1990a, 1990b). Isolated parks will not work, nor will multiple-use management that degrades natural qualities. Identifying ecologically functional regions on the basis of physiography, hydrology, species distributions, population viability, migration routes, watersheds, vegetation patterns, fire regimes, patch dynamics, and other natural criteria is imperative. Then those regions must be managed to perpetuate ecological processes and biodiversity.

Island Biogeography and SLOSS

The biological basis for reserve design has strengthened as conservation science has evolved over the last two decades. The new conservation science of the 1970s was stimulated largely by the equilibrium theory of island biogeography, advanced a decade earlier by MacArthur and Wilson (1963, 1967; see Chapter 2). Recall that this theory considers species diversity on an island to represent a balance between immigration and extinction. Large islands that are close to a source of colonists are predicted to have the highest levels of diversity. Studies of land-bridge islands, isolated by rising sea levels after the Pleistocene, showed an apparent loss of bird species through time (Diamond 1972, Terborgh 1974). The analogy between land-bridge islands and terrestrial habitat patches isolated by development of the surrounding landscape was persuasive. Small isolated reserves, it was predicted, are doomed to lose species. Evidence began to accumulate that this prediction was correct.

Drawing largely from island biogeographic theory, Diamond (1975), Wilson and Willis (1975), and Diamond and May (1976) proposed rules for the design of nature reserves [Fig. 5.4; similar guidelines were suggested by Willis (1974) and Terborgh (1974)]. Later incorporated into the World

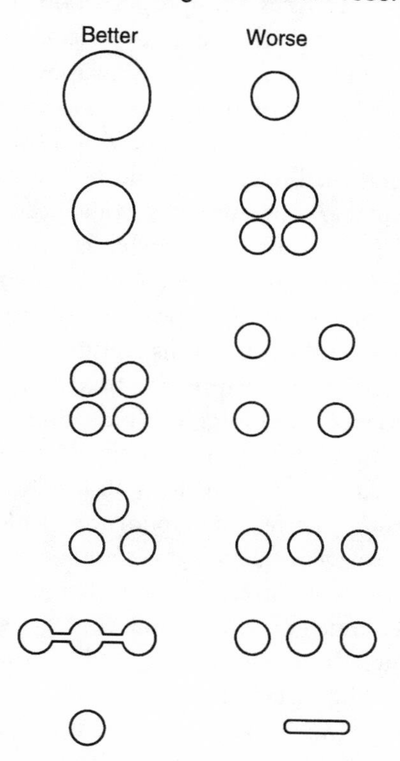

FIGURE 5.4 Principles for design of nature reserves, from Diamond (1975). In each comparison, the reserve design on the left is considered to be better for maintaining species diversity than the design on the right. Based in part on island biogeographic theory, these rules have been controversial. However, most have been validated by experience. Used with permission.

Conservation Strategy (IUCN 1980), the rules state that, all else being equal,

1. Large reserves are better than small reserves.
2. A single large reserve is better than a group of small ones of equivalent total area.
3. Reserves close together are better than reserves far apart.
4. Round reserves are better than long, thin ones.
5. Reserves clustered compactly are better than reserves in a line.
6. Reserves connected by corridors are better than unconnected reserves.

Immediately after these rules were proposed, other scientists challenged them as premature, given the lack of empirical data on island biogeography, problems with the equilibrium theory as proposed by MacArthur and Wilson, and dangers of extrapolating concepts from real islands to habitat islands (Simberloff and Abele 1976). These authors pointed out that in some situations several small reserves would be preferable to a single large reserve of the same total area because they would contain more species than the large reserve. The only criterion for what is better in these rules is the total number of species at equilibrium (Simberloff 1991). As we discussed in Chapter 4, species richness is only one of many important criteria for assessing the conservation values of alternative sites.

The six rules of reserve design appear to have been based as much on the collective field experience and biological intuition of those who proposed them as they were on island biogeographic theory. Following the rules might amount to "making the right decision for the wrong reasons" with regard to reserve design (Abele and Connor 1979). However, it is also possible that incorrect management decisions could be made if the mechanisms underlying the theory are incorrect (Simberloff 1991).

Rule #2 generated much more academic controversy than any other: the suggestion that a single large reserve is preferable to several smaller ones of equivalent total area. The literature on this debate, which came to be known by the acronym SLOSS (single large or several small), is perhaps larger than on any other topic in the history of applied ecology. The SLOSS debate was sometimes acrimonious and made for entertaining reading. It was not just coincidence that scientists arguing for small reserves mostly studied insects or plants, whereas those arguing for large reserves studied birds and mammals. Vertebrates, especially large-bodied species, are less likely than insects or plants to maintain viable populations in small areas. But besides giving academics lots of publications for tenure review and promotion, the SLOSS controversy accomplished little and was finally recognized as a red herring. When would decisions about selecting and designing reserves in the real world ever boil down to choices between single large or several small? In putting the SLOSS controversy to rest (although it still raises its ugly head from time to time), two biologists who had been on opposing sides of the controversy concluded that "bigness" and "multiplicity" are both essential criteria for establishing a system of reserves (Soulé and Simberloff 1986):

> Nature reserves should be as large as possible, and there should be many of them. The question then becomes how large and how many. There is no general answer. For many species, it is likely that

> there must be vast areas, while for others, smaller sites may suffice so long as they are stringently protected and, in most instances, managed. If there is a target species, then the key criterion is habitat suitability. Suitability requires intensive study, especially in taxa that contain species with narrow habitat requirements.

Today, few biologists would disagree that we need big reserves and lots of them. We expect that most would also agree that conservation decisions involve much more than just these two criteria, and that autecology (especially the life histories of species that are highly vulnerable to extinction or pivotal in the ecosystem) must assume a greater role in reserve design decisions (Soulé and Simberloff 1986; Simberloff 1988, 1991; Noss 1992b).

But where does this leave us? Do we need detailed studies of every species that might be sensitive to human activities before we make any recommendations about how to design reserve networks? We think not. In the absence of detailed autecological information, some empirical generalizations for reserve design stand out. In their conservation strategy for the northern spotted owl, Thomas *et al.* (1990) listed five reserve design concepts that they characterized as "widely accepted among specialists in the fields of ecology and conservation biology." We agree, and paraphrase these guidelines below, adding a sixth (from Noss 1992b) that applies to species that are especially sensitive to human disturbance and, therefore, greatly in need of protection.

1. Species well distributed across their native range are less susceptible to extinction than species confined to small portions of their range.
2. Large blocks of habitat containing large populations of a target species are superior to small blocks of habitat containing small populations.
3. Blocks of habitat close together are better than blocks far apart.
4. Habitat in contiguous blocks is better than fragmented habitat.
5. Interconnected blocks of habitat are better than isolated blocks, and dispersing individuals travel more easily through habitat resembling that preferred by the species in question.
6. Blocks of habitat that are roadless or otherwise inaccessible to humans are better than roaded and accessible habitat blocks.

Note that these guidelines are not all that different from those offered by Diamond and others 15 years earlier. They have proven to be extremely robust and are among the best-supported generalizations that conservation biology has to offer (Wilcove and Murphy 1991). The sixth guideline can be shown to apply to most large carnivores, often the most sensitive species in an ecosystem, and also to the desert tortoise (U.S. Fish and Wildlife Service 1993) and other organisms likely to be exploited or persecuted by humans. Although these guidelines are oriented toward target species, they also apply to conservation planning at higher levels. But as we have hinted already and will explore in greater detail later, still other factors should be considered when designing reserve networks.

Biosphere Reserves

The old model of isolated parks has failed. Unless it contains many millions of acres, no reserve can maintain its biodiversity for long. Smaller parks are not only less likely to maintain viable populations, but they are also more heavily assaulted by activities beyond their boundaries. Scientists studying boundary problems around reserves (Schonewald-Cox and Bayless 1986, Buechner 1987, Dasmann 1988, Schonewald-Cox 1988) emphasize the importance of large areas and well-managed buffer zones. None of our national parks is big enough to maintain its diversity over time. Most are becoming more insular as surrounding habitats are modified by logging, grazing, suburban development, and other human activities. Smaller parks are losing species of mammals more rapidly than large ones (Newmark 1985, 1987).

A big step toward better integration of reserves and their surrounding landscapes was the development of the biosphere reserve model as part of UNESCO's "Man and the Biosphere" (MAB) program (UNESCO 1974). A main purpose of the program was to create a global network of protected areas for scientific research and monitoring and for protecting genetic diversity (Hough 1988). The MAB program stimulated development of a global classification system (Udvardy 1975) so that reserves could be selected to represent all of the earth's major biomes or biogeographical provinces.

Another impetus for biosphere reserves was the recognition that economic development ideally should lead to sustainable ways of life and that conservation will not succeed in the long run if it fails to consider the needs of local people (Dasmann 1988). This argument has been particularly cogent for the Third World. In some tropical countries, indigenous tribes that had lived sustainably with their landscape were displaced by new national parks, resulting in disintegration of their cultures and open hostility toward the parks (Dasmann 1988). This hostility was often expressed by vigorous poaching and other destructive activities. However, in the United States, it is not so much *people* who might be displaced by new reserve designations on public lands, but rather certain kinds of human activities. When conducted as intensively as usual, clearcut logging, road building, livestock grazing, and mining are not sustainable ecologically. Off-road vehicle use is so blatantly harmful and frivolous that we wonder why there is even a debate about continuing this use on public lands. Eliminating destructive activities from public lands ultimately enhances both biodiversity and human well-being.

The zoning concept of biosphere reserves holds promise for integrating conservation and human activities. The basic biosphere reserve model (Fig. 5.5) portrays strictly protected core zones surrounded by one or more buffer and transitional zones and often containing areas for research, restoration, monitoring, and compatible human settlements. Unfortunately, this kind of zoning has seldom been implemented. By 1990 almost 280 biosphere reserves had been designated in more than 70 countries, including 43 in the United States (Dyer and Holland 1991). But most biosphere reserves have been superimposed on existing national parks and other protected areas without adding land. In 1983 only 1.6 percent of the

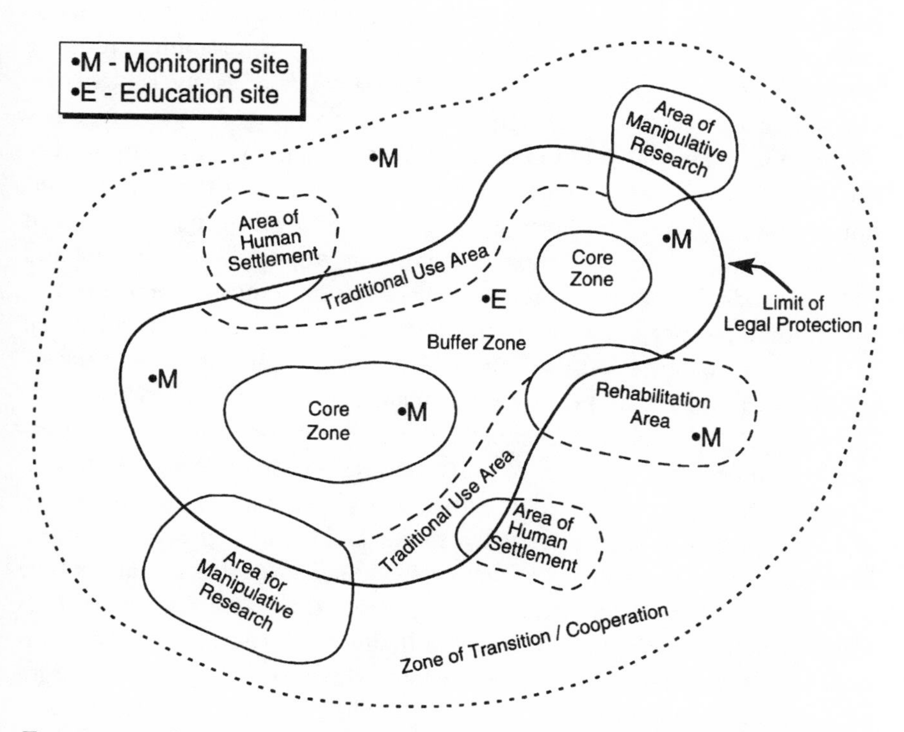

FIGURE 5.5 Conceptual layout of an ideal biosphere reserve (based on Hough 1988).

area in biosphere reserves worldwide was newly protected land, and by 1986 only 101 (52 percent) of 193 biogeographical provinces classified by Udvardy (1975) had been represented. Moreover, as noted above, biosphere reserves in the United States lack buffer zones. The biosphere reserve ideal remains unfulfilled.

Reserve Networks

A level beyond biosphere reserves is the concept of reserve networks. The basic idea is this: If functionally connected, a system of reserves may be united into a whole that is greater than the sum of its parts. Although no single reserve may be able to support a long-term viable population of a species with large area requirements, such as cougar or grizzly bear, reserves linked by corridors or other avenues of movement may do so (Noss and Harris 1986). Thus, whereas individual reserves are unlikely to encompass ecosystems replete with all native species, a well-connected network of reserves just might.

The reserve network concept was promoted by Noss (1983), Harris (1984), and others who looked at conservation opportunities from a landscape or regional perspective and emphasized the need for animals to move between reserves or other areas of favorable habitat. Landscape ecology developed in earnest in North America in the late 1970s and early 1980s (Forman 1981, Forman and Godron 1981) and was accompanied by studies showing that many animals use corridors when traveling through human-dominated landscapes (Wegner and Merriam 1979, Johnson and Adkisson 1985) and that corridors can enhance persistence of populations (Fahrig and Merriam 1985, Henderson *et al.* 1985). These studies, mostly carried out in agricultural landscapes, showed that population dynamics in individual woodlots are of only local importance. We must consider a network of woodlots to understand persistence of many species in these landscapes.

Noss (1983) described a "regional landscape approach to maintain diversity." He urged an expansion of conservation concern beyond local sites, emphasizing protection of old growth and other natural areas wherever they occur, a complex of large and small reserves, and broad corridors of natural habitat connecting reserves. He further suggested that effects of alternative designs and management strategies on diversity should be assessed regionally rather than site by site. Although a fragmented landscape may contain high species diversity, the species favored are mostly weedy and edge-adapted, whereas sensitive species decline. The net effect of these changes on regional biodiversity is negative. In line with this suggestion, an Australian study showed that, though a system of many small reserves maximizes species richness, those species found only in reserves are best served by a system of large reserves (Humphreys and Kitchener 1982). These findings reinforce the advice of Diamond (1976):

> [S]pecies must be weighted, not just counted; the question is not which refuge system contains more total species, but which contains more species that would be doomed to extinction in the absence of refuges. A refuge system that contained many species like starling and house rat while losing only a few species like ivory-billed woodpecker and timber wolf would be a disaster.

Harris (1984) presented a management strategy for old-growth forests in the western Cascades of Oregon in his landmark book, *The Fragmented Forest*. Harris' strategy was based on maintaining the characteristic regional fauna, rather than species richness per se, in line with Diamond's advice. He recommended the conservation of existing old growth, an integrated system of long rotation forest islands surrounded by replacement stands, and a dendritic network of corridors containing many nodes of small and medium-sized forest islands linking larger forest islands. He noted that his approach was designed to integrate conservation and development planning and "should be based on sound principles of multiple use, not preemptive policies of exclusive use. The archipelago approach shifts the emphasis away from any single old-growth habitat island toward a system of islands integrated into the managed forest landscape."

Harris (1984) also introduced the concept of the multiple-use module (MUM), essentially a generalization of the biosphere reserve model to a variety of spatial scales and management scenarios. The MUM concept was further elaborated by Noss and Harris (1986) and Noss (1987a). Its basic feature is a concentric design where intensity of use increases outward from the core and intensity of protection increases inward (Fig. 5.6). As applied to the Oregon Cascades, each old-growth island would be surrounded by a long-rotation buffer zone, thereby increasing effective island size, improving chances of forest persistence over time, and providing a spatial gradation from old growth into young managed forest (Harris 1984). More generally, it is essential that MUMs be located to capture nodes or hot spots of biodiversity and be linked by habitat corridors into a functional regional network (Fig. 5.7). Noss and Harris (1986) stressed that the MUM network strategy was consistent with agency mandates for multiple use because the system as a whole provides for many different uses. However, the most sensitive sites and species are insulated from disturbances arising in the landscape matrix.

In simplified form, then, the regional reserve network model consists of

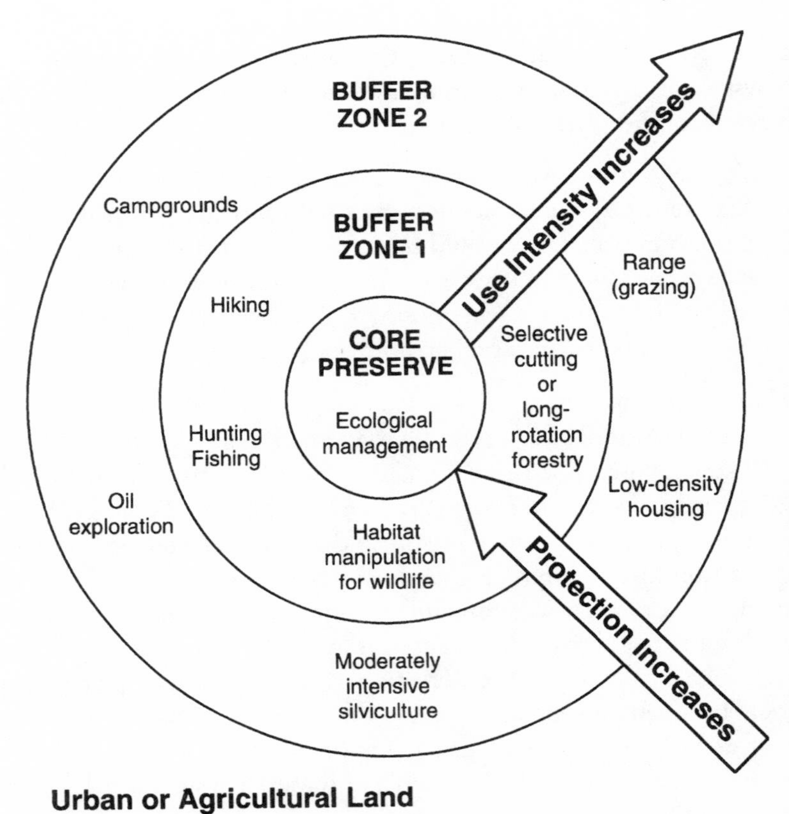

FIGURE 5.6 A multiple-use module (MUM). An inviolate core reserve is surrounded by a gradation of buffer zones, with intensity of human use increasing outward and intensity of protection increasing inward (from Noss 1987a, modified from Harris 1984). Used with permission of the Natural Areas Association.

two or more reserves connected by broad corridors, surrounded by a gradation of buffer zones, and connected to other regions (where biogeographically appropriate) by interregional corridors (Fig. 5.8). Thus, the strategy involves a combination of (1) more reserves, (2) bigger reserves, (3) interconnected reserves, and (4) more sensitive management of multiple-use lands. The strategy rejects resource tradeoffs and insists that we can have the best of all possible worlds if we put our minds to it and are willing to reduce our resource consumption and intensity of land use for the sake of the land. Below, we review some major features and functions of the three essential components of a regional reserve network: core reserves, multiple-use buffer zones, and connectivity.

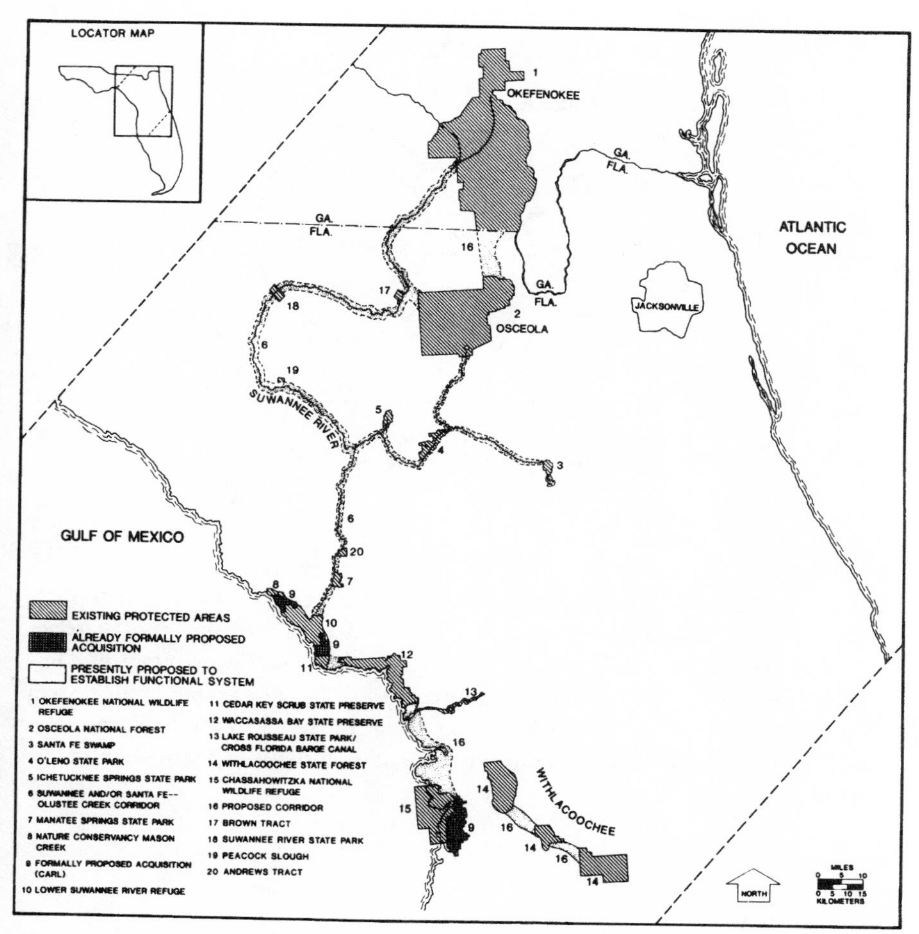

FIGURE 5.7 A MUM network, or regional network of reserves, based on the Suwannee River and its tributaries. From Noss and Harris (1986). Used with permission.

Core Reserves

Core reserves are the backbone of a regional reserve system. Without strictly protected areas representing most of a region's biodiversity, losses are inevitable. Criteria and methods for selecting core reserves were reviewed in Chapter 4. Size and scale issues—how much do we need?—and general management principles will be discussed later in this chapter.

Just what is a core reserve? National parks, wilderness areas, research natural areas, state parks and preserves, BLM areas of critical environmental concern, national wildlife refuges, Nature Conservancy and Audubon

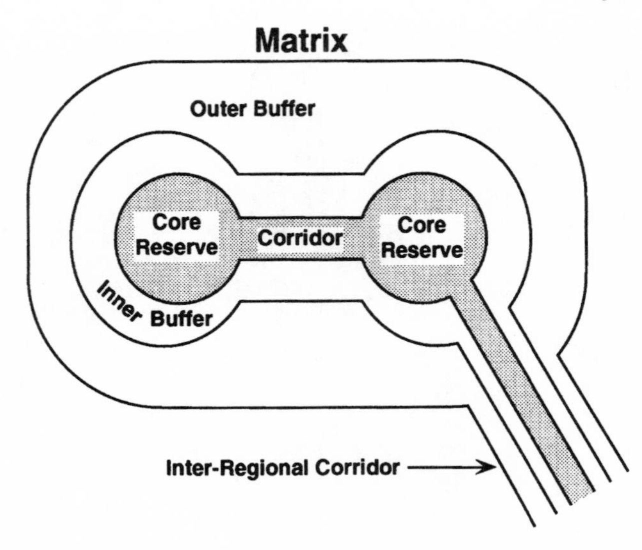

FIGURE 5.8 A model regional reserve network, consisting of core reserves, connecting corridors or linkages, and multiple-use buffer zones. Only two core reserves are shown, but a real system may contain many reserves. Inner buffer zones would be strictly protected, while outer zones would allow a wider range of compatible human uses. In this example, an interregional corridor connects the system to a similar network in another bioregion. "Matrix" refers to the landscape surrounding the reserve network. From Noss (1992b). Used with permission of the Cenozoic Society.

Society preserves, and several other kinds of areas may qualify. Recall that Scott *et al.* (1993) (see Chapter 4) considered four categories of land protection in the gap analysis project. The highest category, what we would call *core reserves,* includes any area "that is maintained in its natural state and within which natural disturbance events are either allowed to proceed without interference or are mimicked through management" (Scott *et al.* 1993). Most national parks (or parts thereof), Nature Conservancy preserves, and Audubon sanctuaries meet these criteria, but only some Forest Service wilderness areas, national wildlife refuges, and research natural areas (the latter often fail particularly because they are too small).

We suggest that changes in management, for example removal of livestock from wilderness areas and wildlife refuges, are necessary to make these areas qualify as the core reserves they should be. Design and management guidelines for core reserves will vary on a case-by-case basis.

Multiple-Use Buffer Zones

In most regions, a system of core reserves will be necessary but not sufficient to maintain biodiversity. They must be complemented by multiple-use lands. Buffer zones also provide an opportunity to find ways to integrate development (and human activities in general) with conservation. Humans are as much a part of the earth as any other species. We just have to relearn how to get along with our nonhuman kin. What better place to relearn these lessons than in seminatural wildlands? Integration has yet to be fully realized, but there are some promising experiments, such as the Sian Ka'an Biosphere Reserve on Mexico's Yucatan Peninsula. This reserve, which contains 1.3 million acres of tropical moist forests, marshes, mangrove swamps, and freshwater and marine systems, is essentially pristine but supports about 800 people in subsistence farming, fishing, and small-scale tourism (Tangley 1988). Promising experiments in the United States are virtually nonexistent. We will have to create them.

A multiple-use buffer zone, as we define it here, is a zone that permits a greater range of human uses than core reserves but is still managed with native biodiversity as a preeminent concern. Because its allowable uses are less intense than in the general landscape matrix, it should serve to shield or insulate core reserves from harmful activities. Because road access is a major threat to sensitive species and creates many other ecological problems (see Chapter 2), road density in buffer zones should be kept low (certainly no more than 0.5 miles per square mile). Allowable activities in buffer zones might include nonmotorized recreation (including fishing and hunting, unless they pose a threat to sensitive species), selection forestry, light grazing (for grassland types that are adapted to grazing), and small-scale subsistence agriculture.

Ecologically, buffer zones serve a number of potential functions (Noss 1992b). Especially in the case of small reserves, they may ameliorate edge effects that would otherwise be intense near reserve boundaries. Wind, sun, exotic weeds, agricultural chemicals, noise, and opportunistic predators that thrive in suburban landscapes might all be filtered out by well-managed buffer zones. For large reserves, such edge effects are not expected to be as much of a problem, but a buffer from intensive logging and other commodity production will help protect sensitive species.

Ideally, buffer zones enlarge the effective size of a reserve and provide some temporal stability to the landscape. As noted earlier, we can expect many species to be able to persist in multiple-use landscapes. Reserves are needed mainly for the more sensitive species and to provide a buffer against our ignorance about the conditions that species need to survive. Because all environments are dynamic, a buffer zone may have to take on the functions of a reserve if disturbances temporarily make reserve habitats unsuitable. A lightly used buffer zone will be easier to convert to reserve functions than would a massive industrial clearcut, soybean field, or housing subdivision. Buffer zones can also provide connectivity between reserves, allowing animals to move long distances without molestation or a high chance of mortality on busy roads. Even if buffer zones are technically population sinks for some species (i.e., areas where death rates exceed birth rates), they can still contribute to overall metapopulation peristence by at least temporarily supporting resident individuals while serving as connections between source habitats (see Pulliam 1988, Howe *et al.* 1991).

Opportunities for creating buffer zones appear most promising in regions where vast areas of national forest or BLM lands surround national parks, wilderness areas, and other reserves. Reducing road density and eliminating or scaling down harmful activities such as logging, mining, livestock grazing, and off-road vehicle use on these lands are needed. Politically, such changes will not be easy. In many cases the Forest Service has vigorously opposed the idea of buffer zones. The Park Service and Forest Service have maintained an antagonistic relationship throughout their histories (Grumbine 1990a, 1991). The national forests will not serve as buffer zones for national parks without sweeping changes in national leadership and agency organization (for example, moving the Forest Service from the Department of Agriculture back to the Department of Interior, as is often proposed).

Where reserves are surrounded by private lands, buffer zones may seem more difficult to establish. However, creative solutions are possible. Farmers, for example, might be paid easements to maintain perennial cover (such as pasture) on their lands that border reserves. County planning boards can zone for low-density developments (e.g., agricultural zoning) and limit road construction in areas surrounding reserves. High-density developments, although desirable for other environmental reasons (they save space and energy), should be kept away from reserves and other sensitive sites, for they will be sources of noise, chemicals, housecats, and opportunistic predators like raccoons and human trespassers.

Connectivity

Connectivity is fundamental to our concept of regional reserve networks. Biological functions of connectivity have been discussed in some detail (e.g., Harris and Gallagher 1989; Hudson 1991; Saunders and Hobbs 1991; Noss 1992b, 1993b) and have been vigorously debated (Soulé and Simberloff 1986, Noss 1987c, Simberloff and Cox 1987, Noss 1991c, Simberloff *et al.* 1992).

Despite uncertainty about optimal width of corridors, mortality risks, tradeoffs with other uses of conservation dollars, and other issues, the fundamental need for populations of many species to be connected in order to be viable is widely recognized. Determining the best ways to provide connectivity, however, is a tremendous challenge.

Connectivity is essentially the opposite of fragmentation. Instead of breaking landscapes into pieces, we are seeking ways to preserve existing connections and restore severed connections. The connectivity of interest to biologists and conservationists is *functional connectivity,* usually measured according to the potential for movement and population interchange of target species. Many factors determine the degree of functional connectivity between habitat patches in a landscape (Table 5.1). Variation in the quality of linkages affects their use by organisms (Henein and Merriam 1990).

Connectivity is not just corridors. For species that disperse in more or less random directions, such as the northern spotted owl (Thomas *et al.* 1990) or cabbage butterfly (Fahrig and Paloheimo 1988), connectivity is affected more by the suitability of the overall landscape matrix than by the presence or absence of discrete corridors. Multiple-use buffer zones with low road density and minimal human disturbance, as described earlier, should provide adequate connectivity for most organisms. Scale must be specified in discussing connectivity. A multiple-use landscape 20 or even 50 miles wide that lies between two national parks can be considered a corridor at a regional scale, if in fact it functions as such. Biogeographers discuss corridors as broad, heterogeneous zones, such as the Bering Land Bridge, that permit migration of species from one region to another over long periods of time (Brown and Gibson 1983). For conservation planning, connectivity

TABLE 5.1 Determinants of Functional Connectivity

Mobility or dispersal characteristics of the target species
  Species-specific habitat preferences for movement
  Dispersal distance or scale of resource utilization
  Rate of movement or dispersal (through various types of habitats)

Other autecological characteristics of the target species (e.g., preference for particular plant species or structural features of the habitat; feeding and nesting requirements; mortality risks)
Landscape context: Structural characteristics and spatial pattern of landscape (patch, corridors, matrix, mosaics)
Distance between patches of suitable habitat
Presence of barriers to movement (e.g., rivers, roads)
Interference from humans, predators, etc.

should be evaluated at several spatial and temporal scales, ranging from daily movements within home ranges to long-distance dispersal events connecting populations once every generation or two. Critical planning and management questions exist at each of these scales.

For corridors or other habitat linkages to serve conservation goals, their functions must be stated explicitly and analyzed carefully (Soulé 1991b). The scientific literature on connectivity has concentrated quite narrowly on discrete habitat corridors and specifically on a conduit function: allowing individuals of a target species to move from one place to another. But a habitat linkage in a real landscape may have several functions and affect many species. Although a particular target species may be the main concern in a corridor plan, the net effect of alternative landscape designs on a whole suite of species and ecological processes should be considered, whenever possible. Two major roles of landscape linkages (defined here as specific pieces of land that provide a connectivity function) in biological conservation are to: (1) provide dwelling habitat for plants and animals and (2) serve as a conduit for movement. The conduit role can be further subdivided into several functions: (a) permitting daily and seasonal movements of animals; (b) facilitating dispersal, consequent gene flow between populations, and rescue of small populations from extinction; and (c) allowing long-distance range shifts of species, such as in response to climate change. These functions have been discussed by Noss (1993b) and are summarized below.

Linkages as habitat. Some types of linkages, such as riparian forests, are distinct in the natural landscape. Riparian forests have many ecological values, including rich alluvial soils and an associated high biological productivity; microclimates moderated by a dependable source of water; abundant insects and plant foods such as woody browse or mast for vertebrates; and many tree cavities and substrates to serve as homes or nests for birds and mammals (Harris 1989). Riparian forests and other naturally linear habitats would be important to protect even if they served little as movement corridors (Harris and Gallagher 1989).

Wide protected linkages are basically extensions of core reserves. The width of corridor needed to contain an adequate amount of interior habitat and minimize edge effects and mortality rates is uncertain and depends on habitat type and quality both within and outside the corridor (Noss 1983, 1987c, 1993b) and on the target species and the mortality risks it faces (Soulé 1991b, Soulé and Gilpin 1991). Another consideration for determining optimal linkage width is the territory or home range size of target species, particularly when the length of the linkage exceeds normal dispersal distances

(Bennett 1990; Harrison 1992; Noss 1992b, 1993b). This issue will be discussed later under "linkages for dispersal."

Linkages for movements within home ranges. Movement of animals between reserves or other suitable habitat patches is the function most often associated with corridors (Soulé and Gilpin 1991). A core reserve may not encompass a single annual or even a daily home range of a large mammal. Maintaining safe travel opportunities for wide-ranging species is largely a matter of protecting them from human predation and road mortality. Every species faces mortality risks when moving about its home range on a daily or seasonal basis. Minimizing sources of mortality—human activity, internal fragmentation of corridors, ambush sites for predators—is a major consideration in corridor design for target species (Noss 1987c, Simberloff and Cox 1987, Soulé 1991b, Soulé and Gilpin 1991).

Vertebrates, especially certain ungulates, often use traditional migration routes between summer and winter range. Elk, for example, generally use forested travel lanes, if possible, for migratory movements (Adams 1982). Mule deer also commonly use distinct corridors for travel between winter and summer range (Thomas and Irby 1990). Tunnels or underpasses have been built under highways to facilitate deer movements in several regions (Reed *et al.* 1975, Reed 1981).

Highways can be significant mortality sinks for many animals. In southern California, seven of thirty-five radio-collared cougars were killed by automobiles in the first two years of a study (Beier and Barrett 1991). Several animals have been documented using corridors, especially canyon bottoms with dense vegetation. Cougars used canyon corridors 0.5 to 1.0 km wide and 6 km long and crossed under interstate highways by way of culverts. Some animals that failed to use culverts were killed on the highway (Beier 1993, Harrison 1992). But other studies found that cougars used bridge underpasses but not culverts; animals crossed roads rather than pass through culverts (P.J. Mock, personal communication). Similar to the situation in southern California, the greatest source of mortality for Florida panthers is roadkill (Harris and Gallagher 1989). The Florida panther uses wooded strands and other natural corridors in the south Florida landscape for movement (Maehr 1990). Panthers and dozens of other vertebrate species regularly use wildlife underpasses built on Interstate 75 across the Big Cypress Swamp (Foster and Humphrey 1991).

Animals such as amphibians that use different habitats in a landscape for different life history stages also depend on connectivity. Small "toad tunnels" have been used in several European countries for decades to help migrating amphibians avoid the hazards of road crossings (Langton 1989). Aquatic turtles may migrate hundreds of meters from rivers or ponds to find sandy substrates in which to lay their eggs. Conversely, in times of drought, many upland animals move downslope to riparian areas (Brown *et al.* 1987). If habitat in between is fragmented (for example by a highway) they may not be able to make such movements safely.

Linkages for dispersal. Dispersal refers to movement of organisms away from their place of origin, such as the movement of subadult animals out of the parental home range. Dispersal can potentially counteract the isolating effects of habitat fragmentation, but only if adequate dispersal habitat remains. For a regional metapopulation of a species to persist, enough individuals must move between patches to balance extirpation from local patches (den Boer 1981, 1990). Preserving natural linkages between existing populations may increase the chance of metapopulation persistence. Dispersal linkages are most important for late-successional species (which commonly have poor dispersal capacities), other habitat specialists, and for species such as large carnivores that risk being killed by humans or vehicles in developed landscapes. Dispersal is more likely to be successful when habitat in a linkage is similar to the habitat in which a species lives (Wiens 1989). But desert-dwelling mountain sheep will move across basins from one mountain range to another if mountainous corridors are lacking (Bleich *et al.* 1990).

Linkages that support resident populations of animals may be more likely to function as long-distance dispersal conduits for those species. Genes might then flow in both directions, filtering through resident breeding animals, and minimum corridor widths can be based on average home range or territory diameters of target species (Bennett 1990; Harrison 1992; Noss 1992b, 1993b). The notion of maintaining dispersal corridors wide enough to support resident individuals or pairs of target species is controversial. A model by Soulé and Gilpin (1991) predicts that animals in wide corridors may spend time wandering around rather than reaching their goal, but such models may not be accurate for most vertebrates (Noss 1993b).

The wide-corridor strategy has been proposed specifically for cases where the distance between population centers exceeds normal dispersal distances for the target species. A wide corridor of suitable habitat appears to be optimal under such circumstances. Logically, the ideal connectivity is continuity. Unless resident individuals of a target species exist within linkages or within a series of stepping-stone habitats—none separated by impenetrable barriers or distances greater than those commonly traversed—the populations will be isolated from one another. Interchange necessary to maintain the metapopulation must then take place through translocations, an expensive and uncertain long-term undertaking, especially when applied to multiple species.

When long-distance dispersal corridors are designed, all of an animal's life-history requirements should be appraised (Beier and Loe 1993). A culvert or underpass will not suffice. Consider the cougar, with an average female monthly home range of 43 km^2 (10,625 acres) and an average annual home range of 155 km^2 (38,301 acres, range 24,463–87,227 acres) in southern California; male annual home ranges average about 450 km^2 or 111,120 acres (P. Beier, personal communication). To accommodate a female monthly home range twice as long as it is wide (based simply on the observation that ranges are commonly rectangular or elliptical), a linkage would need to be about 3.1 miles wide (Harrison 1992). Linkages would need to be wider to accommodate the larger, nonoverlapping home ranges of males. Because male cougars disperse longer distances and make longer intrarange movements (Anderson 1983), one male may be able to mate with several females.

Although intuitively attractive and suggested independently by several authors, the idea of basing dispersal corridor widths on home range diameters is not supported by much empirical evidence. Nevertheless, a prudent strategy would be to maintain or restore wide habitat corridors whenever the intent is to link areas farther apart than normal juvenile dispersal distances. It is expected that inhabited linkages will be internally heterogeneous and might contain many bottlenecks (for example, highway underpasses) without seriously limiting their effectiveness. Interregional dispersal corridors for large carnivores, such as those proposed for grizzly bears in the northern Rocky Mountains which link population centers up to 150–200 miles apart (see Noss 1992b, Shaffer 1992), must necessarily consist of landscapes many miles wide with low road density in order to protect animals from poachers.

To summarize, a lesson of dispersal studies for reserve network design is that reserves or population centers of target species should be connected by linkages of suitable habitat. If no corridors are possible, reserves should be very close together and not separated by insurmountable barriers (Diamond 1975, Thomas *et al.* 1990). Reserves separated by distances longer than normal dispersal distances of target species should contain resident individuals or populations between them, either distributed more or less continuously or in stepping stone habitats. For species that refuse to cross even a few meters of unsuitable habitat, such as a road (Mader 1984, Swihart and Slade 1984, Mader *et al.* 1990), continuous habitat linkages are needed both for movements within home ranges and for dispersal. Even some birds, despite their ability to fly, will not cross areas of unsuitable habitat (Diamond 1975). In all cases, the scale of planning must correspond to the scale at which organisms use the environment.

Linkages for long-distance range shifts. Another potential function of connectivity is to provide for long-distance migration of species in response to climate change. Many temperate plant species have migrated hundreds of miles northward or hundreds to thousands of feet upward in elevation since the Pleistocene (e.g., Davis 1981). Models of global warming predict dramatic shifts in habitat conditions in most regions over the next few decades. But human activities have imposed a new set of barriers on the landscape that, in addition to natural barriers, may interfere with long-distance movements. If rates of global warming in the next few decades are as fast as predicted, many species will be unable to migrate quickly enough, even along ideal corridors. Species with short and rapid life histories, such as introduced weeds, will probably adjust well to climate change, as will broadly distributed species. But species with limited and discontinuous distributions or poor dispersal capacities are at high risk of extirpation (Peters and Darling 1985, Peters 1988, Peters and Lovejoy 1992).

Mountainous regions with broad elevational spans provide better opportunities for adaptation to climate change than flatter regions. A 3° C rise in temperature (as is predicted to occur with greenhouse warming) translates to a latitudinal range shift of roughly 250 km (155 miles), but an elevational range shift of only 500 m (1640 feet) (MacArthur 1972). Perhaps the best way to facilitate adaptive migration in response to climate change is to maintain intact environmental gradients. Complete, unfragmented elevational gradients will offer opportunities for species to migrate upslope in response to global warming. The full spectrum of topoclimates and substrates should also be maintained in each landscape, so that species can adjust distributions to changes in temperature and soil moisture conditions, as controlled by topography, aspect, and soil characteristics. Ultimately, habitat continuity on a continental scale will be needed to facilitate movement of entire floras and faunas in response to climate change.

From Regional to Continental Networks

Reserve networks with the components described above—core reserves, buffer zones, and connectivity—can be implemented at many scales, from counties to continents. We emphasize a regional scale of planning for reasons expressed throughout this book: regions are often physiographically or biogeographically distinct, they provide a convenient scale for mapping and analysis, and they inspire a sense of belonging and protectiveness in their more enlightened human inhabitants.

The Wildlands Project, mentioned earlier, is an effort to stimulate and promote bioregional conservation planning throughout North America. The Wildlands Project is a continental-scale venture built from a collection of regional reserve networks. Many regional conservation plans are being developed by local people. Examples of proposed regional networks are shown for Florida (Fig. 5.9), the southern Appalachians (Fig. 5.10), and the northern Rocky Mountains (Fig. 5.11). Another example, for the Oregon Coast Range (Noss 1992c, 1993a), was provided in Chapter 4 (Fig. 4.4). We emphasize that all of these proposals are preliminary and need refinement. But this is the strategy of The Wildlands Project and of regional network planning in general: Put forth a bold vision of what it might take to maintain all of biodiversity in a region and then work out the details later. The vision will provide direction and motivation for all subsequent work.

Size and Scale of Networks

What does it take to maintain in perpetuity all of biodiversity in a region? This is a complex question. Though it has occupied the minds of conservation biologists since the SLOSS controversy and, in some cases, much earlier (Shelford 1933, Kendeigh *et al.* 1950–51), there is no general answer. Usually the question has been framed for single reserves and one group of organisms. For instance, how big a reserve do we need to maintain populations of neotropical migrant birds? But we now know that single-reserve analyses are incomplete, especially when species are distributed as metapopulations. For birds, the regional abundance and spatial pattern of forest, not just the size of individual forest tracts, are necessary considerations because dispersal of birds from other forests can be important in maintaining populations (Askins *et al.* 1987). Unless suitable habitats are truly isolated, with no interchange between them, we must consider size and scale in a regional context.

Some general considerations for determining optimal size and scale of reserve networks have been presented. They include all the criteria for capturing hot spots and representing ecosystems discussed in Chapter 4 and criteria for maintaining viable populations of those species with the largest area requirements (typically large carnivores). They also include allowing for the operation of natural disturbances and other key processes and maintaining opportunities for organisms to migrate and otherwise adjust their distributions to changing environments.

We recommend that a draft reserve system first be designed to meet the representation objectives discussed in Chapter 4. This draft system would include at least one representative of each major habitat or vegetation type,

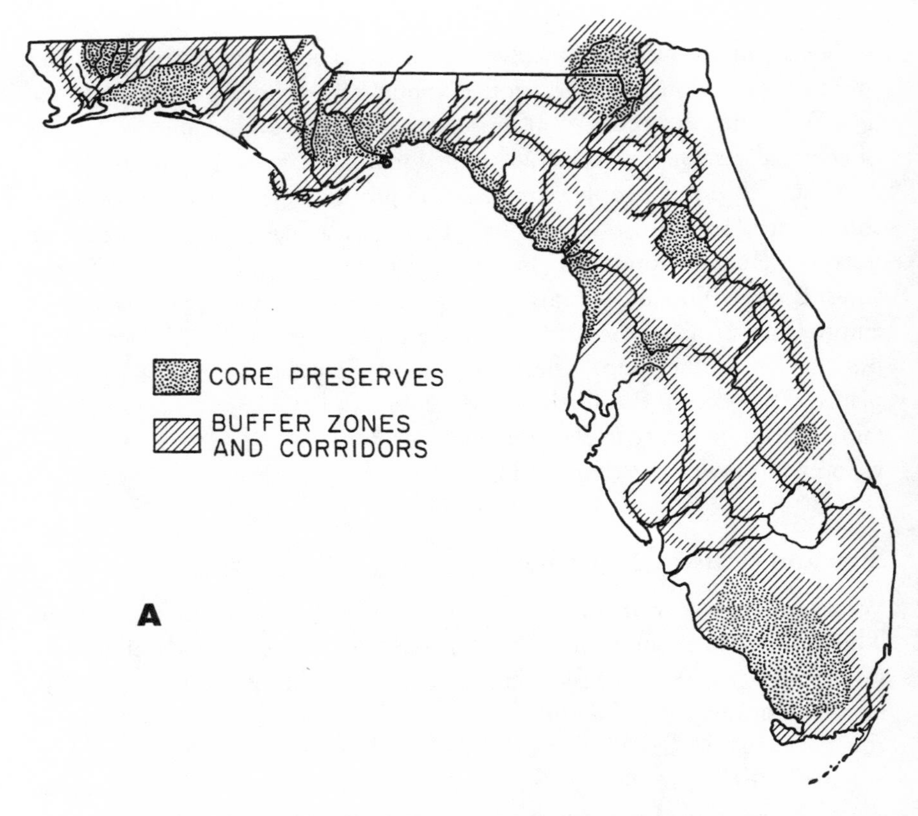

FIGURE 5.9 (A) A proposed reserve network for Florida. The system was designed to capture hot spots of endemism and large areas with low road density, and to provide for movements of wide-ranging species, especially Florida panther and black bear (modified from Noss 1987a). (B) Ecological Resource Conservation Areas, showing existing public lands, private reserves, proposed state acquisitions, and areas of conservation interest, as determined by a 1991 workshop sponsored by The Nature Conservancy, Florida Audubon Society, and Florida Department of Natural Resources. Used by permission of the Florida Natural Areas Inventory.

and, for those taxa with distributional data available, at least one population of every extant species native to the region. Depending on the biogeography of the region, the draft system based on these objectives might include some 10 to 50 percent of the total land area, or in rare cases much more (Ryti 1992).

The next step is to identify the species with the largest area requirements and figure out what it takes in area and connectivity to provide a high probability for maintaining their populations for several hundred years. This step might involve formal population viability analyses (see Soulé 1987,

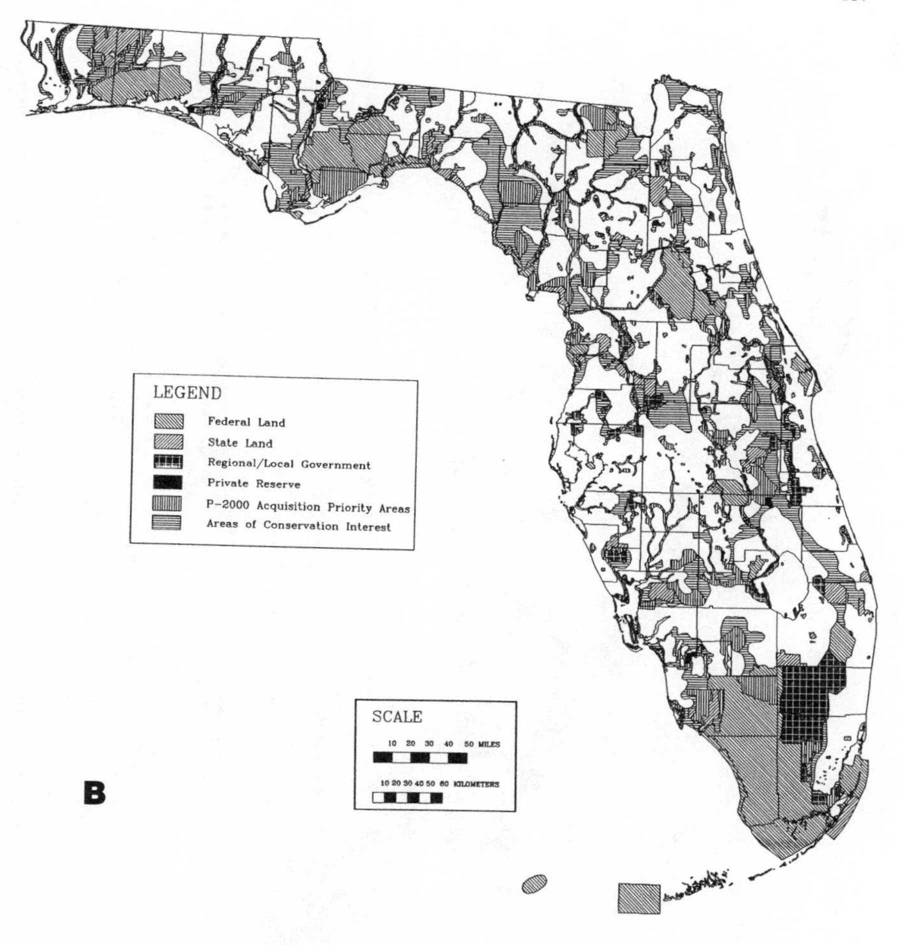

Boyce 1992), or if not enough demographic data exist (as is often the case), area requirements should be estimated on the basis of the best available information and professional judgment of conservation biologists. Finally, one should distinguish the processes necessary to maintain healthy ecosystems in the study region, such as hydrological and disturbance–recovery regimes, and determine how much more area is needed for these processes to operate effectively.

We hasten to note that strictly protected core reserves are not needed to serve all these functions. Most species and processes will probably persist

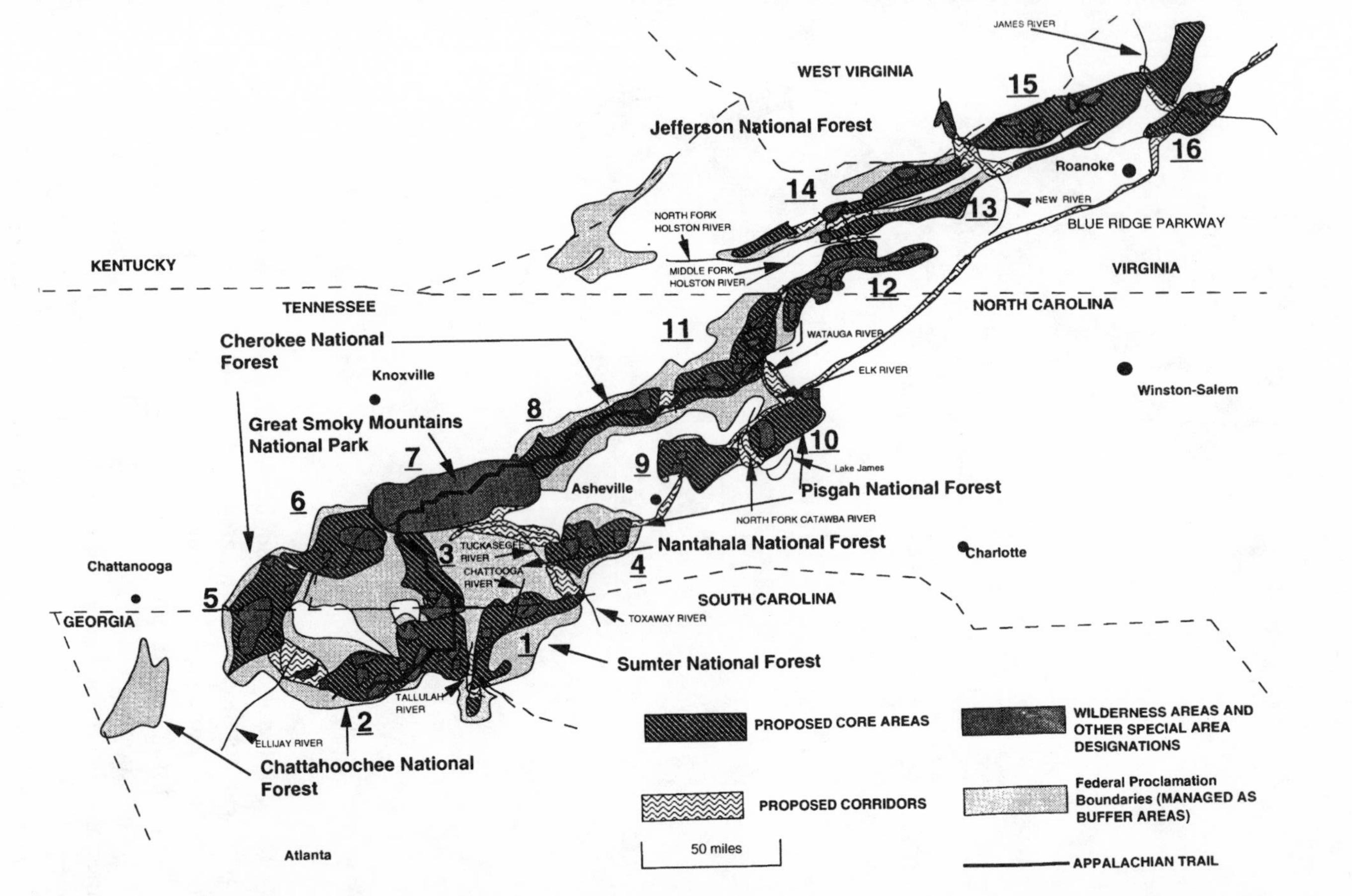

FIGURE 5.10 Proposed core and corridor areas in the Blue Ridge Province of the Southern Appalachian Bioregion (from Newman *et al.* 1992). Used with permission of the authors and the Cenozoic Society.

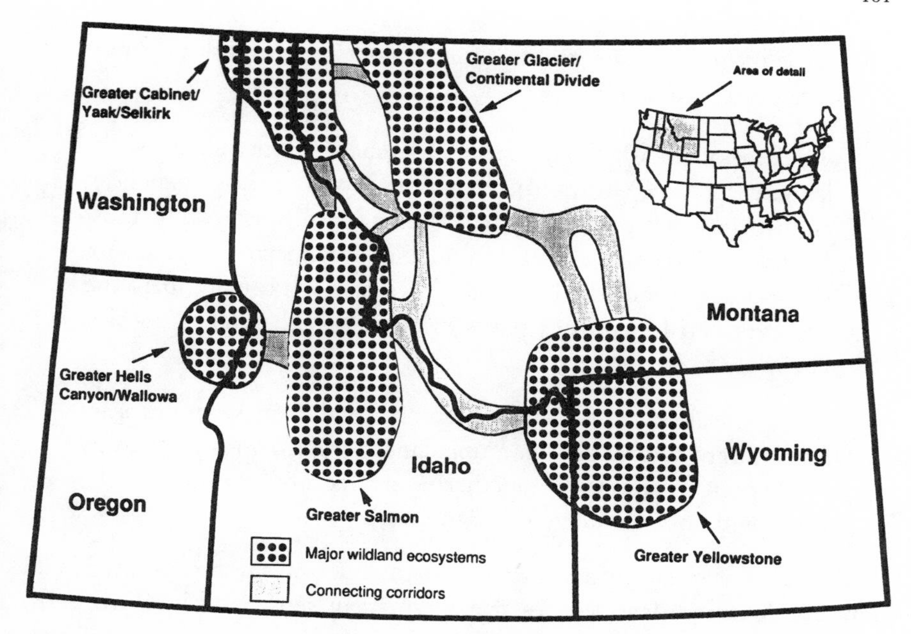

FIGURE 5.11 Conceptual plan for protecting and linking major wilderness areas in the northern Rocky Mountains, as proposed by the Alliance for the Wild Rockies. Used with permission.

in well-managed buffer zones. However, a conservative approach would represent each ecosystem type at least once in core reserves and create a secure network of reserves for large carnivores and other species that are especially sensitive to human activity.

Area Requirements of Large Carnivores

Regional conservation plans, such as those being developed in cooperation with The Wildlands Project, often focus on large carnivores. Species that need a large amount of secure, wild habitat to maintain viable populations should receive priority attention in a regional plan. Otherwise, they will not be around for long. Often, they have already been extirpated and decades of wilderness recovery will be needed to bring them back. Many American conservationists have recognized the significance of large carnivores. The following quotes from prominent ecologists reinforce the point that carnivores deserve special attention:

> The animals requiring first and most careful consideration are the carnivores, likely to be unpopular with the agricultural (broad

> sense including game culture) interests outside the park or forest. (Shelford 1933: 244)

> The National Parks do not suffice as a means of perpetuating the larger carnivores; witness the precarious status of the grizzly bear, and the fact that the park system is already wolfless . . . The most feasible way to enlarge the area available for wilderness fauna is for the wilder parts of the National Forests, which usually surround the Parks, to function as parks in respect to threatened species. (Leopold 1949: 276–277)

> It is in the absence of the large predators that many sanctuaries are not entirely natural and have unbalanced populations of the various species. Very large sanctuaries are required to contain the large predators. (Kendeigh *et al.* 1950–51)

The decline of large carnivores in North America is unprecedented in its rapidity. Within a few decades after permanent settlement by Europeans, cougars and wolves were essentially gone from the East and grizzly bears were absent from most of the West. The extirpation of these species was a consequence of deliberate extermination, promoted by bounties. The Florida panther, virtually extinct, is the last remnant cougar subspecies east of the Mississippi River, though scattered reports from further north are common and may involve western cougars released from captivity. The red wolf was extinct in the wild until recently reintroduced to two areas in the Southeast. Jaguars are gone from the Southwest, except occasional strays that may be trying to escape habitat destruction or hunters in Mexico. The grizzly bear and gray wolf have both been eliminated from over 90 percent of their ranges in the lower 48 states. The same appears to be true for the elusive wolverine. Even black bears have suffered a severe range contraction (Schoen 1990).

Aldo Leopold, especially in his later years, emphasized large carnivores in his writings because he recognized that these predators provide a critical test of society's commitment to conservation (Meine 1988, 1992). It is relatively easy to save a patch of wildflowers or even a few thousand acres of woods for songbirds. No one will call such actions radical. But to advocate protecting millions of acres of roadless habitat for creatures capable of maiming and eating people, even though they do so rarely, requires an extraordinary sense of humility, ecological conscience, and courage.

Leopold also knew from his field observations that a vigorous population of large carnivores is a sign of healthy land. Carnivores at the top of the food chain are correctly considered keystone species, umbrella species, flagship species, and supreme indicators of success in conservation. In some situations, say Long Island or central Illinois, opportunities for predator recovery are unlikely for many years to come. But across much of the United States, including heavily populated areas such as Florida (Noss 1987a) and southern California (Beier 1993), short-term viable populations of at least one large carnivore—cougar—could be reestablished without radical changes in land use. Long-term population viability for these species, however, will require more drastic changes. In any case, a regional conservation plan that fails to create conditions favorable for native carnivores is incomplete.

What would it take to restore long-term viable populations of large carnivores? The estimates may alarm some people. Probably tens of millions of acres of wild habitat are needed for long-term persistence. Schonewald-Cox (1983) estimated that large carnivores require reserves on the order of 1–10 million hectares (2.5–25 million acres). Assuming a short-term minimum viable population of 50, Hummel (1990) estimated that grizzly bear populations in Canada require an average of 49,000 km^2 (12.1 million acres), wolverines about 42,000 km^2 (10.4 million acres), and wolves about 20,250 km^2 (5 million acres).

Grizzly bear densities in the northern Rockies of the United States are somewhat higher than the Canadian average. Assuming four bears per 100 square miles, Metzgar and Bader (1992) calculated that 50,000 square miles (32 million acres) of wild habitat are needed to maintain 2000 grizzly bears in the northern Rockies. This population corresponds to an effective population of only 500 bears, the official U.S. Fish and Wildlife Service recovery goal for the species. As for many mammals and birds, an effective breeding population of grizzlies is about one-quarter of the total population, largely because individuals do not contribute equal amounts of genetic material to the next generation (Allendorf *et al.* 1991). The required area of 32 million acres for a long-term viable population is roughly 60 percent of the U.S. northern Rockies region.

An analysis of recovery requirements for the Florida panther, which once ranged throughout the Southeast, came to similar conclusions (Noss 1991f). A genetically effective population of 50 panthers, which would have a reasonable chance of short-term survival, translates to an actual adult population of 100–200 (Ballou *et al.* 1989). For calculating reserve area, it is useful to focus on male panthers because they have larger and nonoverlapping home ranges that usually encompass the overlapping ranges of one or more females. Assuming an adult sex ratio near unity (generally true for cougar populations; Anderson 1983), a reserve network for short-term viability of panthers should provide for 50–100 males. Given an average male home range of 558 km^2 (137,600 acres) (Maehr 1990), each reserve network would have to encompass about 6.9 to 13.8 million acres. To provide a margin of safety, this estimate was rounded upward to 10 to 15 million acres (Noss 1991f). But again, this is only the area needed for short-term viability. For long-term persistence, the U.S. Fish and Wildlife Service (Jordan 1991) again calls for an effective population of 500, which (they fail to note) translates to 1000–2000 adult panthers requiring a total of 100-150 million acres of wild land (Noss 1991f). This is about the size of the entire states of Florida, Georgia, South Carolina, and Alabama combined, or roughly 60–70 percent of the original range of the Florida panther in the Southeast. Whether or not this much land is really needed for long-term persistance of the panther is uncertain, of course. But given the available data, it is a credible hypothesis.

Single reserves of 100–150 million acres are highly unlikely anywhere in temperate North America in the near future. But the problem becomes less daunting when we recognize that viable populations of panthers and other large carnivores need not be contained within single reserves. Rather, they can be distributed over a much larger area, but in much smaller units, as a metapopulation. This is where the concept of regional reserve networks comes into play.

From what we can gather from historical records, populations of many large carnivores were virtually continuous across much of their presettlement ranges in North America. Most of these species are habitat generalists, limited mainly by food supply, human persecution, or social behavior (territoriality or mutual avoidance). Areas of unsuitable habitat, such as desert playas and high peaks, were only partial barriers to movement. Cougars commonly disperse distances of 100 miles or more (Anderson 1983) and usually seek cover in riparian vegetation or other shelter when moving across broad areas (Young 1946). Therefore, populations and even classified subspecies of cougars, wolves, bears, and other carnivores were probably genetically connected by occasional dispersal over immense areas. Population viability of these species should not be analyzed within single parks, watersheds, or even physiographic regions, but instead across a scale of habitat that corresponds to the genetic and demographic structure of the species. This often will mean several interconnected regions, for example, the entire southeastern United States for the Florida panther and the entire Rocky Mountains for the grizzly bear.

Noss (1991f) suggested that the recovery goal of an effective population of 500 panthers, or 1000–2000 adults, could be met by establishing 10 smaller populations of 100–200 panthers, each requiring about 10–15 million acres (or perhaps less, if areas where panthers are reestablished support higher prey densities than in the South Florida habitats where home range sizes were calculated). The system could be managed as a "metapopulation of metapopulations" in reserve networks. Each regional reserve network would consist of several core reserves and buffer zones linked by wide habitat linkages that would permit dispersal movements. How land is apportioned among reserves would depend on the distribution of suitable and restorable wildland relative to developed areas. These small metapopulations would be connected by restored long-distance linkages, which can be developed over the next few decades, to enhance long-term persistence of the larger metapopulation of 1000–2000 adults across the Southeast. If necessary to prevent genetic problems before interregional linkages are restored, individual animals can be translocated from one population to another (translocations for this purpose have now been approved for recovery of the Florida panther).

Area Requirements for Natural Disturbance Regimes

A small reserve is vulnerable even to natural events. One lightning strike, for example, could result in a fire that destroys the last stand of old-growth forest in an eastern state. For a forest type, such as ponderosa or longleaf pine, that requires frequent fire to persist, the low chance of lightning striking a small, isolated reserve is just as great a threat. Without fire, fire-sensitive species will invade and radically change community structure. For these kinds of reasons, small reserves usually require more intensive management to maintain the conditions for which they were set aside (Pyle 1980, White and Bratton 1980).

How big does a reserve need to be to maintain a natural disturbance regime? First, it is helpful to visualize a landscape as a "shifting mosaic" of patches in various stages of recovery from disturbance (Cooper 1913, Watt 1947, Bormann and Likens 1979). Shifting mosaics in forest ecosystems will be considered in more detail in Chapter 6. For now, we need only recognize that reserves that are small relative to the spatial scale (patch size) of disturbance may experience dramatic fluctuations in the proportions of different seral stages over time, which in turn may threaten populations that depend on certain stages. If a core reserve is to maintain a reasonably stable mix of seral stages and species, it must be large enough that only a relatively small part of it is disturbed at any one time. Another requirement is that a source of colonists (that is, a reproducing population of the same species) exists within the reserve or within dispersal distance so that populations can be reestablished on disturbed sites. Optimal reserve size can therefore be estimated by knowing something about the scale of natural disturbance and the landscape context in which the reserve exists.

Pickett and Thompson (1978) defined a "minimum dynamic area" as "the smallest area with a natural disturbance regime, which maintains internal recolonization sources, and hence minimizes extinction." In theory, a minimum dynamic area should be able to manage itself and maintain habitat diversity and associated native species with no human intervention. Shugart and West (1981) estimated that landscapes must be some 50–100 times larger than average disturbance patches to maintain a relative steady state ("quasi-equilibrium") of habitats. In a steady-state landscape, the proportions of seral stages in the overall landscape would be relatively constant over time, even though the sites occupied by different stages would change.

A limitation of the Shugart and West model is that all disturbances are assumed to be the same size, whereas real disturbance regimes create a range of patch sizes, increasing diversity (Baker 1992a). Furthermore, a steady state may not be a natural condition in some ecosystem types, such as those regularly experiencing large, catastrophic fires (Baker 1989). But the concept is still useful because landscapes that are close to steady state will experience less radical fluctuations in seral stages and associated species populations than smaller areas and consequently should experience fewer extinctions (Shugart and Seagle 1985). The greatest stability occurs when the disturbance interval is long compared with recovery time and only a small portion of the region is affected. Although no reserve size can guarantee stability, larger reserves have a lower probability of major shifts in landscape dynamics caused by rare disturbance events (Turner *et al.* 1993).

Keep in mind that, although absolute stability or equilibrium is not the way of nature (Botkin 1990), extreme fluctuation is also abnormal in most ecosystems and, when caused by human activity, is often what threatens biodiversity. As pointed out by Pickett *et al.* (1992):

> The new paradigm in ecology can, like so much scientific knowledge, be misused. If nature is a shifting mosaic or in essentially continuous flux, then some people may wrongly conclude that whatever people or societies choose to do in or to the natural world is fine. The question can be stated as, "If the state of nature is flux, then is any human-generated change okay?" . . . The answer to this question is a resounding "No!" And the resonance is provided by the contemporary paradigm and the ecological knowledge that underwrites it. Human-generated changes must be constrained because nature has *functional, historical,* and *evolutionary*

> limits. Nature has a range of ways to be, but there is a limit to those ways, and therefore, human changes must be within those limits.

Baker (1992a) proposed that perpetuation of natural disturbance regimes should be a fundamental design goal of nature reserves. Attributes of disturbance regimes and associated landscape patterns are essential, higher order expressions of biodiversity. One need not defer to species-level concerns, as did Pickett and Thompson (1978), in defining minimum dynamic areas, but can look to the landscape directly (Baker 1992a). For example, a certain disturbance-produced mosaic of seral stages might be needed to maintain viable populations of moose in a region, but that mosaic is also an expression of landscape diversity in its own right.

Although no one can specify exactly how large a reserve should be relative to natural disturbance regimes, large reserves will minimize management problems because (1) disturbances will not affect an entire reserve at once, leaving it open to invasion by weedy species; (2) disturbances will be less likely to spread from the reserve into adjacent human-occupied lands; and (3) the natural size distribution of disturbances will not be truncated by suppression at reserve boundaries (Baker 1992a). Besides size considerations, locating a reserve so that disturbance initiation and export zones are all contained within its boundaries is sensible. Disturbances can then be allowed to run their course without suppression. Buffer zones may also help control the spread of disturbances (Baker 1992a).

Other Criteria for Scale of Wildlands

Noss (1991b, 1991d, 1992b) recently hypothesized that about 50 percent of an average region needs to be protected as wilderness, or equivalent core reserves and lightly used buffer zones, to restore populations of large carnivores and functional disturbance regimes and meet other well-accepted conservation goals. Noss (1992c, 1993a) subsequently calculated the acreage in proposed Class I and Class II Reserves in the Oregon Coast Range plan (see Chapter 4), as determined from a variety of criteria, and it added up to almost precisely 50 percent of this region of 5.3 million acres. It was recognized that connections to other regions (the Siskiyou Mountains to the south and Cascades to the east) would be necessary to restore viable populations of large carnivores.

Other ecologists and conservationists have converged on similar estimates of the optimal amount of wild habitat for a region. Philosopher Arne Naess proposed an ideal balance of one-third wilderness with no human habitation, one-third "free nature" (containing mixed communities of humans and other species living in largely nondomesticated environments), and one-third intensive human land uses (cited in Sessions 1992). Recently ecologist Malcolm Hunter has called for a similar "triad" of reserves, lightly managed forest, and intensive tree farms for forest regions (Hunter and Calhoun 1994). Ecologist Paul Sears called decades ago for 25 percent of the United States to be preserved as wilderness. More recently, authors Edward Abbey and Dave Foreman independently suggested 50 percent as a reasonable compromise (D. Foreman, personal communication).

Considering sustainability of ecosystem processes and quality of life for people, Eugene Odum (1970) recommended that 40 percent of the state of Georgia remain as natural area, 10 percent in urban-industrial systems, 30 percent in food production, and 20 percent in fiber production. Odum pointed out that Georgia is an excellent microcosm for the United States as a whole, as it has close to the mean human population density and land use patterns. Similarly, Odum and Odum (1972) proposed that half natural and half cultural land use in southern Florida was optimal ecologically, economically, and culturally. As discussed in Chapter 4, calculations of the area necessary to represent all species and ecosystem types in a region can run as high as 99 percent, but are usually in the range of 25 to 75 percent. The range of estimates is large because of inherent differences in distribution patterns of organisms among regions.

The amount of wild habitat needed to meet the conservation goals espoused in this book will vary from region to region in response to numerous factors (Table 5.2). The optimal amount of wild land will depend, in part, on the kinds and intensities of human uses on nonreserved lands and on what we consider an acceptable quality of life for ourselves as well as other creatures. However, the convergence of estimates cited above and summarized in Table 5.3 leads us to suggest that most regions will require protection of some 25 to 75 percent of their total land area in core reserves and inner buffer zones, assuming that this acreage is distributed optimally with regard to representation of biodiversity and viability of species and well connected within the region and to other reserve networks in neighboring regions. Protection should not imply a "lock-up," as many core reserves and buffer zones can accommodate a variety of human uses, so long as they are compatible with conservation objectives.

Our estimate of area needed in reserves is an order of magnitude beyond what is currently protected in most regions. We harbor no illusions about our vision being easy to accomplish. A smaller and better-educated human population is ultimately required. During the many years it will take to establish reserve networks of this scale, ways of managing land for human commodities and biodiversity may be improved. But because success in

Table 5.2. Factors That Influence Estimates of the Proportion of a Region That Must Be Protected in Reserves to Meet Conservation Goals

Factor	Type of influence
Size of region	The size of region considered will affect estimates of protected area needed. A small wild region containing communities not represented elsewhere may need to be protected entirely. Many regions are too small to maintain viable populations of carnivores without connection to other regions.
Heterogeneity	A highly heterogeneous region, with many habitats and associated species assemblages, will require more area to meet representation goals than will a more homogeneous region.
Classification	The system used to classify vegetation or other community types will determine the area needed to represent each type. Systems that recognize more types, or lower levels in a hierarchical classification, will lead to larger area estimates.
Replication	The number of sites in which a species or community type must be protected to meet representation goals will influence the total area required.
Unit size	The size of individual reserves in which a species or community is represented will influence estimates of total area required.
Area requirements	According to home range size and other aspects of autecology, species require different areas to maintain viable populations or metapopulations.
Population viability criteria	The length of time considered and specified probability of persistence will influence target population size and area required. Models with different assumptions or parameter values will lead to different estimates.
Habitat quality	Quality of habitat will, in part, determine population density of target species and therefore area required for persistence.
Human uses inside	Human activities inside reserves, for example livestock grazing or hunting, will affect habitat quality and survivorship of target species. If uses have a negative impact, larger areas will be required.
Human uses outside	Human activities outside reserves will determine amount of habitat needed in reserves. If multiple-use lands support populations of target species, reserve area may not need to be as large.
Natural disturbances	Regions with natural disturbance regimes characterized by large stand-replacing disturbances (e.g., northern Rocky Mountains conifer forests) will require more area in reserves than areas with small patch disturbances (e.g., eastern deciduous forest).
Connectivity	Connectivity of reserves and other suitable habitats, both within and among regions, will affect area estimates. If reserves are well connected, the total area needed in reserves may be smaller.
Human quality of life	People differ in their need for natural areas to provide a high quality of life. People with more tolerance for domesticated environments will be biased toward smaller area estimates for reserves.
Sustainability	The ability of a region to maintain optimally functioning ecosystems (in terms of services to human society) and be energetically self-sufficient depends on some high percentage of the region maintained as natural area (Odum and Odum 1972).
Politics	A desire to be politically reasonable or expedient will bias estimates of needed protection downward, often well below estimates based on biological criteria.

TABLE 5.3 Estimates of the Proportion of a Given Region Needed to Accomplish Conservation Goals

Region and authors	Goal	Proportion needed
Australian wetlands (Margules *et al.* 1988)	Represent each plant species at least once	4.6% of total number of wetlands, but 44.9% of total wetland area
	Represent all wetland types and all plant species at least once	75.3% of total wetland area
Islands in Gulf of California (Ryti 1992)	Represent all bird, mammal, reptile, and plant species at least once	99.7% of total area
Canyons (habitat islands) in San Diego County (Ryti 1992)	Represent all bird, mammal, and plant species at least once	62.5% of total area
State of Idaho (Scott *et al.* in press)	Represent all vertebrate species at least once	4.6% of total area[a]
	Represent all endangered, threatened, and candidate vertebrates and plants at least once	7.7% of total area
	Represent all 119 vegetation types at least once	8% of total area
Northern Rocky Mountains of United States (Metzgar and Bader 1992)	Maintain an effective population of 500 grizzly bears (actual population = 2000)	32 million acres, or roughly 60% of region

Southeastern United States (Noss 1991f)	Maintain an effective population of 500 Florida panthers (actual population = 1000–2000)	100–150 million acres, or roughly 60–70% of original range
Oregon Coast Range (Noss 1992c, 1993a)	Capture all clusters of rare species and community occurrences, protect all remaining primary forest, provide for large carnivore recovery	About 25% of region within each of two categories of reserves and an additional 25% in buffer zones
Average region in the United States (Noss 1992b)	Maintain viable populations of large carnivores and sustain natural disturbance regimes	Roughly 50% of region
Average region (A. Naess, cited in Sessions 1992)	Optimize human and nonhuman well-being	1/3 wilderness, 1/3 mixed communities of humans and other species, 1/3 intensive human use
State of Georgia (Odum 1970)	Optimize ecosystem services and human quality of life in self-sufficient system	40% natural, 10% urban–industrial, 30% food production, 20% fiber production
South Florida (Odum and Odum 1972)	Optimize ecosystem services and economic and cultural well-being	50% natural, 50% developed

[a]Area as represented by 389 equal-area hexagons of 635 km^2 each.

multiple-use management is uncertain, and because wild places are valuable for their own sake, the prudent course is to risk erring on the side of protection.

How Does Our Present Reserve System Compare?

According to the report *America's Biodiversity Strategy: Actions to Conserve Species and Habitats* (USDI–USDA, 1992), 10 percent of the land area in the United States is in "specially-protected areas"; 90 percent of this in national parks, wildlife refuges, wilderness areas, and wild and scenic rivers. The report claims that "[l]ittle or no development or human activity, other than controlled recreation, is allowed in these specially-protected areas,"—failing to mention that 35 percent of designated wilderness areas are open to livestock grazing (Reed *et al.* 1989); that national wildlife refuges are seriously threatened by livestock, logging, exotic species, trespassing, energy development, pollution, heavy sport hunting, and other incompatible activities; and that many national parks are overdeveloped by private concessionaires. The true figure for the proportion of land strictly protected in the United States is unknown, but is probably less than 3 percent.

Among the conspicuous oversights in the USDI–USDA report is the lack of any recognition of the inadequate size, connectivity, and buffering of existing reserves, or of their failure to represent ecosystem diversity. Sizes of existing reserves are far too small to meet most conservation objectives (Fig. 5.12). Research natural areas, designated for scientific and ecological values, are tiny—93 percent are smaller than 1000 ha (2500 acres), and the remaining 7 percent are smaller than 5000 ha (Noss 1990b). National parks are also dominated by small areas, with 55 percent of units smaller than 1000 ha (Schonewald-Cox 1983). Wilderness areas are somewhat larger, with most between 1000 and 100,000 ha, but only 12 percent are larger than 100,000 ha and only 1 percent larger than 1 million ha (Noss 1990b).

As of 1989, only about 1.8 percent of the lower 48 states was designated wilderness, or 4 percent of the United States including Alaska (Watkins 1989). Although the National Wilderness Preservation System has grown over the years, the amount of land that qualifies as wilderness in the United States has declined precipitously. In 1936 there were no designated wilderness areas, but an estimated 150 million acres of roadless lands in the lower 48 states qualified as wilderness. When the Wilderness Act was passed in 1964, 9.5 million acres were designated and 100 million acres of de facto wilderness remained. By 1985 the statutory wilderness system had grown to 32 million acres, but total wilderness (designated and de facto) totaled only 86 million acres (Wolke 1991). Wilderness is dwindling, not increasing.

Moreover, designated wilderness areas do a poor job of representing

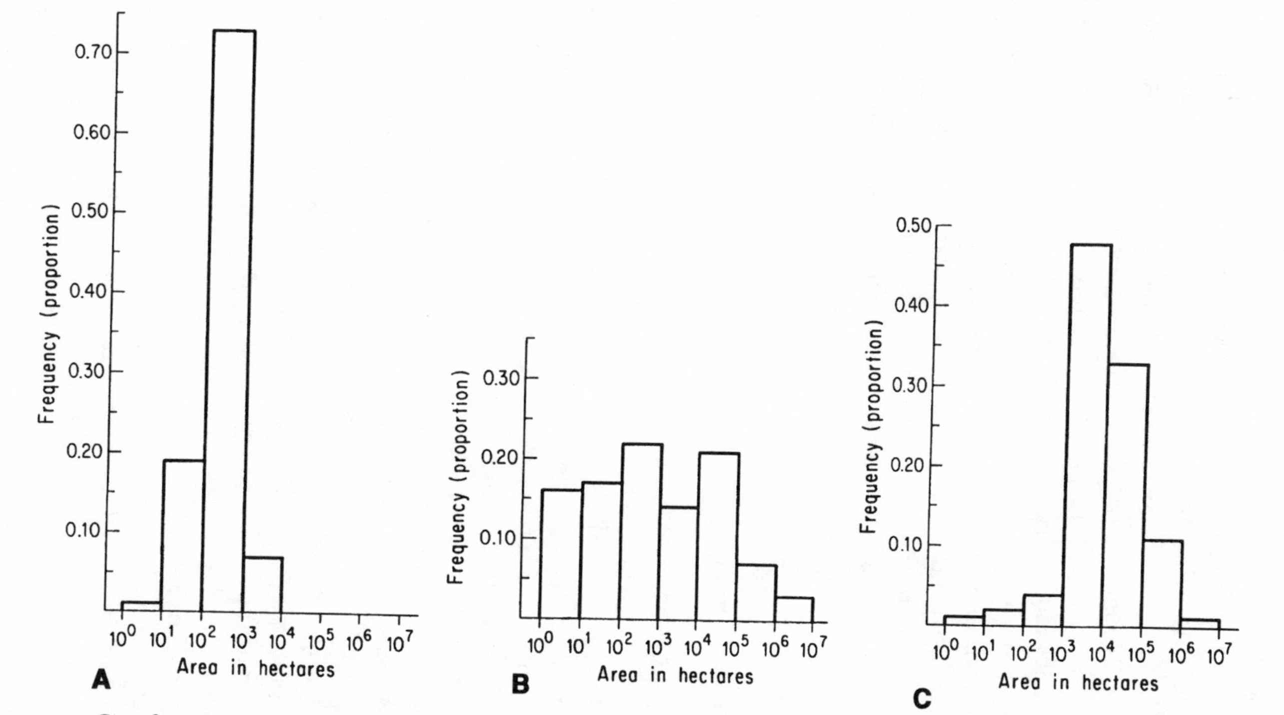

FIGURE 5.12 Size frequency distribution of (A) 213 Forest Service research natural areas; (B) 320 units in the national park system; and (C) 474 designated wilderness areas. From Noss (1990b).

ecosystem diversity. Many are truly rocks and ice (Forman and Wolke 1989, Wolke 1991). Davis (1988) found that 104 (40 percent) of 261 major terrestrial ecosystems in the United States and Puerto Rico (Bailey-Kuchler ecosystems; see Davis 1988, Noss 1990b) were not represented at all in wilderness areas. Applying a modest size criterion of 100,000 ha, only 50 (19 percent) of Bailey-Kuchler ecosystems were found in wilderness areas. East of the Rocky Mountains, only four ecosystem types are represented in wilderness areas in 100,000-ha units. Using a more ecologically reasonable size criterion of 1 million ha, which might be sufficient for short-term survival of large carnivores, only five (2 percent) of the 261 Bailey-Kuchler types were represented, all of them in Alaska (Noss 1990b).

The present system of protected areas in the United States is deficient. A major conservation priority is to use the criteria reviewed in Chapter 4 and in this chapter to expand dramatically the amount and quality of land protected throughout the nation.

Management Considerations

Establishing a reserve network is a necessary step toward maintaining biodiversity in a region, but it is not sufficient. The system of reserves, buffer zones, and linkages must then be managed. We discuss details of management for forests, rangelands, and aquatic ecosystems in forthcoming chapters. Here we offer only a few general comments.

Perhaps the key principle for managing landscapes for biodiversity is prudence: be cautious, move slowly, stay out of sensitive areas, avoid overmanipulation of habitats. Someday we may know more about how to manage land for all its values. For now we must concentrate on protecting the most sensitive components of the ecosystem. Prudence does not imply hands-off management in most cases. Indeed, much human labor will be required to protect a reserve from harm, to substitute for natural processes that have been disrupted, and to restore damaged habitats and recover populations. But we need to be careful.

In the case of the reserve system proposed for the Oregon Coast Range (see Chapter 4), management guidelines were offered for each of three land protection categories and for private lands outside the reserve network (Table 5.4). Similar guidelines would apply to other landscapes, particularly forested ones, where reserve networks are developed. The Coast Range strategy, like most regional case studies of The Wildlands Project, is a "100-year plan." Some elements of the strategy, such as carnivore reintroduction, might not be fully implemented for decades. But other actions are needed

TABLE 5.4 Protection, Management, and Restoration Guidelines for Three Categories of Land Protection and Undesignated Lands in the Oregan Coast Range[a]

Class I Reserves

No logging in primary forests of any age. No other timber cutting except thinnings and other silvicultural manipulations designed to restore plantations to natural structure and composition.

No new road construction or reconstruction.

Prompt closure of all roads, except major highways and other roads necessary to access private property or to conduct restoration activities. Obliterate and revegetate roadbeds. Reduce overall road density to well under 0.5 miles/square mile.

Limited trail systems and other access (follow typical wilderness area standards).

Initiate land acquisition programs to consolidate federal and state holdings by purchase of inholdings and other private lands. Some private lands can be protected by conservation easements and some may house reserve managers and researchers.

No grazing of domestic livestock.

No horses.

No mineral, oil, or gas exploration; no mining.

No collection of plants or other natural objects for commercial purposes.

Eliminate exotic (introduced) species, as feasible.

No control of native insects or diseases.

Fire suppression to be determined on a case by case basis, but generally discouraged (particularly after private inholdings have been purchased).

Reintroduce extirpated species (e.g., large carnivores) after road density sufficiently reduced, private lands or easements acquired, and feasibility studies suggest high probabilities for survival.

No off-road vehicles or other motorized equipment or mountain bikes.

Hunting permitted only as necessary to control herbivores.

Hiking, primitive camping, nature study, environmental education, and nonmanipulative research encouraged (manipulative research in the form of restoration experiments also encouraged on human-disturbed sites).

Class II Reserves

No logging in primary forests of any age. No other timber cutting except thinnings and other silvicultural manipulations designed to restore plantations to natural structure and composition or to accelerate development of old-growth characteristics in second growth.

No new road construction or reconstruction.

Prompt closure of unnecessary roads on public lands, with obliteration and revegetation of roadbeds. Gradual reduction of overall road density to no more than 0.5 miles/square mile.

[a] From Noss 1993a.

(Continued)

TABLE 5.4 (*Continued*)

Class II Reserves (*Continued*)
Trail system more extensive than in Class I Reserves, but limited enough to provide security to sensitive species.
Pursue acquisition of private lands or conservation easements, but can be more gradual than in Class I Reserves. In some cases, management agreements with landowners will suffice.
No off-road vehicles (except mountain bikes) or other motorized equipment on public lands.
No grazing of domestic livestock on public lands.
No mineral, oil, or gas exploration; no mining.
No collection of plants or other natural objects for commercial purposes.
Eliminate exotic (introduced) species, as feasible.
No control of native insects or diseases.
Fire suppression to be determined on a case by case basis, but generally discouraged (particularly after private inholdings have been purchased).
Reintroduce extirpated species (e.g., large carnivores) after road density sufficiently reduced, private lands or easements acquired, and feasibility studies conducted.
Hiking, horseback riding, mountain biking, legal hunting, primitive camping, nature study, environmental education, and nonmanipulative research encouraged (manipulative research in the form of restoration experiments also encouraged on human-disturbed sites).

Multiple-Use Buffer Zones
No logging in primary forests of any age. Timber management in plantations and some second growth permitted, but emphasizing long rotations, "New Forestry," selection logging, and other silvicultural systems that seek to simulate natural disturbance–recovery regimes. Restoration forestry and sustainable forestry experiments encouraged.
No new road construction or reconstruction.
Gradual reduction of overall road density to no more than 1.0 mile/square mile, except where higher densities are necessary to access private property.
Pursue conservation easements and management agreements with private landowners.
No motorized off-road vehicles on public lands.
Eliminate exotic (introduced) species, as feasible, on public lands.

Undesignated Lands
Practice sustainable resource production.
Protect riparian zones and other sensitive sites.

immediately. Otherwise, future conservation options will be limited. For the Coast Range, several priority conservation actions were proposed:

- A moratorium on logging of old growth and other virgin forests;
- A rapid phasing out of road construction, with existing roadless areas protected;
- Closure of unnecessary roads;
- A moratorium on development in all natural and near-natural habitats, instead channeling development into areas already manipulated or degraded;
- Initiation of restoration projects, both short-term and long-term; and
- Public education about what it takes to fully restore the richness of life in this remarkable region

These proposed short-term actions are not uncontroversial, but land managing agencies in the Coast Range region have shown a sincere interest in the plan. We feel that bold conservation plans will generally be taken seriously if they are scientifically defensible and have strong support from at least some local people. In the Coast Range, a regional group with several hundred members (the Coast Range Association) is vigorously advocating the plan and has gained support from other conservationists. It is too early to say that the Coast Range strategy or similar plans in other regions will be successful, but clearly the combination of science and citizen involvement can powerfully influence the future of a region. We hope to see this influence increase in all regions.

CHAPTER SIX

MANAGING FORESTS

> In many respects the attitude of the forester toward a forest is radically opposed to that of the ecologist. To the former it represents merely the means to an end, to the latter it is the end in itself. The fundamental idea . . . in the forester's treatment of the forest is utility. He estimates the value of a tract of woodland in board feet. His chief ambition is to secure a maximum yield per acre of the most desirable lumber. He regards the sawmill as the logical destination of every healthy tree. To him an over mature stand of heart-rotted veterans is an eyesore—they should be felled without delay to provide space for younger generations. The ecologist, on the other hand, sees in such a group of trees the glorious consummation of long centuries of slow upbuilding on the part of Mother Nature . . . To precipitate their downfall with the axe seems little short of desecration.
>
> G.E. Nichols (1913)

The point of view expressed throughout this book is decidedly that of the ecologist in the dichotomy depicted above. Whether forestry and ecology are closer in understanding today than when George Nichols wrote his article in 1913 is uncertain. Change in the forestry profession has been slow. The maximum-yield philosophy persists in forestry schools, land-managing agencies, and industry. On the other hand, today many foresters are also ecologists and have contributed much to our understanding of forest ecosystems. Although utility remains a primary concern in forestry—we all use wood and paper—many forest ecologists now recognize noncommodity values in forests and are seeking ways to obtain products from forests without jeopardizing these values. A few forest ecologists have been in the forefront of efforts to inform their scientific peers and the public about the values of natural forests (Franklin *et al.* 1981, Maser 1989, Perry *et al.* 1989a, Hunter 1990, Norse 1990, Zahner 1992) and in educating Congress and other policymakers of the ecological risks involved in forest management decisions (Thomas *et al.* 1990, Johnson *et al.* 1991, Thomas *et al.* 1993).

In this chapter, we offer an overview of forest management and conservation issues in the United States. We first review the ecology of forests and distinctions between natural forests and plantations. We then discuss fragmentation and other threats to forests, reserve management, and multiple-use management.

Ecological History and Principles

Forests of some type or other have occupied the earth for perhaps 400 million years, but their composition has changed continuously throughout this period. Early forests of the Carboniferous period contained *Lepidodendron, Calamites, Sigillaria,* and other ancestors of today's diminutive clubmosses, spikemosses, and horsetails. Then came tree ferns and gymnosperms, including the cycads, ginkgos, and conifers. Lastly, about 140 million years ago, arose the angiosperms, the flowering plants that include hardwood trees. In most regions, forests have more species than other kinds of ecosystems because their large aboveground and belowground biomass, multilayered structure, and horizontal patchiness induced by site factors and disturbances provide niches for many types of organisms (Hunter 1990).

Today forests and woodlands occupy roughly 30 percent of the world's land surface and 30 percent of the United States (Haynes 1990, World Resources Institute 1990). At the time of European settlement, roughly half of the conterminous United States was forested. Forest area declined from the late 1700s until about World War I, stabilized at less than half of the presettlement value, then recovered somewhat due to abandonment of cropland, chiefly in the Piedmont of the Southeast, the Northeast, and the upper Great Lakes states (Clawson 1979). But the forest we see today differs in many ways from the primeval forest. About 85 percent of the primary forests in the United States, and 95 to 98 percent of those in the conterminous 48 states, have been destroyed. Most of what remains is on public lands in the Rocky Mountains, California, Pacific Northwest, and Alaska (Findley 1990, Postel and Ryan 1991).

Although people view remnants of virgin forest as primeval, the associations of species in these communities are often quite young. The "ancient" Douglas-fir forests of the Pacific Northwest took shape only about 6000 years ago when modern climatic conditions were established (Brubaker 1991). However, some of the genera that compose these forests date back to the Arcto-Tertiary flora of at least 25 million years ago. More recent is the Big Woods vegetation of southern Minnesota. Although described as the regional climax, this forest of elm, basswood, sugar maple, and other mesic hardwoods came into being only about 300 years ago, when the climate became cooler and wetter during the Little Ice Age, and the fires that had maintained oak woodland and savanna in this region declined in frequency (Grimm 1984). Then, even more quickly, the Big Woods was all but eliminated by agriculture.

Environmental Gradients

Forests vary in species composition, structure, and ecological processes from place to place, largely due to the response of species to environmental gradients in space and time (Gleason 1926). Variation in microclimate with topography and elevation is a major control of species distributions within forest landscapes. In mountainous regions, the pattern of different forest types and other communities often corresponds predictably to elevation and topographic–moisture gradients, as shown by Whittaker (1956) in his work in the Great Smoky Mountains (Fig. 6.1). Later workers built similar "gradient mosaic" diagrams for other regions (e.g., Fig. 6.2). These diagrams are useful for showing the spatial relationships of vegetation to environment and showing how different communities fit together into a larger landscape pattern.

The response of species to environmental gradients is sometimes controlled by strong affinities to particular soil types. In the Siskiyou Mountains of southern Oregon and northern California, Whittaker (1960) found a vegetation pattern on serpentine (an unusual metamorphic rock rich in magnesium and some other minerals but deficient in calcium and others) very different from that on the more common diorite (a coarse-grained igneous rock). In addition, competition and other biotic factors often limit a species to a smaller portion of the environmental gradient than it might be able to grow in alone. For instance, many plants associated with serpentine actually grow better on more common substrates but are restricted from them by superior competitors.

Although species distributions along environmental gradients might be diagramed most conveniently for mountainous areas, analogous patterns occur on flatter terrain. The vegetation of Florida, for example, provides strong evidence for subtle changes in elevation and soil moisture profoundly affecting plant distribution. Natural pine flatwoods in Florida are dominated by longleaf pine on mesic sites; by slash pine in lower, wetter areas, near the coast, and in south Florida; and by pond pine on some of the wettest sites (Abrahamson and Hartnett 1990). Still wetter areas are typically shrub swamps, cypress ponds, or other wetlands. Ridges in these landscapes, sometimes only a few feet higher than the surrounding flatwoods,

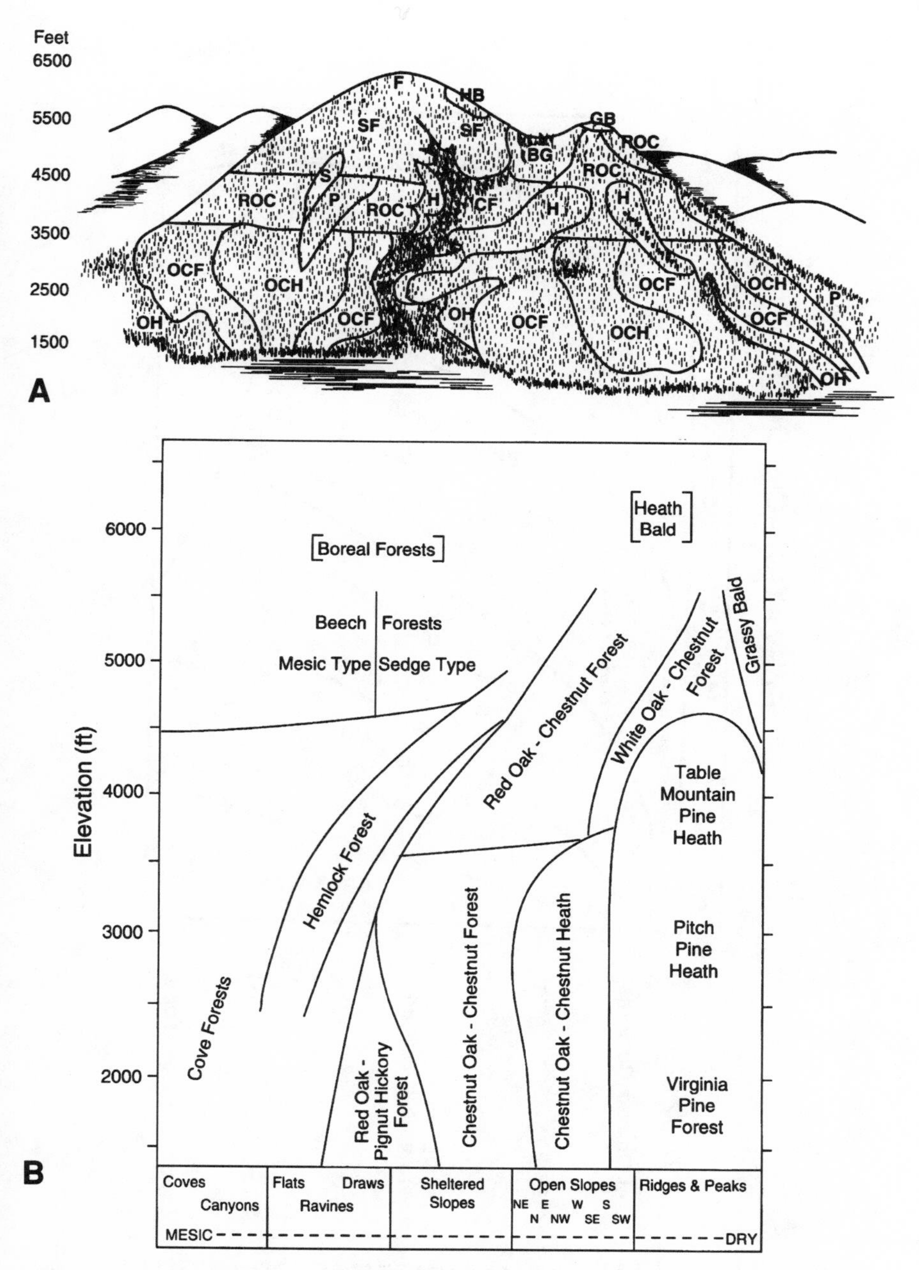

FIGURE 6.1 (A) Topographic distribution of vegetation types on an idealized west-facing slope in the Great Smoky Mountains, USA. Vegetation types: BG, beech gap; CF, cove forest; F, Fraser fir; GB, grassy bald; H, hemlock; HB, heath bald; OCF, chestnut oak-chestnut; OCH, chestnut oak-chestnut heath; OH, oak-hickory; P, pine forest and heath; ROC, red oak-chestnut oak; S, spruce; SF, spruce-fir. (B) Nonboreal vegetation types in the same area, displayed with respect to elevation and topographic–moisture gradients. Both from Whittaker (1956). Copyright: the Ecological Society of America.

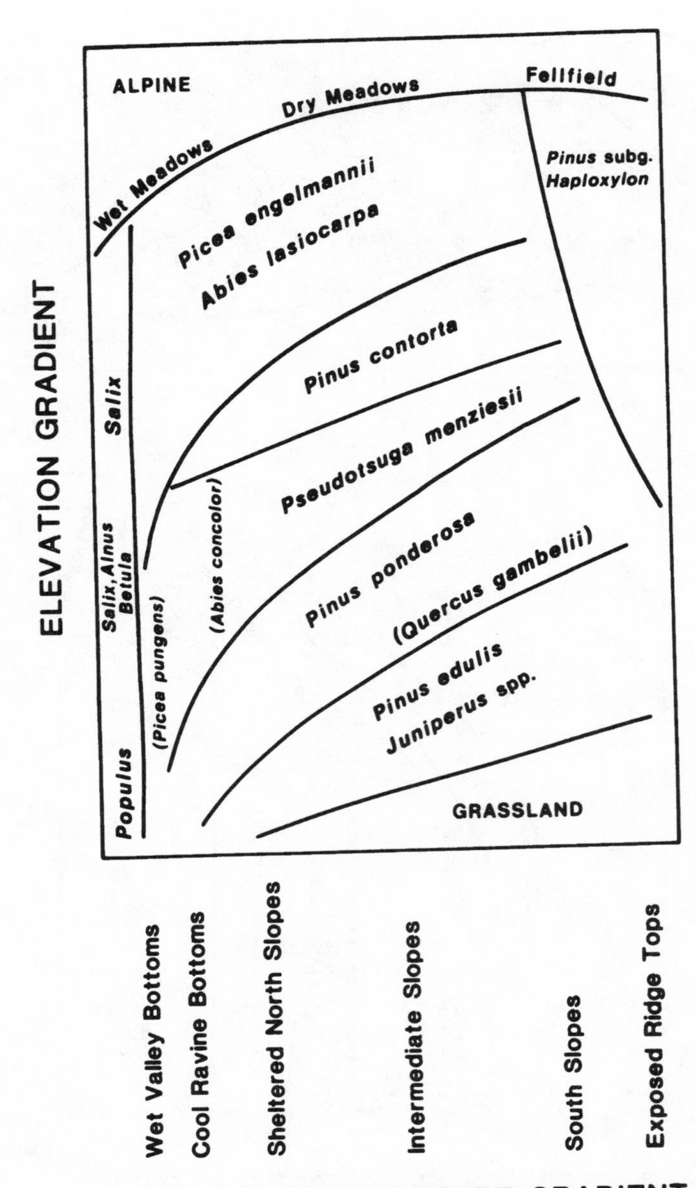

FIGURE 6.2 Major vegetation zones of the central Rocky Mountains, as related to elevation and topographic–moisture gradients. From Peet (1988). Used with permission of the author and Cambridge University Press.

contain longleaf pine and turkey oak, often with islands of scrub dominated by sand pine and various scrub oaks (Myers 1985, 1990; Noss 1988).

Disturbances and Heterogeneity

The distribution of vegetation across a landscape cannot be explained entirely by responses to environmental gradients. Typically, disturbances impose one or more scales of pattern on top of the underlying pattern determined by the physical environment. Canopy gaps, caused by death or fall of trees, and other small disturbances create a fine-grained pattern, whereas stand-replacing fires, major windstorms, and other large disturbances create a coarse-grained pattern (Fig. 6.3). Disturbances are typically patchy in time and space, so that new disturbances occur in some portions of the stand or landscape while previously disturbed areas are recovering. Thus the mosaic of seral stages may appear to shift across the landscape over time (Bormann and Likens 1979). Although many ecologists have only recently come to appreciate the contribution of disturbance and patch dynamics to biodiversity, as evidenced by the dramatic increase in interest in the topic in the late 1970s and 1980s (e.g., White 1979, Pickett and White 1985), scientific recognition of these phenomena dates back at least to the early twentieth century. For example, W.S. Cooper observed the following in his studies of the virgin forest on Isle Royale, Lake Superior:

> The forest is a complex of windfall areas of differing ages, the youngest made up of dense clumps of small trees, and the oldest containing a few mature trees with little young growth beneath ... The result [of disturbance–recovery processes] in the forest in general is a mosaic or patchwork which is in a state of continual change. The forest as a whole remains the same, the changes in various parts balancing each other. (Cooper 1913)

The significance of canopy gaps to forest ecology and biodiversity is now well appreciated (Bray 1956, Williamson 1975, Hartshorn 1978, White 1979, Runkle 1981, Brokaw 1985). In many different forest types across the world, about 0.5 to 2.0 percent of the forest canopy is opened up by new treefalls each year (Runkle 1985). If all areas of the forest were equally susceptible to gaps, these disturbance rates would result in complete turnover of the forest every 50 to 200 years. Complete turnover generally does not happen because some sites are better protected than others, some trees escape mortality simply by chance, and many dominant forest trees grow slowly for many years under a closed forest canopy. Thus, dominant trees across most of the earth have lifespans on the order of 100 to 1000 years.

One function of gaps is to enhance opportunities for regeneration and

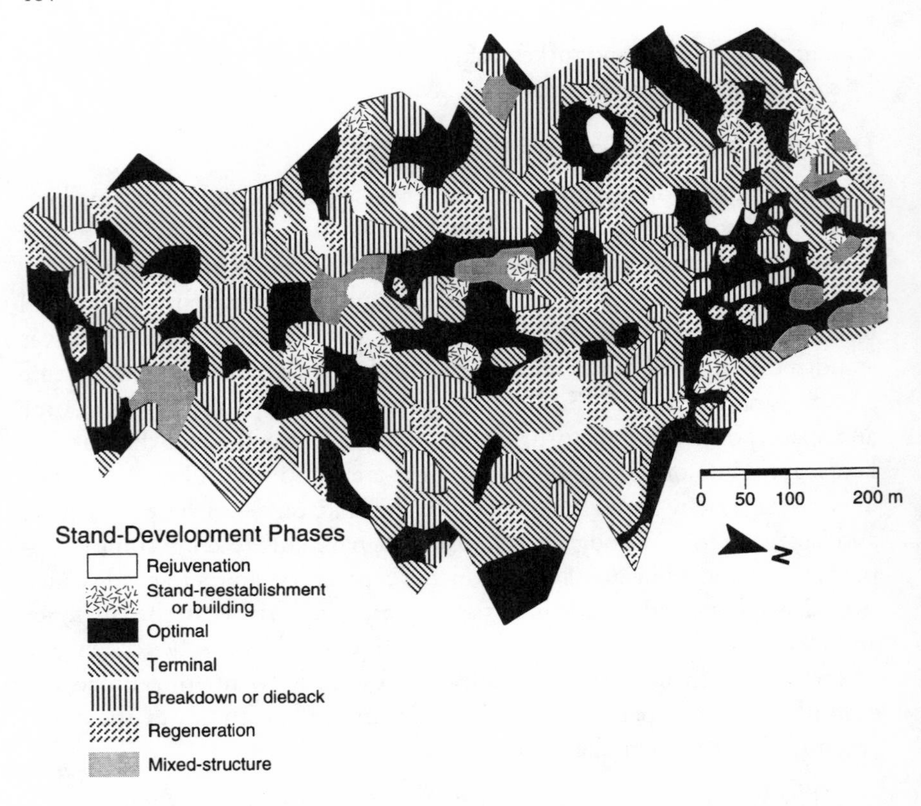

A

FIGURE 6.3 (A) Stand development phases in a 1-km wide section of virgin forest, Yugoslavia. Patches average about one-half ha in size. From Mueller-Dombois (1987), copyright American Institute of Biological Sciences. (B) Fire mortality patches for 1800–1900 in the Cook–Quentin study area, Willamette National Forest, Oregon. Note scale is about 10 km from left to right. From Morrison and Swanson (1990). Reprinted with permission of Peter Morrison.

B
High-mortality patch
Medium-mortality patch
Low-mortality patch
N
Scale: 1 km

growth of trees and other plants. Even the shade-tolerant trees of old-growth forests, while able to persist in a suppressed state for many years under a closed canopy, may require occasional gap episodes in their vicinities to get enough sunlight to grow into the canopy. Trees that fit this description include beech and maple in the eastern deciduous forest and western hemlock in the Pacific Northwest. Even small openings in the canopy can enhance growth for these species (Canham 1989). Other species require larger gaps. For instance, tuliptree (yellow poplar) and yellow birch seldom regenerate in openings smaller than 400 m^2 (Runkle 1985) and some other species highly intolerant of shade require much larger openings.

An important consequence of gap dynamics in forests is the creation of vertical and horizontal heterogeneity. Because of their tall, uneven canopies and older trees, old-growth forests have higher rates of gap formation than younger stands (Clebsch and Busing 1989, Lorimer 1989). This patchiness largely explains why old-growth forests are so diverse (Meslow *et al.* 1981). Horizontal patchiness of vegetation is one of the best predictors of the number of species, especially birds, at both stand and landscape scales (MacArthur *et al.* 1962, Roth 1976, Boecklen 1986, Freemark and Merriam 1986). Furthermore, the snags, down logs, pits, and mounds created by death and eventual fall of trees provide many different microhabitats and regeneration niches for plants and animals. Because old-growth forests possess a larger spectrum of tree sizes than young forests, a broader array of fallen logs is maintained in various sizes and stages of decomposition. Many of these structures persist after a stand-replacing disturbance through later seral stages, but their replacement depends on redevelopment of old growth.

Often, more bird species occur in gaps than in undisturbed forest understory. Researchers in Illinois mist-netted more birds, especially foliage-gleaning insectivores, granivore–omnivores, and migratory frugivores, in gaps than under a closed canopy (Blake and Hoppes 1986, Martin and Karr 1986). Densities of most bird species in a mixed hardwood forest in Florida were higher in plots with abundant gaps. Some birds were gap specialists (Noss 1991g). These studies suggest that naturally patchy old-growth forests often contain more species of birds and other wildlife than the more homogenous second-growth forests.

Stand-replacing disturbances also enhance biodiversity, but they must be viewed at a broader spatial scale. An event that is destructive at a local scale, perhaps killing most of the aboveground biomass on a site, can contribute to diversity at a landscape scale by creating different habitats and opportunities for regeneration. For example, three related species of pine—lodgepole pine in the West, jack pine in the upper midwest, and sand pine in Florida—usually require stand-replacing disturbance, generally fire, to regenerate. All three species have races with serotinous cones that open only under intense heat. Without fire, these short-lived pines are replaced by other species. Lodgepole pine in the Rocky Mountains is successionally replaced by shade-tolerant subalpine fir and Engelmann spruce. Windstorm or insect damage to lodgepole pine will hasten replacement by fir (Fahey and Knight 1986, Peet 1988), an interesting case in which disturbance accelerates, rather than sets back, succession.

The diversifying effects of large disturbances can be attributed largely to the uneven way in which each disturbance affects the landscape. Few natural disturbances wipe out everything in their paths. Islands of surviving vegetation may persist in a lava flow or flood. Most stand-replacing fires, even those as large as the Yellowstone fires of 1988, which burned 400,000 ha (nearly 1 million acres), are not monotonous scorches but a mosaic of many patches of varying fire intensity (Christensen *et al.* 1989). Some patches are untouched, others are burned to bare ground, and a broad assortment fall somewhere in between. For two fire history study areas in the western Cascades of Oregon, one (Fig. 6.3b) experienced fairly frequent, low to medium severity fires that occasionally created even-aged patches, but more commonly thinned preexisting stands. The other area experienced small fires of low to medium severity at higher elevations and infrequent, large stand-replacing fires (Morrison and Swanson 1990). Much of the heterogeneity of large disturbances can be traced to the underlying diversity of topography and microclimate in the landscape. Riparian zones and other moist, protected sites rarely burn, whereas exposed southern and western slopes may burn often.

A fascinating example of how biodiversity can be enhanced by fire interacting with environmental gradients can be found in the Apalachicola lowlands of Florida and some other areas in the Gulf coastal plain. In these landscapes, the dominant vegetation is a fire-dependent ecosystem with longleaf pine in the overstory and wiregrass in the groundcover. The natural disturbance regime is low-intensity ground fires recurring at intervals of 2 to 5 years. As fires burn downslope along elevational gradients usually no more than a few feet high, they prune back wetland shrubs that would otherwise encroach upslope. In the process, they maintain an open herb bog community of pitcher plants, sundews, orchids, and many other unusual plants (Fig. 6.4) (Noss 1988, 1989, Wolfe *et al.* 1988, Noss and Harris 1990). Longleaf pine flatwoods and adjacent herb bogs are incredibly rich in species, often with over 100 herbaceous plants in sites no larger than an acre (Clewell 1989). At least 191 rare and endemic plant taxa are associated with these and similar wiregrass-dominated communities in the Southeast (Hardin and White 1989). The richest aspect of biodiversity in these

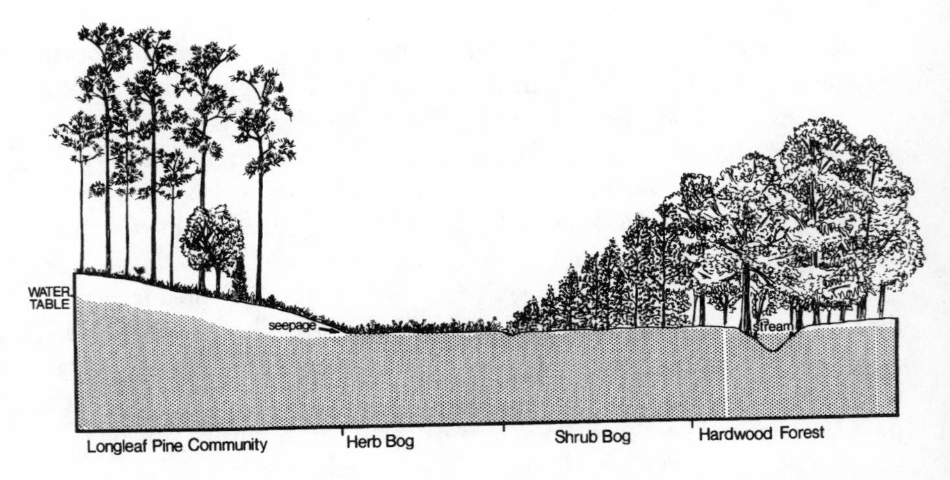

FIGURE 6.4 Open seepage bogs in pine flatwoods of the Florida Panhandle are maintained by frequent fires, which prune back wetland shrubs that would otherwise encroach upslope. From Wolfe *et al.* (1988). Used with permission of Steve Wolfe.

landscapes is beta diversity, the change in species composition along the slope-moisture gradient from sandhills to wetlands. If this gradient is disrupted by fire suppression, plowing of fire lanes, or mechanical site preparation for tree-planting, biodiversity declines markedly (Clewell 1989, Noss and Harris 1990).

Thus, environmental gradients and disturbances of various types and scales are essential to the creation and maintenance of biodiversity in forest landscapes. Without sufficient disturbance, communities are expected to be dominated by a few superior competitors. Conversely, with frequent or intense disturbance, only those few species that can tolerate such extremes will persist. Because the species in any area have adapted to a certain disturbance regime through natural selection, altering that regime—such as by substituting a natural gap replacement process by clearcutting—is bound to affect biodiversity in a negative way.

From Natural Forests to Plantations

For conservation of biodiversity, we need to distinguish between natural forests and those heavily modified or planted by humans. Unfortunately, concepts of natural, virgin, old-growth, ancient, primary, secondary, managed, and unmanaged forests are confused and intertwined, especially in the popular literature. We offer a few definitions to clarify these concepts.

We must first define what we mean by natural. Though one of the most ambiguous in the conservation lexicon, the term *natural* persists because it signifies something of great esthetic and spiritual importance to many people. Naturalness may also have scientific values, not the least of which is that it provides a buffer for our ignorance. That is, natural forests have ecological qualities that we have barely begun to fathom and have no idea of how to replicate. Although generally considered an intangible concept, naturalness has some aspects that can be quantified. Anderson (1991) suggested three criteria for assessing an area's naturalness: (1) the amount of cultural energy required to maintain the system in its present state; (2) the extent to which the system would change if humans were removed from the scene; and (3) the proportion of the fauna and flora comprising native versus non-native species. Forests that can take care of themselves, that would change little if we left them alone, and that are made up of native species are more natural.

We define *primary natural* forests (Table 6.1) as virgin forests of any age, arising through natural regeneration (primary or secondary succession) after natural disturbance. Virgin forests are considered to be essentially uninfluenced by human activities (Peterken 1981, Whitney 1987). *Old growth* has been defined in many ways, but usually refers to older seral stages of natural forests, generally more than 150 years old for most forest types. Most old-growth forests in the United States are primary natural forests, though some are old second growth. In these days of long-range transport of air pollutants, acid deposition, thinning of the ozone layer, and global warming, no forest is entirely free of human influence, but primary natural forests have not experienced substantial logging or other direct human modification. Forests with a history of light selection cutting, signs of which have been largely obliterated by natural decomposition and regrowth, may qualify as primary natural. Nationally, private lands may contain about 5 million acres of primary natural forests; state lands outside of Alaska have less than 5 million acres; and the BLM has less than one-half million in western Oregon. The remainder, about 20 million acres, occurs on national forests, almost entirely in the West (Graecen 1991).

Secondary natural forests, often called second growth, are defined here as those that have regenerated naturally on sites formerly logged, farmed, or otherwise cleared by humans. The time needed to regain naturalness or old-growth characteristics will vary, depending on forest type, severity of past disturbance, climate, soil fertility, and other factors, but may exceed 200 years for most types. Many forests of the eastern and southern United States are secondary natural and some are beginning to regain old-growth characteristics. Second-growth forests in the southern Appalachians have regained most of their natural qualities after decades of "benign neglect"

TABLE 6.1 Generalized Qualities of Primary Natural (Virgin) Forests, Secondary Natural Forests, and Plantations

Characteristic	Primary natural	Secondary natural	Plantation
Origin	After natural disturbance (fire, windstorm, flood, landslide, lava flow, etc.)	After timber harvest, agricultural, mining, or some other human disturbance that involved clearing of preexisting vegetation	Usually same as secondary natural but also may be established on existing natural vegetation (e.g., trees planted in native prairie)
Regeneration	Natural reseeding, recolonization, germination from seed bank, or vegetative propagation from remnant structures	Same as primary natural	Artificial regeneration (planting of seeds, seedlings, etc.)
Within-stand structural diversity	High in many forest types, with abundant standing and down dead wood in various size classes (but less dead wood in forests with frequent fire) and high vertical complexity (multilayered)	Initially low, but gradually increasing with stand age and elevated rates of gap formation	Usually low throughout cycle unless rotations are unusually long, in which case diversity increases as stand ages and disturbance, mortality, injury, and self-pruning occur
Within-stand horizontal patchiness	Usually high due to heterogeneity in physical site conditions (soil type, microtopography, etc.), canopy gap dynamics (especially in old growth), and horizontal variation in disturbance intensity	Variable but initially lower and gradually increasing with stand age and formation of gaps and pit/mound microtopography	Usually low, except in cases of long rotations, where patchiness will increase over time with disturbance and differential mortality
Landscape diversity (between-stand patchiness)	Variable, but generally high due to responses of species composition and structure to environmental gradients and effects of natural disturbances of different types, sizes, and intensities	Expected to be intermediate between primary natural forests and plantations, but increasing with time since abandonment of intensive human activities	Superficially high if variety of age classes present, but because stands are structurally similar and monotonous in pattern, overall landscape diversity is relatively low

Functional diversity (processes)	High diversity of natural disturbances with variable intensity and frequency; diverse food webs and energy/nutrient pathways; diverse biotic interactions	Probably low initially but converging on primary natural condition with recovery from human domination	Usually low due to simplified structure and low species richness in monocultural, chemically manipulated stands
Species diversity	Varies among forest types and latitudes, but generally is maximum for region, especially in early and late seral stages (often lower in mid-succession)	May be high in early seral stages with influx of weedy colonists; often low in mid-succession, high in old growth	May be high in early stages due to abundant weeds; lower thereafter unless long rotation permits colonization by more species as stand ages and thins
Animal population density	Variable but probably higher for most species than in plantations	Probably intermediate, converging on primary natural condition with stand age	May be episodically high for some species (e.g., herbivorous insects); otherwise generally lower than in natural forests
Stability (resistance and resilience)	Variable, but probably maximum for region	Expected to be intermediate, converging on primary natural condition with stand age	Low, often requiring continual, expensive inputs of energy and chemicals (e.g., pesticides) to maintain
Esthetics	Increasingly regarded as beautiful, sublime, and inspirational by modern cultures	Can be almost as highly regarded as virgin forests, especially when old and few virgin forests remain in region	Generally and increasingly viewed as dull, monotonous, and ugly, although perhaps appreciated by silviculturists
Conservation value	Extremely high because rare and depleted in most regions; supports species sensitive to human disturbance; high diversity	Moderate to high; have increase in some regions (e.g., northeastern United States and Piedmont of Southeast United States) over last century; high diversity	Low except in highly altered landscapes, where these may be the only forests available; otherwise of low or marginal value to wildlife; common

(Zahner 1992), although the time required for complete recovery of herbaceous flora after clearcutting may exceed several decades or even centuries (Duffy and Meier 1992; but see rebuttals in *Conservation Biology*, June 1993). Ecological recovery may not result in convergence on the same species composition as before disturbance, as climate and other regeneration conditions differ today. In the central hardwood region of the eastern United States, available data show little difference between primary old growth and second growth 100–150 years old (Parker 1989). Second-growth longleaf pine in the southeastern coastal plain, recovering after extensive clearcutting in the late nineteenth and early twentieth centuries, has regained most of the fauna characteristic of old-growth stands, including the red-cockaded woodpecker in some areas (Jackson 1986). For longleaf pine forests, the major requirement for recovery seems to be that fire is not excluded for too long and that the native groundcover remains intact (Clewell 1981, 1989).

In some regions, such as the northeastern United States, managed forests are mostly naturally regenerated rather than planted. These secondary natural forests are largely on private lands managed for timber production, yet they vary tremendously in biodiversity, depending on how they are managed (Hunter 1990 and personal communication). Intelligent management of secondary natural forests can make a tremendous contribution to biodiversity in regions where these forests are abundant.

Plantations, defined as stands of trees planted by humans, are easy to recognize. They are usually dense, monospecific stands of conifers in straight rows. They are seldom allowed to age beyond a few decades before harvest. The species planted may be either native or exotic to the region. Even when native species are used, genotypes may be foreign, or the species may be planted on sites where it would not normally occur, a process called type conversion (Norse *et al.* 1986). For example, slash pine or loblolly pine are often planted on sites formerly occupied by longleaf pine in the Southeast. Confusingly, plantations are sometimes called second growth, although they usually differ dramatically from naturally regenerated stands.

Consequences of Forest Conversion for Biodiversity

Conversion of forests to anthropogenic habitats is the most significant threat to forest biodiversity in the United States and worldwide. Replacement of natural forests by cropland, pasture, or urban developments is obviously destructive—the trees are gone. But the conversion of natural forests to plantations is nearly as harmful from the standpoint of biodiversity, and is more insidious because it is invisible to many people. Somewhat less extreme, but still significant, is the conversion of primary forests to second growth (Table 6.1).

Consequences at the Site Level

How do plantations and other intensively managed forests differ from natural forests? We can begin by considering the ways in which stands of trees originate. Although a few tree species regenerate and grow in deep shade, most trees regenerate and grow after treefalls have created small gaps in the canopy, or after fire, windstorm, flood, or other disturbances have created larger openings. Because natural disturbances usually leave abundant dead (and often living) wood behind, they tend to enhance biodiversity. In contrast, modern logging and silvicultural methods create barren patches mostly devoid of natural structure. The practice of removing harvested trees from a site, called "skidding," also damages sensitive habitats, especially when logs are dragged across streams or wetlands (Norse *et al.* 1986).

After the trees on a site have been cut and skidded away, some form of site preparation is usually applied to remove competing vegetation and logging residues and prepare a seedbed for planting of new trees. Site preparation is also used in many "natural regeneration" systems such as seed tree or shelterwood silviculture where trees from the former stand are left to reseed harvested areas. Site preparation techniques include yarding and piling unmerchantable materials, burning, chopping, disking, bedding, windrowing, and applying herbicides (the latter are often used to control competing vegetation after the desired trees have been established on a site, a practice called "release"). The more intensive forms of site preparation remove virtually all wood and other structural materials from sites, increase erosion and losses of nutrients, and can have dramatic effects on biodiversity (Hunter 1990).

Southern loblolly and slash pine plantations, most of them on former longleaf pine sites, are among the most intensively managed ecosystems in the world. Intensive mechanical site preparation of these plantations, though profitable in the short term in many cases, is a threat to their long-term sustainability. For example, site preparation practices that remove residual organic matter from a site can cause significant losses of nitrogen, the availability of which often limits tree growth (Vitousek and Matson 1984), and may therefore reduce productivity. Disking (using a harrow) and bedding (forming soil into low ridges) can result in severe soil erosion. Leaching of nutrients from beds can lead to eventual infertility (Schulz and Wilhite 1975).

Windrowing, where displaced groundcover vegetation, debris, and topsoil are piled into long rows that are sometimes several feet high, is perhaps the most severe of all site preparation techniques. Nutrients displaced into windrows can amount to more than 10 percent of a site's nutrient reserves (Morris *et al.* 1983). Mechanical site preparation also decreases litter insulation and reduces transpiration, resulting in greater temperature extremes, high soil temperatures during the growing season, higher water tables, and extreme nutrient flushes (Pritchett and Morris 1982). Although pine growth often responds positively to these ecological changes in the short term, after three or four rotations production may decline dramatically.

Species respond in different ways to site preparation. The overall effect of intensive site preparation on vegetation in southern pine plantations is to increase the abundance and diversity of weedy herbaceous species (Moore *et al.* 1982), usually at the expense of more sensitive native herbaceous plants. Woody plants are virtually eliminated in the short term. Responses of animals to site preparation are variable. Mobile animals often escape immediate impacts by fleeing a site or escaping down burrows, such as those of gopher tortoises (where they still occur). Such escape responses have obvious adaptive value in ecosystems having frequent fires. However, reptiles and amphibians are not able to evade all the impacts of mechanical site preparation and associated soil compaction and burrow destruction. Clearcutting and site preparation reduce amphibian numbers and cause a significant decline in both abundance and species richness of reptiles; the more intense the treatment, the greater the impact (Enge and Marion 1986). Rare soil-dwelling species may be particularly threatened. As the negative effects of intensive site preparation become more widely appreciated, many ecologists and foresters in the South are calling for a return to gentler methods, such as light prescribed fires. Fire is often more than adequate to expose mineral soil for pine regeneration and control competing hardwoods, and generally benefits native biodiversity (Noss 1988, Hunter 1990).

Negative effects of site preparation and other vegetation management are not limited to southern pine forests. In many forest types, removal of woody debris and use of herbicides to control competing vegetation may reduce productivity, sometimes immediately and other times after several rotations. The specific causes of reduced productivity are still elusive in most cases. Loss of mycorrhizal fungi associations, complex changes in soil biology, and changes in soil structure seem to be responsible for regeneration failures of planted conifers in some areas of the West. Some clearcut sites at high elevations in Oregon have been replanted several times without success, even though adjacent natural forests are quite productive (Perry *et al.* 1989b).

If a clearcut site regenerates naturally and is allowed to recover over many decades or centuries, structural diversity will gradually accumulate as trees die, self-prune, create snags and gaps, contribute logs to the forest floor, and develop multilayered canopies. This diversification will occur to some extent in aging plantations but will be cut short by harvest. In Douglas-fir plantations under 100-year rotations, snag and log abundance declines to about 30 percent that of the preharvest natural forest by the end of the first rotation and 6 percent after the second rotation (Spies and Cline 1988). Plantations managed under conventional rotations will never produce large snags and logs. Whatever structural legacies are left from the natural forests that once occupied these sites will inevitably disappear over time.

Loss of structural diversity in plantations and other managed stands results in lower species diversity and reduced abundance of many species (Norse 1990). Cavity-nesting birds and mammals, bats, and other animals that depend on holes, loose bark, or other features of standing dead trees will be obvious victims of any management strategy that fails to retain and recruit large snags. Large down logs also contribute to diversity, particularly of invertebrates, nonvascular plants, algae, and bacteria, but also of small vertebrates, some of which are prey for larger vertebrates such as the spotted owl (Franklin *et al.* 1981, Maser and Trappe 1984, Maser *et al.* 1988). Many species of animals and some plants require old-growth forests, with the best documentation from Douglas-fir forests of the Pacific Northwest (Carey 1989, Norse 1990, Ruggiero *et al.* 1991, Spies 1991, Thomas *et al.* 1993).

Natural forests of all ages in many parts of the world are more diverse than plantations. In southern Sweden, the breeding bird community in a natural forest was more diverse than in managed forests of similar or younger age. The natural forest had a density of birds three times greater than a managed forest of the same age and nearly nine times greater than a young spruce plantation (Nilsson 1979). In a comparison of bird communities in exotic conifer plantations and native *Nothofagus* forests in New Zealand, native birds were more diverse and dense in natural forest, whereas several introduced species preferred plantations (Clout and Gaze 1984). Breeding birds, small mammals, and amphibians in the Oregon Coast Range were more diverse in a natural mature (not old-growth) forest than in a plantation. Although differences in species richness were not statistically significant, total abundances of birds and amphibians were significantly higher in the natural forest (Hansen *et al.* 1991). In Florida, a secondary natural longleaf pine forest had significantly higher densities, species richness, species diversity, and biomass of breeding birds than a slash pine plantation (Repenning and Labisky 1985).

Diversity differences between natural forests and plantations appear to be even greater among invertebrates. Old-growth Douglas-fir and western hemlock stands in Oregon contained five times as many species (75) and twice as many functional groups (6) of canopy arthropods as did Douglas-fir plantations (Schowalter 1990). The long-term stability of old growth apparently favors the accumulation of arthropod diversity, including species with poor dispersal capacities (Lattin and Moldenke 1990). This conclusion probably also applies to younger stages of natural forest in which structural legacies remain. The usual practice of clearcutting followed by slash burning in the Northwest reduces total arthropods in the soil by about 90 percent (Moldenke and Lattin 1990). Spiders also decline with clearcutting and site preparation; recovery to a typical forest spider species composition takes at least 30 years on wet sites and considerably longer on dry sites (McIver *et al.* 1990).

Foresters have often claimed that older natural forests are biological deserts and that clearcutting enhances diversity. It is true that total species richness of plants and several other taxa is often highest in early stages of natural forest succession (Schoonmaker and McKee 1988). But species associated with early seral habitats are generally at little risk of extinction (Harris 1984, Franklin 1992). For example, Swindel *et al.* (1983) reported that management of slash pine plantations by clearcutting and site preparation increased plant diversity, but the species responsible for high diversity in these plantations were largely weeds, including many exotics (Harris 1984). Although it has not been well studied, we can predict that native species richness will be higher in natural early seral habitats than in clearcuts or young plantations, because naturally disturbed areas have greater habitat diversity. In any case, these examples show that species richness by itself is a poor measure of biodiversity.

Associations of wildlife with different seral stages are well known. But curiously, recent studies in the Pacific Northwest show that younger stages of primary forests often differ little from old growth in species composition and ecological processes. In the Oregon Coast Range, Hansen *et al.* (1991) found "relatively few strong differences in plant and animal communities between unmanaged young (25–80 years), mature (80–200 years), and old-growth (more than 200 years) forests." The lack of strong differences probably relates mostly to the large down logs (coarse woody debris), snags, large live trees, and other "biological legacies" that persist through the seral stages that follow natural disturbance of an old-growth stand. Spotted owls and many of the other 482 plant and animal species associated with late seral stages in the Pacific Northwest (Thomas *et al.* 1993) may also be found in younger forests that retain structures of old growth. Indeed, the most important ecological distinction in the Northwest and many other regions seems to be between plantations and natural forests of any age.

Differences in diversity probably contribute to greater stability of natural forests relative to plantations, both in resistance (ability to maintain themselves against stresses) and resilience (ability to bounce back after a disturbance). Although the diversity–stability relationship has been argued for decades among theoretical ecologists, empirical evidence shows that the loss of diversity caused by the conversion of natural ecosystems to artificial systems is often accompanied by a loss of stability. Most tree plantations require much energy and labor to maintain their productive simplicity against the natural processes that would diversify them and decrease their timber value. The higher diversity of natural forests, particularly in predaceous arthropods and other insectivores, appears to help these forests resist pest outbreaks (Schowalter 1990). Disease problems are generally more of a problem in plantations than in natural forests (Franklin *et al.* 1989). Furthermore, a diversity of plants, microbes, and other organisms helps maintain healthy linkages between plants and soil, including positive feedback mechanisms that may facilitate adjustment to climate change and other unpredictable stresses (Perry *et al.* 1989b, 1990).

Fragmentation and Edge Effects

Outright transformation of forests to plantations or nonforest habitats is not the only way that forest biodiversity is threatened. Other threats accompany the deforestation process or arise later and jeopardize the integrity of remnant forest patches in human-altered landscapes. Fragmentation of forests and other habitats is considered one of the greatest causes of biotic impoverishment worldwide (see Chapter 2). Disentangling the ecological effects of fragmentation and habitat loss can be difficult, for the two usually occur together. Many landscapes on federal as well as state and private lands have been heavily fragmented over the last few decades. For example, in the Olympic National Forest, Washington, more than 87 percent of the old growth in 1940 was in patches larger than 4000 ha. In 1988, only one patch larger than 4000 ha remained and 60 percent of what remained of the old growth was in patches smaller than 40 ha. Forty-one percent of the remaining old growth in 1988 was within 170 m of edge and therefore vulnerable to blowdown and other edge effects (Morrison 1990).

The problem of edge effects is a relatively new discovery in ecology. Traditionally, wildlife managers encouraged foresters to create checkerboard patterns on the landscape (Fig. 6.5) because such patterns maximize the amount of habitat interspersion and edge. Many wildlife areas today are intentionally managed for maximum habitat interspersion. In his classic text on game management, Leopold (1933) discussed the tendency of edges to be rich in wildlife, particularly species that require more than one habitat type

FIGURE 6.5 A checkerboard harvest system near Geraldton, Ontario. Wildlife managers often advocated this type of habitat interspersion because it maximizes edge effects and favors many game species. However, these patterns also maximize forest fragmentation. Photo by Reed Noss.

to meet life history needs. If forage areas, cover areas, and nesting or denning sites for a species are close to each other, an area should be able to support a higher population density than more homogeneous areas. Indeed, most terrestrial game animals are edge adapted because they depend on habitat mosaics. Edges are also likely to contain species characteristic of each of the abutting habitats, as well as species that actually specialize on edge (Johnston 1947, Yahner 1988). Managing for edge may increase beta diversity, the between-habitat component of species diversity (see Chapter 1). Game biologists have urged managers to create as much edge as possible (Yoakum and Dasmann 1971) with scant concern for the effects these actions may have on nontarget species (Noss 1983).

Research over the past two decades has shown that habitat edge is not benign. Furthermore, an edge is not linear, but rather is a zone that varies in width depending on what is measured. The physical environment near edges differs from that in forest interiors, with a microclimate generally drier, brighter, and more windy. Vegetation structure and species composition respond to these physical factors, often creating a distinct edge vegetation comprising xeric, shade-intolerant species (Ranney *et al.* 1981). These vegetational changes may have greater long-term effects than the more severe but brief disturbance of logging (Whitney and Runkle 1981). A study in Washington found increases in diurnal temperature and decreases in humidity up to 180 m into old-growth and mature Douglas-fir forests from a clearcut. On hot, windy days, these physical edge effects may extend more than 240 m into a forest (Chen and Franklin 1990). Regeneration, growth, density, and mortality of trees are affected by distance from edge, with some effects extending 137 m (Chen *et al.* 1992). Tree blowdown rates are usually greater near edges, up to three tree heights into a forest (Harris 1984, Franklin and Forman 1987).

Differences in environmental conditions and vegetation near edge seem to be responsible for the association of certain animal species with edge (Lay 1938, Johnston and Odum 1956). Again, this phenomenon has been best studied for birds. Each species seems to respond somewhat differently to the edge–interior gradient. In eastern deciduous forests, species such as the gray catbird, rufous-sided towhee, and indigo bunting are well-known edge specialists. But even among forest birds, some species are attracted to edge, some avoid edge, and others are seemingly indifferent (Fig. 6.6; Noss 1991g). From the perspective of a forest interior bird, a small forest patch may be entirely edge, and therefore uninhabitable (Temple 1986).

The first concerns about potential negative effects of edge were voiced by biologists interested in applying island biogeographic theory to reserve design (see Chapters 2 and 5). These biologists emphasized the need for large, intact blocks of habitat, an idea diametrically opposed to that of maximizing habitat interspersion. They also noted that most edge species are widespread and weedy, and at little risk of extinction in human-dominated landscapes (Diamond 1976, Terborgh 1976, Whitcomb *et al.* 1976). Forest interior birds, including many neotropical migrants, are generally of greater conservation concern than edge species because so much of many regions is now edge habitat (Noon *et al.* 1979, Robbins 1979, Whitcomb *et al.* 1981). The best management strategy for these sensitive species is clearly to maximize the amount of unfragmented forest in the landscape.

Dangers more directly associated with edges, rather than with habitat area, were uncovered by studies of predation and cowbird parasitism on birds' nests. Birds were attracted to a forest-field edge in Michigan, but those nesting near the edge suffered reduced reproductive success due to high rates of predation and cowbird parasitism (Fig. 6.7; Gates and Gysel 1978). Thus, edge may be an ecological trap, enticing birds to nest but giving them little chance to fledge young. Other studies confirmed that brown-headed cowbird parasitism is intense near forest edges, declining with distance into the forest (Brittingham and Temple 1983). Nest predation rates, as measured in experiments employing artificial nests, show similar trends

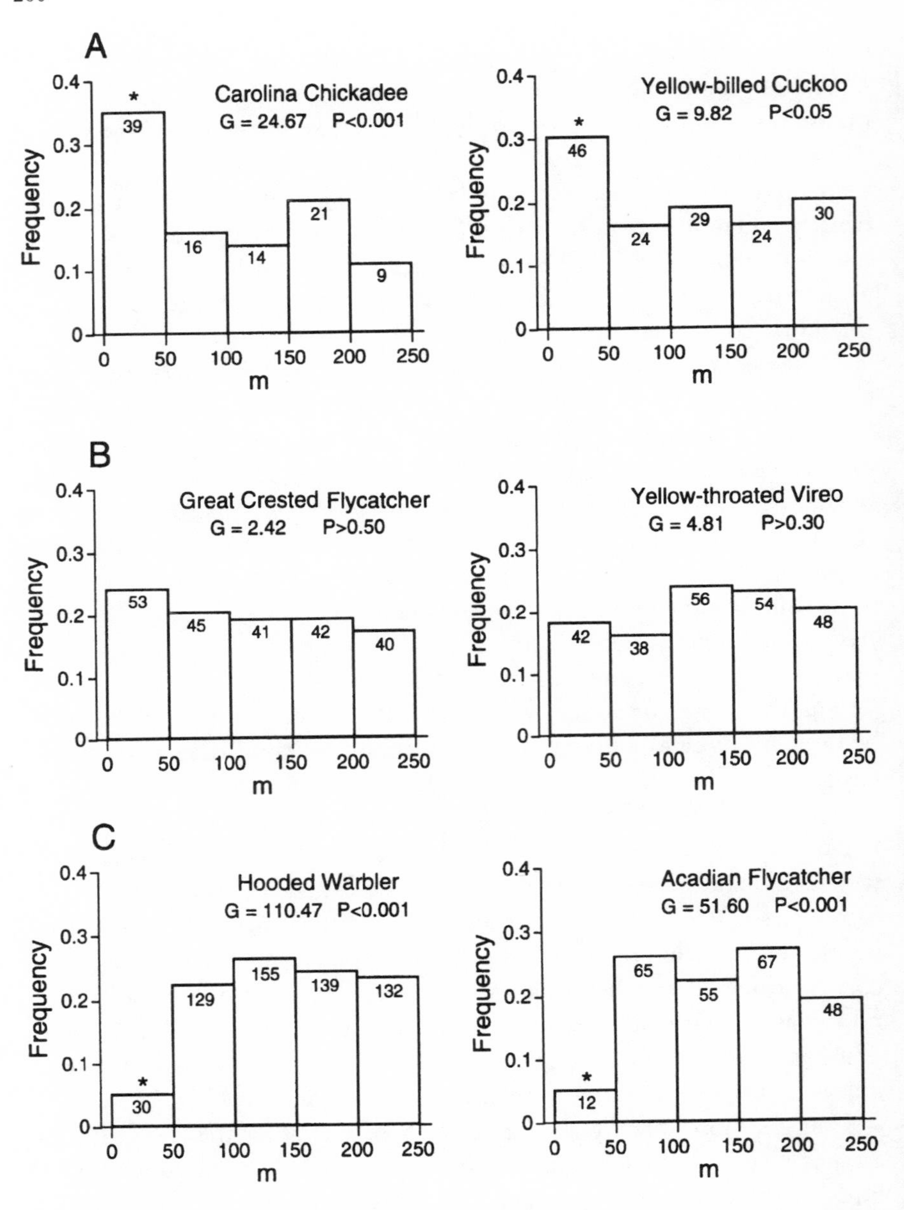

FIGURE 6.6 Examples of birds that are (A) attracted to edge, (B) seemingly indifferent to edge, and (C) repelled by edge in a Florida hardwood forest. The horizontal scale is distance from edge in meters. Numbers within bars indicate bird registrations in each distance category. Asterisks indicate significant differences. From Noss (1991g). Used with permission of the Natural Areas Association.

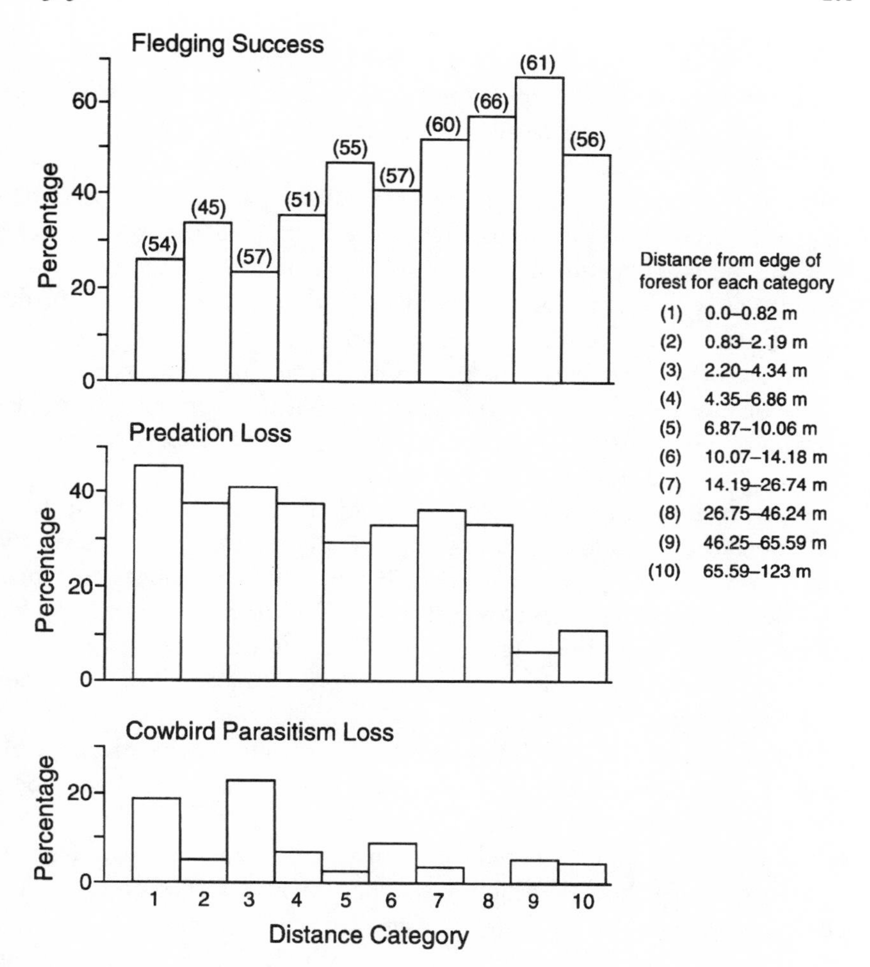

FIGURE 6.7 Nesting success, predation losses, and losses to cowbird parasitism of songbirds as a function of distance from forest edge, based on a study in Michigan. Sample sizes are in parentheses and distance categories are shown at right. From Wilcove *et al.* (1986) as modified by Gates and Gysel (1978). Used with permission of both sets of authors, Sinauer Associates, and Ecological Society of America.

(Fig. 6.8; Wilcove 1985, Wilcove *et al.* 1986, Andren and Angelstam 1988, Small and Hunter 1988). Some other studies failed to confirm the ecological trap hypothesis with regard to nest predation, but nests near abrupt edges may be more heavily preyed upon than those near more natural feathered edges (Ratti and Reese 1988, but see Yahner *et al.* 1989). Despite discrepancies between studies, in part related to regional differences in vegetation and faunas (Hansen and Urban 1992), it is becoming clear that many regions where forest interior habitats have declined greatly have difficulty retaining viable populations of forest birds due to high levels of nest predation and cowbird parasitism (Bohning-Gaese *et al.* 1993). Such regions may be population sinks for many species, their continued presence being dependent on immigration from regions with large intact forests (Temple and Cary 1988, Robinson 1992).

Problems related to edge effects and habitat interspersion may seem to contradict the ecological rule that disturbance creates habitat diversity, which in turn fosters species diversity. But not all disturbances are equal. We cannot expect all species to adjust to an altered disturbance regime. Disturbances that create abrupt forest edges are rare in nature; even a stand-replacing wildfire leaves dead trees in place, producing much less contrast with the adjacent forest than does a clearcut. Generally, the greater the structural contrast between adjacent stands, the more intense the edge effects (Thomas *et al.* 1979, Harris 1984, Noss 1991g). However, Janzen (1983) warned that disturbed successional habitats around reserves can be a source of weedy invaders and suggested that it may be better to surround reserves by intensive agriculture. Within forests, the size of disturbed area is an important consideration. Openings smaller than 0.2 ha in eastern forests may not attract cowbirds or produce other harmful edge effects (Brittingham and Temple 1983, Noss 1991g).

Curiously, most edge effect studies showing deleterious biotic effects of edge, such as predation or parasitism, have been conducted in relatively homogeneous second-growth forests that exist as small fragments in a sea of developed land. In such cases, species may be drawn to edges because they contain structural cues, such as dense shrub growth, similar to those of the patchy natural forest in which they evolved (Gates and Gysel 1978). In regions where forest is still the dominant cover type, edge effects may be less pronounced. For example, a study of nest predation in a forested landscape in Maine found no effect of distance from clearcut edge on predation rates; clearcuts actually had lower predation rates than forest sites (Rudnicky and Hunter 1993).

Old-growth forests may have fewer problems with biotic edge effects than do second-growth stands. A Florida study showed that birds were at-

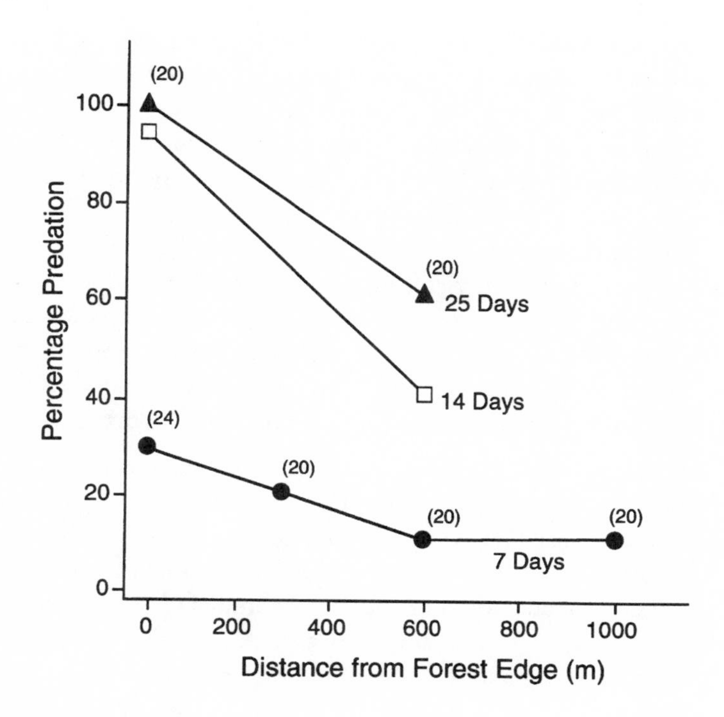

FIGURE 6.8 Percentage of experimental nests preyed upon as a function of distance from forest edge. Graph shows losses after 7 days (bottom), 14 days, and 25 days. From Wilcove *et al.* (1986). Used with permission of author and Sinauer Associates.

tracted less to edge in study plots containing treefall gaps and other shrubby patches than in younger, more homogeneous plots (Noss 1991g). Some species, particularly the hooded warbler, avoided artificial edges but were strongly attracted to natural gaps. Management that allows young stands (secondary natural forests or plantations) to mature to old growth might benefit species dependent on habitat heterogeneity without the dangers of edge effects.

Weeds, Herbivores, and Pests

The fragmentation and road building that accompany intensive forest management increase opportunities for invasion of forests by weedy plants and animals. Some of these problem species are exotic, others are native. Many weedy plants, including herbs (e.g., garlic mustard, Johnson grass, myrtle), vines (e.g., Japanese honeysuckle, climbing euonymus, kudzu), and shrubs (e.g., bush honeysuckles, glossy and common buckthorns), are Eurasian in origin and have outcompeted native species in small forested nature reserves, particularly in the East (Natural Areas Association 1993). Exotic birds such as the European starling often become abundant in agricultural and suburban landscapes and can exclude native cavity-nesters under certain conditions. The brown-headed cowbird, on the other hand, is native to North America but has expanded its range and numbers dramatically due to forest fragmentation, agriculture development, and livestock grazing (Brittingham and Temple 1983). Whereas large, intact forests would be expected to have very few, if any, cowbirds, fragmented landscapes with abundant agricultural food sources may be so saturated with cowbirds that parasitism rates are high even in 2000-ha forest blocks (Robinson 1992).

Another native species that has benefited from fragmentation, particularly in the eastern and midwestern states, is the white-tailed deer. Deer numbers have increased far beyond presettlement abundances because they thrive on habitat interspersion and probably also because their native predators, chiefly cougar and wolves, have been eliminated or greatly reduced. Deer densities in many regions today are at historic highs, leading to severe impacts from browsing and grazing on plant regeneration, abundance, and distribution (Miller *et al.* 1992). A 20-year study in the Allegheny National Forest, Pennsylvania, documented serious damage by deer to the understory of a virgin forest (Hough 1965). Because woody species differ greatly in palatability and sensitivity to browsing, even moderate levels of browsing can greatly affect species composition (Strole and Anderson 1991). Throughout the Allegheny region, the failure of mixed-species hardwood forests to regenerate after clearcutting has sometimes led to nearly pure stands of black cherry, a rapidly growing species unpalatable to deer (Tilghman 1989).

A shift in dominance from eastern hemlock to sugar maple in northern Michigan can be attributed to intense browsing by deer preventing regeneration of hemlock (Frelich and Lorimer 1985). In northern Wisconsin, eastern hemlock, Canada yew, and eastern white cedar—all favored deer food—are scarcely regenerating in many areas due to heavy deer browsing. Yew are mostly restricted to rocky outcrops inaccessible to deer (Alverson *et al.* 1988). Herbaceous species, an important part of the spring and summer diet of deer, are also vulnerable to deer grazing. Wildlife management to support a huntable deer herd may threaten rare plants in adjacent natural areas (Bratton and White 1980). A recent survey determined that at least 98 species of threatened or endangered plants in the United States are being disturbed by deer (Miller *et al.* 1992).

The problem of elevated predation on bird eggs and nestlings near the edges of forests, or even throughout forest patches in severely fragmented landscapes, is often due to population excesses of opportunistic birds and mammals that thrive in human-dominated landscapes. As with deer, these "mesopredators" (medium-sized predators) may be overly abundant because their natural predators (e.g., wolves and cougar) have been extirpated in most fragmented landscapes. Indeed, when large predators are lacking, populations of small and medium-sized omnivores and predators may be up to ten times more abundant than normal (see citations in Soulé *et al.* 1988). In southern California chaparral, predation on bird nests is lower in sites where coyotes are present, apparently because coyotes prey on foxes, raccoons, skunks, opossums, and domestic cats, all of which are important bird predators (Soulé *et al.* 1988). Another reason these opportunists are so common in human-dominated landscapes is that they are subsidized by human garbage. Wilcove (1985) found that predation rates are significantly higher in forest fragments surrounded by suburbs than in those surrounded by agricultural land, apparently because subsurbs provide more food subsidies to mesopredators.

Invasion of forests by pests and diseases, especially along road systems, is a severe problem in some regions. A well-known example involves the decline of the Port Orford cedar across its range in northern California and southern Oregon. Spores of the Port Orford cedar root rot fungus seem to be carried largely in mud in the tires of logging trucks. Other pests that disperse along road systems in the West include black stain root disease fungus, gypsy moth, and spotted knapweed (Schowalter 1988). In many cases the worst pests of all are humans, whose firewood collecting, poaching, arson, off-road vehicle use, and other activities inimical to biodiversity are all made possible through road access. By contributing large amounts of sediment to streams, logging roads also threaten aquatic habitats and survival of fish populations, particularly where slopes are steep and soils unstable (Swanson *et al.* 1990).

Recommendations for Forest Reserve Management

Forest conservation has become more sophisticated with the realization that whole forest landscapes, not just a few set-asides, must be managed for biodiversity. As discussed in Chapter 5, the reserve strategy by itself will not fulfill all the conservation goals espoused in this book. In the near term, few regions are likely to have reserves large and numerous enough to maintain the full spectrum of native species and ecological and evolutionary processes. But without reserves we might as well throw in the towel. Our multiple-use experiments are, so far, not very encouraging. Many managers are sincerely trying to improve management practices and simulate natural processes on multiple-use lands, but it is too early to claim success. Besides, people appreciate reserves for more than just biodiversity. They appreciate—and even need—natural beauty, opportunities for solitude and contemplation, and knowing that other species are living their lives independent of human control. For these reasons, we believe that reserves should be emphasized more than they are now in discussions of ecosystem management.

In developed landscapes, most remaining forest patches and other natural areas will be too small to represent their nominal community type. Is a 20-acre oak–hickory forest really an oak–hickory forest if it is missing many of the birds and mammals characteristic of this community and the groundcover is dominated by exotic weeds? Perhaps not, but in a soybean–corn landscape there may be little else to work with. On the other hand, a large forested landscape, say in the Rocky Mountains, may appear wild, yet it has lost grizzly bears and wolves due to human persecution. Fire suppression, even across huge wild landscapes such as the Boundary Waters Canoe Area in Minnesota, can significantly affect many aspects of biodiversity at several temporal scales (Heinselman 1973, Wright 1974, Baker 1989, 1992b). The only way to understand how human activity affects biodiversity is to have reserves large enough to be essentially self-managing. That is, we need true control areas for our management experiments.

Because reserve management differs substantially between highly fragmented landscapes and wildland landscapes, we divide our recommendations accordingly.

Guidelines for Highly Fragmented Landscapes

1. **Control exotic plants and animals, and other pests, ideally before they get out of hand.** Populations of many species will be unbalanced in fragmented landscapes due to missing predators and competitors and a landscape matrix that favors opportunists. Small reserves, with their high perimeter-to-area ratios, are extremely vulnerable to exotic invasions and other cross-boundary problems, as are disturbed habitats in general. Although eliminating all exotic species is often a goal in reserve management, control activities need to be planned intelligently and undertaken in a cautious and experimental fashion. Otherwise, the "cure" of exotic removal might turn out worse than the disease (Westman 1990), especially if exotics play functional roles in the ecosystem formerly filled by native species that have been extirpated. In certain cases, not all exotics need be removed. Even the noxious bush

honeysuckles and buckthorns, major pests in eastern nature reserves, serve a function. If these shrubs are eliminated too rapidly, before native shrubs can grow to take their place, shrub-nesting birds may decline (Whelan and Dilger 1992). In any case, exotic species vary tremendously in the threat they pose to native communities. Also, some exotics are virtually impossible to eliminate. Control priorities should be based on ecological information indicating the degree of threat each exotic species represents in a specific management situation.

2. **Control overabundant herbivores and omnivores.** When populations of deer, other herbivores, or opportunistic omnivores such as raccoons and opossums are overabundant (often the case in fragmented landscapes with plentiful edge habitat and no large predators), do not hesitate to use hunting and trapping as management tools (see Diamond 1992). Opening a nature reserve to public hunting and trapping may be risky, as licensed hunters in the United States are not required to know anything about wildlife and cannot be counted on to select the correct prey; visitors may even be in danger. In such cases, managers themselves or hired sharpshooters and trappers may need to do the killing.

3. **Control visitor activities.** Especially in small reserves close to human population centers, overuse by visitors can be a major threat, leading to soil compaction and erosion, harassment of wildlife, littering, dispersal of weeds (for instance, from burs on pantlegs), and other problems. In all cases, keep trails away from sensitive sites, such as small springs and wetlands, rare plant populations, wading bird rookeries, and raptor nests. As wide trails can create edge effects (Noss 1981), including reduced reproductive success of forest interior birds (Hickman 1990), keep trails narrow. Use interpretive signs, public programs, law enforcement, and other means to encourage respectful behavior within reserves.

4. **Do not fragment or disturb a reserve internally by overmanipulating habitats.** Many nature reserve managers try to maximize habitat diversity within reserves, more for esthetic and interpretive than for biological reasons, a practice that can have many undesirable consequences (Noss 1981, 1983; Noss and Harris 1986). If habitat manipulation is required to maintain early successional habitats for species of interest, it is best to manage for such habitats around the edges of a forest reserve, rather than fragmenting the interior. However, gaps might be created to simulate treefalls by felling trees in young forests with low rates of natural gap creation, which may help restore patchiness.

5. **Simulate natural disturbances, as necessary.** For fire-dependent forest types, prescribed burns may be needed, as natural ignitions will not occur often enough in small isolated areas. Regular burning is essential for communities, such as longleaf and ponderosa pine, that require frequent, low-intensity fires. Prescribed burns should be conducted within the same season as natural fires and follow natural patterns of frequency. If fire exclusion has resulted in large fuel accumulations, these may have to be reduced by hand (for example, clearing small trees and raking litter from around bases of large trees) before fire is reintroduced.

6. **Enlarge reserves.** One of the most beneficial actions that can be taken for a small forest reserve is to enlarge reserve area by acquiring surrounding lands and reestablishing native trees on them. A larger reserve means larger populations of area-sensitive forest species, a smaller perimeter–area ratio and therefore fewer harmful edge effects, and often less expensive management per unit area (Noss 1983). For a park agency considering acquisition options, it may often be wiser to enlarge the area of existing reserves than to acquire more small reserves (but see the many considerations for conservation evaluation in Chapter 4).

7. **Establish buffer zones.** Where possible, establish buffer zones of low-intensity land use around reserves. Planted conifers, for example, might help reduce edge effects related to altered microclimate around reserve boundaries (Ranney *et al.* 1981). Some hardwood species buffer conifers from fire (D. Perry, personal communication). Conservation easements, landowner agreements, and tax incentives can be used to encourage proper land use on private lands surrounding reserves.

8. **Link reserves together.** Examine opportunities to link a reserve with other forests by restoring wooded corridors. As reviewed in Chapter 5, a narrow, edge-dominated corridor may do more harm than good if it favors weedy generalists and encourages invasion of reserves by opportunistic species. However, in a landscape where no deep forest interior remains, wooded fencerows and other narrow corridors may provide habitat and dispersal routes for many native species that might otherwise go extinct (Merriam 1988). Whenever possible, maintain or restore corridors wide enough (ideally miles wide) to serve sensitive interior species (Noss 1993b). The ideal situation may be an interconnected landscape matrix in which early successional or anthropogenic habitats exist as patches within a mosaic of older natural forest, rather than vice versa.

Guidelines for Wildlands

1. **Keep them wild.** A forest landscape large enough to manage itself with a natural disturbance regime, generally on the order of millions of acres (see Chapter 5), should be kept as wild as possible by limiting human uses. These rare, true wilderness landscapes offer our only real controls for comparison with managed landscapes and for monitoring effects of global threats such as climate change (Noss 1990a, 1990b; Baker 1992b).
2. **Manage disturbances.** For reserves too small to manage themselves with a natural disturbance regime, use prescribed fire (or, very cautiously, silvicultural manipulations) as needed to simulate natural conditions in fire-dependent communities. Fires in many coniferous forest types often covered hundreds of thousands of acres. Even our largest national parks may be too small to sustain a relatively balanced mix of seral stages without prescribed fires and other disturbance management.
3. **Enlarge reserves.** As for small reserves in fragmented landscapes, our large wildland parks, wilderness areas, and other reserves would benefit greatly from further enlargement. The ideal size (the goal) should be based on consideration of the scale of natural disturbance, the requirements of large carnivores and other area-dependent species, connections to other reserves, and other aspects of landscape context (see Chapter 7).
4. **Avoid internal fragmentation by roads, campgrounds, concessions, and other developments.** If such developments are deemed necessary, they should be placed on the periphery of the reserve and away from sensitive habitats. As a case in point, building campgrounds and concessions at Fishing Bridge (an area heavily used by grizzly bears) and other sensitive sites within Yellowstone National Park was a grand mistake with lethal results for grizzly bears. Developments already constructed in sensitive sites should be removed and roads closed and obliterated.

Recommendations for Multiple-Use Management

Within the next few decades, at least, more forest land is likely to remain in multiple-use management than is converted to strict reserve status. Therefore, the ways in which these lands are managed will have a tremendous bearing on the status of biodiversity across the country. The concept of multiple-use management is undergoing vigorous debate and revision, both within and outside land-managing agencies. Within agencies such as the

Forest Service, scientifically trained ecologists and other staff are at odds with a previous generation that was trained and motivated to produce timber. Both camps may believe sincerely in multiple-use management on federal lands, but they disagree about ways in which forests should be managed, as well as about the primacy of timber and other hard commodities in the multiple-use spectrum.

The technical aspects of a credible position on multiple-use forest management are unclear. We are in the midst of a period of vigorous experimentation (and even more vigorous speculation) with new silvicultural techniques, cutting patterns, and regeneration strategies. The intent and framework of this book do not permit a detailed examination of these techniques. They have been discussed by other authors (e.g., Norse *et al.* 1986, Perry *et al.* 1989a, Hunter 1990). However, we will draw from current research in making our recommendations.

New Forestry, New Perspectives, and Ecosystem Management

Many of the new approaches to multiple-use forest management fall under the rubric of "New Forestry," a concept largely developed and popularized by Jerry Franklin of the University of Washington (formerly with the Forest Service) and his many colleagues in Oregon, Washington, and other regions (Franklin 1989, 1992; Perry *et al.* 1989a; Hopwood 1991, Franklin and Spies 1991; Swanson and Franklin 1992). New Forestry is essentially an attempt to emulate nature in forest management. As such, its basic goals are highly laudable. As we have seen, past management has resulted in fragmentation, edge effects, and structurally simplified plantations that differ profoundly from stands arising after natural disturbance (Fig. 6.9). New Forestry emphasizes aggregation of cutting units to reduce fragmentation, minimal road building, and retention of coarse woody debris and scattered live trees or clumps of trees in harvest areas. Thus, a site harvested under New Forestry prescriptions might resemble an area after a natural fire more than it resembles a clearcut (Fig. 6.10).

New Forestry has been claimed to represent a shift in natural resource management philosophy from regulation of undesirable uses to sustained yield management (emphasizing wood and other commodities) to sustained ecosystem management (Salwasser 1990, Swanson and Franklin 1992). However, the related Forest Service programs of "New Perspectives" and "Ecosystem Management" (Salwasser 1990, 1991, 1992; Kessler *et al.* 1992) continue to embrace the anthropocentric notion that forests are first and foremost for human use (Frissell *et al.* 1992; Grumbine 1992, 1994; Lawrence and Murphy 1992). Despite millions of dollars spent on GIS

FIGURE 6.9 Two kinds of landscape mosaics: (A) one produced by fire in Yellowstone National Park (photo by George Wuerthner); and (B) one produced by clearcutting in the Willamette National Forest, Oregon (photo by Reed Noss). The mosaic produced by clearcutting differs from the natural mosaic in having grossly simplified structure within disturbance patches (i.e., little wood remaining), sharp boundaries and edge effects between adjacent patches, and the presence of roads.

FIGURE 6.10 Differences between a traditional clearcut (A) and a New Forestry-style partial or retention cut (B) in the Willamette National Forest, Oregon. Photos by Reed Noss.

technology and glossy brochures, these programs so far have not delivered what they promised in terms of an ecosystem approach. Many New Perspectives and Ecosystem Management projects seem to have been public-relations campaigns by the Forest Service to disguise the negative effects of logging and road building in old growth and roadless areas (Noss 1991b, Frissell *et al.* 1992, Lawrence and Murphy 1992).

Much more than new slogans and fancy technology is needed to achieve true sustainability in forest management. But although we remain skeptical about the political use of New Forestry, New Perspectives, and Ecosystem Management as justifications for logging of natural forests, the techniques of New Forestry hold considerable promise for managing plantations and, in some cases, second-growth forests. Partial or retention cuts, because they open the canopy and stimulate tree growth, may hasten the development of old-growth characteristics, a more diverse groundcover, and denser shrub layer in plantations and second growth. Precommercial and commercial thinnings, however, would do the same, yet retain more live trees. The clustering of harvest units proposed by New Forestry will allow a greater proportion of the landscape to remain in large, unfragmented blocks and reduce the need for extensive road systems (Franklin 1992). Under conventional silvicultural systems, road density averages about 5 miles per square mile on commercial forest lands in the Pacific Northwest (Norse 1990), a level too high to sustain sensitive elements of biodiversity (see Chapter 2). In sum, New Forestry holds promise from a silvicultural perspective, but is no replacement for protection of natural forests.

Selection Forestry

New Forestry silviculture was developed for coniferous forests and shade-intolerant trees, particularly Douglas-fir. For forest communities characterized by shade tolerance and gap-phase replacement, single tree and group selection cuts (uneven-age management) are arguably more suitable because they more closely mimic the natural disturbance regime. However, selection forestry, often advocated by environmentalists who have little scientific understanding, also has its dangers. A common kind of selection called high-grading, where the highest quality trees in a stand are removed, is suspected of causing genetic impoverishment and contributing to forest health problems in ponderosa pine and other forests where it has been conducted.

High-grading is not the only kind of selection forestry. The opposite approach of "natural selection" forestry removes unhealthy, broken, and dying trees from a stand. Although this selection is claimed to simulate natural mortality patterns, it actually depletes a stand of the structures most critical for wildlife. Furthermore, selection forestry requires a road density at

What Do We Do with Sick Forests?

Healthy stands of natural forest are usually best left alone. But in some cases entire forest landscapes, even whole regions, are suffering from mismanagement. A classic case is the forests east of the Cascades crest in the Pacific Northwest, which have been declared in a "state of ecological collapse" by the U.S. Forest Service (Gast *et al.* 1991, Wickman 1991).

The forest health crisis in the eastside of the Northwest and other regions (such as the Sierra Nevada) is a result of poor management. Most damaging of all has been fire suppression. For at least seven decades, fires were suppressed in ponderosa pine forests that naturally experience low-intensity ground fires every five to twenty-five years, depending on site conditions. Fire suppression allowed stands to grow unnaturally dense and be invaded by Douglas-fir and white and grand firs. Passive suppression resulting from roads, clearcuts, and other artificial firebreaks also reduced fires in many areas. In central Oregon, early foresters recorded 10–30 trees per acre with average diameters over 17 inches. Today, those same areas have over 300 trees per acre and diameters less than 10 inches (Daniel 1990). These denser forests are more vulnerable to insect infestations, fungal diseases, dwarf mistletoe parasitism, and drought (due to competition among trees for scarce water). Livestock grazing adds to the problem by decreasing herbaceous biomass and therefore the frequency of ground fires. If a fire starts in a forest that has gone too long without a burn, it may be catastrophic due to a heavy accumulation of fuels that "ladder" up to the canopy.

Another management mistake in ponderosa pine forests was selective cutting of the largest and healthiest individual trees, a practice known as high-grading, which resulted in loss of trees that were most important ecologically and genetically. Probably many ponderosa pine stands today are genetically impoverished due to selective cutting, a practice that also encouraged invasion of firs and other trees from off site.

Under pressure to supply wood to the mills and make up for reduced cutting in the "spotted owl forests" west of the Cascades crest, the Forest Service is planning mas-

least as high as even-age management, and frequent entries that may disturb sensitive species. Perhaps most dangerous, selection forestry is not nearly as ugly as clearcutting; from a distance it may be invisible. Hence, it creates a green illusion of unbroken forest pleasing to the eye, but beneath the canopy, biodiversity may be seriously impoverished. Selection forestry, though it has a place in small-scale forestry operations, is no solution to the problems of modern forest management.

sive salvage logging of dead and dying trees in Eastside forests. The salvage timber sales will be sweetened by including many of the remaining healthy ponderosa pines.

The "cure" of widespread salvage logging on the Eastside could be much worse than the disease of declining forest health. Instead of rushing in with a single solution motivated by politics, forest restoration on the Eastside should be cautious and experimental. Several restoration forestry treatments could be applied (Noss 1992d) including:

1. Control areas with no management (wildfires would be allowed to burn).
2. No management except fire suppression.
3. No management except prescribed burning (various treatments).
4. Light noncommercial thinning of small (e.g., less than 11 inches dbh) live and dead trees that have invaded since fire suppression, followed by periodic prescribed burning that emulates the natural fire regime (several treatments possible).
5. Salvage of dead trees in various amounts, proportions, and size classes (several different treatments), but with live trees untouched. Salvage would be followed by prescribed burning on a regular basis.
6. Noncommercial thinning of small live trees, plus salvage of dead trees in various amounts, proportions, and size classes (several different treatments, including prescribed burning combinations).

All treatments must be properly replicated and controlled. A cautious approach would emphasize the less intrusive treatments (#1–4) in terms of land area affected. Those few natural and near-natural stands that remain in relatively good health should be completely protected, as they provide the benchmarks, blueprints, and ingredients for restoration of degraded areas. Intrusive treatments should be confined to accessible stands, avoid roadless areas, and harm no mature or old-growth trees.

Conservation and Restoration

Even-age (clearcutting and similar techniques) and uneven-age (selection) management both threaten biodiversity if practiced intensively. To the best of current knowledge, large or even moderate amounts of wood simply cannot be extracted from an ecosystem without damaging some of its components. The damage may be temporary or long lasting. In most cases we do not know enough about the ecosystem to accurately predict long-term effects, which might not show up for decades.

No harvest method or silvicultural regime is best for all circumstances. Partial retention cuts, where scattered trees or clumps of trees are left in logged areas, and other New Forestry approaches are suitable for some forest types, such as those characterized by stand-replacing disturbances. On the other hand, selection forestry is almost certainly better for forests characterized by gap-phase replacement (though a variety of gap sizes are probably necessary to maintain a diversity of regeneration conditions). Neither is without dangers, but probably both can be practiced sustainably if in moderation. We emphasize moderation.

Apparently the only way to protect biodiversity while practicing forestry is to reduce the intensity of forest management for commodities. This in turn requires that society—American and global—reduce its consumption of wood products. Credible estimates (e.g., Postel and Ryan 1991) suggest that Americans could reduce their demand for raw wood by 50 percent through aggressive recycling and conservation programs. This will eliminate the need to harvest timber (except for restoration) on public lands and will ease the pressure on private forests. Of course, reducing wood consumption will not contribute to overall environmental health if consumers turn to other regions (such as Siberia) for wood products or to other products, such as concrete and steel, that require greater use of fossil fuels during production (Koch 1991, Salwasser *et al.* 1992). Higher taxes on fossil fuels will help prevent that shift. Conservation practice must ultimately encompass all resources and nations, but it has to start somewhere.

Probably the most useful concept for management of multiple-use forests today is ecological restoration. Most forests have been damaged to some degree by human activities and require healing. Restoration should not imply returning to some pristine natural condition, an elusive and perhaps impossible goal. Rather, restoration forestry means, among other things: (1) managing the landscape for older forests by preserving existing old growth and other late-successional stands, allowing many second-growth stands to mature, and placing managed stands on longer rotations; (2) retaining structural diversity, including snags and down logs, in managed stands; (3) retaining and restoring large, intact patches of forest unfragmented by roads, clearcuts, or other openings; (4) retaining and restoring corridors and other linkages between forests and providing connectivity across the regional landscape as a whole; (5) allowing natural fires to burn, using prescribed fire, or applying silvicultural manipulations that simulate fire and other disturbances, in order to maintain a full spectrum of natural seral stages and structures; (6) stopping road construction and reconstruction, and obliterating and revegetating most existing roads; and (7) recovering viable populations of rare species and reintroducing extirpated species (Noss 1993c).

Simply put, restoration forestry means *reversing* the landscape changes that have been associated with loss of biodiversity. Importantly, a management program based on restoration, a labor-intensive activity, can probably employ at least as many workers as one based on extraction. True, restoration produces no immediate revenue, but one would be hard pressed to find a more worthy program for public funding. The long-term payoff—ecological, economic, and emotional—will be considerable.

The following recommendations summarize conservation and restoration priorities for multiple-use forests.

1. **Save virgin forests.** Do not subject primary natural forests of any age to commercial forestry. Virgin forests are severely depleted nationwide and worldwide and should be protected wherever they remain. However, prescribed burning and sometimes thinning may be needed in some situations to return fire-suppressed stands to natural structure, particularly for fire-dependent forest types such as longleaf and ponderosa pine. These forests are naturally open and parklike due to frequent, low-intensity fires but have often suffered from years or decades of fire suppression.

2. **Protect many regenerated forests.** In regions where little virgin forest remains, secondary natural forests and remnant old trees assume high conservation value. In such cases, these forests should be protected on public lands and protected or harvested sustainably on private lands.

3. **Consider landscape ecology.** Use knowledge of landscape context to inform decisions on the spatial pattern of cutting. Schedule harvests in parts of the landscape that are least sensitive. The selection process for core reserves and corridors (see Chapters 4 and 5) should precede multiple-use planning, so that harvest units can be located in sites that conflict minimally with conservation objectives.

4. **Reduce road networks.** Minimize road construction by restricting harvests to sites already accessible by road. Generally any site far enough from a road to be inaccessible for timber operations is valuable as a refuge and should remain roadless. Reduce road density as much as possible by closing, obliterating, and revegetating roads.

5. **Stop clearcutting and intensive site preparation. Apply ecological silviculture.** Clearcuts and intensive site preparation mimic no known natural disturbance; we need not experiment further to learn that they

are destructive. Immediate effects of these activities range from mild to severe, depending on site conditions, but the long-term consequences are probably always undesirable. Partial or retention cuts substituted for clearcutting for shade-intolerant species better simulate natural disturbances (though not perfectly) and leave more habitat structure through subsequent rotations. Single or group selection better mimics natural disturbances for forests dominated by shade-tolerant trees. For fire-dependent forest types, and tree species that require bare mineral soil to regenerate, use prescribed fire, supplemented by light chopping as necessary, to prepare sites. In all cases, use knowledge of forest ecology to select the appropriate harvest and regeneration system for each forest type and site condition.

6. **Protect sensitive sites.** Protect riparian zones, wetlands, and other sensitive habitats with buffer zones wide enough to prevent direct and indirect impacts and edge effects.

7. **Minimize mechanical means.** Use horses, oxen, human labor, and other nonmechanical means to cut and haul timber, reducing the need for heavy equipment for skidding and associated soil disturbance and use of fossil fuels. Whenever possible, use narrow trails instead of haul roads. But, we do not advocate a return to log drives down rivers!

8. **Regenerate naturally.** Whenever possible, rely on natural colonization and regeneration of vegetation. If planting is necessary, use native species and local genotypes of trees. Plant a mixture of tree species in the same general proportions as were in the predisturbance forest, and encourage growth of native vegetation and development of natural structure in all vertical strata.

9. **Schedule long rotations.** Base rotation length on the life span of the tree species involved rather than on culmination of mean annual growth increment or other economically derived measures. Among other benefits, long rotations will contribute structure to a stand and may help foster genetic diversity.

10. **Encourage alternative uses.** Encourage sustainable alternative uses of forests, including *light* harvest of mushrooms or nuts, nonmotorized recreation, nonmanipulative scientific research, environmental education, and nature study.

11. **Emphasize restoration forestry and adaptive management.** Most commercial forest lands, public and private, have been damaged by past mismanagement. Restoring native biodiversity on these lands will re-

quire decades or centuries of healing. In some cases recovery might be best accomplished by hands-off protection; in other cases it can be expedited by human labor. Test various restoration techniques cautiously, employing rigorous standards of experimental design, and adjust management on the basis of new information.

12. **Reduce consumption of wood products.** Especially at a time when the human population is growing rapidly, forests can only be conserved by dramatic reductions in the per capita demand for paper and wood products of all types. When world population stabilizes and then (hopefully) declines, continued reductions in per capita consumption will allow for widespread recovery of forests and other natural ecosystems.

CHAPTER SEVEN

MANAGING RANGELANDS

> Human management, even the most intelligent and enlightened, is not as effective at facilitating species preservation at multiple trophic levels and maintaining sustained levels of productivity as are the mechanisms produced by 50 million years of evolution, including coevolution, of grasses, the herbivores that feed on them, and other members of natural grassland trophic webs.
>
> S.J. McNaughton (1993)

Until recently, rangelands have received scant attention from conservationists. Conservation efforts have focused on forests, and issues such as habitat fragmentation have been described in terms of forest islands in a sea of agriculture. Yet rangelands, defined here as wildland landscapes in which the dominant plants are not trees, make up roughly 70 percent of the terrestrial surface of the earth (Holechek *et al.* 1988). In the western United States, rangelands comprise 70 percent of the land in the 11 states west of the 100th meridian. Rangelands include a wide variety of plant communities ranging from grasslands and deserts to alpine and arctic tundra. Without rangelands, the biodiversity of the world and of North America would be quite incomplete.

A primary human use of rangelands has historically been livestock grazing. Most people probably picture cows and cowboys when they think about western rangelands, even though the popular song "Home on the Range" contains not one word about livestock (we thank Andy Kerr of the Oregon Natural Resources Council for bringing this notable fact to our attention). In North America, rangelands are increasingly in demand for such uses as recreation and mineral extraction, but are also being appreciated for their intrinsic values as wild habitat.

Many investigations and reports have testified to the continuing degradation of the western rangelands, as noted in Chapter 2. These reports, many of them by agencies with major rangeland management responsibilities, including the U.S. Bureau of Land Management, confirm that few arid or semiarid lands in the United States have not been degraded or desertified. Desertification is "the impoverishment of arid, semiarid, and some subhumid ecosystems by the combined impact of man's activities and drought. It is the process of change in these ecosystems that can be measured by reduced productivity of desirable plants, alterations in the biomass and diversity of the micro and macro fauna and flora, accelerated soil deterioration, and increased hazards for human occupancy" (Dregne 1977). Loss of biodiversity is an inevitable accompaniment of desertification. Thus, reports of at least moderate desertification on 98 percent of the arid lands of the United States (Dregne 1977) are cause for concern.

Compared with forests, rangelands are poorly understood ecologically. Four important factors contribute to this situation. First, far less scientific research has been done on rangeland biodiversity than on forest biodiversity. What research has been done has concentrated on narrow aspects of herbivore ecology or on effects of livestock management on a few selected wildlife species such as deer and elk.

Second, many important ecological processes on rangelands are not easily observable. For example, a key process in maintaining healthy perennial grasses is the annual restoration of carbohydrate root reserves. This restoration cannot easily be observed aboveground until chronic depletion of root reserves is manifested in decadent or dead plants.

Third, ecological processes may take a long time to express themselves, and many depend on rare or unpredictable events (Westoby *et al.* 1989). For example, establishing a new stand of a "climax" grass like Indian ricegrass on a degraded range may require a combination of temperature and soil moisture that only occurs once every 20 years on a particular site. The long-term, erratic nature of many processes on rangelands limits the effectiveness of science based upon 3- to 5-year studies.

Finally, livestock grazing—the most severe and insidious of the impacts on rangelands—is a classic cumulative effect. It literally occurs one bite at a time. One bite, one day, or even one year of grazing may have a negligible effect on rangeland biodiversity. But the cumulative effect of more than 100 years of unregulated grazing has been tremendous. Some of the changes have come so slowly that they are almost imperceptible over a span of a few years. Furthermore, long-term changes due to climate may mask or confound impacts due to grazing (e.g., Archer and Smeins 1992, West 1992). Science based on short-duration studies may not effectively detect such changes or determine their causes.

We desperately need a "new range management" to parallel the best aspects of "new forestry" described in Chapter 6, but no such movement is underway. In fact, the pool of scientists and managers studying rangeland ecosystems holistically may be too small to develop such a movement. The analysis and recommendations in this chapter are therefore based on limited scientific background and societal experience in managing rangelands for biodiversity. They represent an attempt to summarize and integrate rudimentary scientific knowledge using extrapolation, extension of knowledge from other ecosystems, and a heavy dose of intuition. We hope that our guidelines will be tested scientifically and that this experience will form the basis for better management in the future.

This chapter focuses on the grasslands and shrublands of the western United States because this region has been better studied than some other rangeland types, because it is heavily affected by human activity, and because a long-standing controversy exists over management of these lands. Hopefully, some of the thoughts and guidelines for these areas can be modified and adapted for other rangeland areas such as arctic and alpine tundra and grasslands in eastern states.

Rangeland Ecology

Given the great diversity of ecosystems that qualify as rangelands, a summary of ecological relationships must be cursory at best. We review here a few key ecological characteristics of rangelands as background for management recommendations. For a more thorough, less selective review see West (1983).

Rangelands are characterized by low or erratic precipitation or by low or erratic moisture availability during the growing season. Many rangelands also have a limited growing season due to temperature. As a result of limited moisture during the growing season, rangelands develop their characteristic treeless vegetation of lichens, grasses, herbs, shrubs, and cacti. Compared with forests, rangelands have little vertical structure. But, on a landscape scale, most rangelands include areas with large shrubs or trees, typically along permanent or seasonal watercourses or scarps where moisture is available for a longer time. Many rangeland organisms depend upon the structure provided by trees. Thus, inclusions of large woody vegetation in rangelands are critical to maintaining native biodiversity.

Because conditions for vigorous plant growth are limited and often unpredictable, many rangeland plants and animals have adopted an opportunistic way of life. In deserts, for example, spring rains are usually followed by large blooms of herbaceous annual plants. Herbivores such as the desert tortoise obtain much of their annual nutrient intake during these blooms. Large grazing animals such as bison often move long distances to find favorable food, cover, or water. Many organisms have evolved ways to survive long periods with limited water or nutrient intake. Thus most grasses have a dormant period, and animals such as prairie dogs hibernate or estivate. Perennial plants must store enough reserves of carbohydrates and other nutrients to survive the dormant period, regrow, and set seed during the next growing season. Similarly, prairie dogs and many other herbivores must store enough fat to survive a steady weight loss during the dormant season.

Soil and Water Relationships

Much of the life of the rangeland goes on beneath the surface of the ground despite the fact that rangeland soils are typically thin and poorly developed. Although some rangelands have deeper soils, most notably the tall-grass prairie that once flourished in the Midwest, most of these areas have been converted to agriculture and no longer support natural communities. On most rangelands, maintenance of biodiversity depends highly upon maintaining a precariously thin layer of soil.

Lack of available moisture has often been noted as a factor limiting rangeland production. Vegetation cover is critical in intercepting moisture and allowing it to infiltrate into the soil where it can be used by plants rather than running off overground. Writers at the turn of the century pointed out that overgrazing was evidently responsible for the denudation of mountain slopes, for increased surface runoff and erosion, and in some instances for the conversion of mountain slopes to barren wastes (Gifford 1984). However, short bursts of intensive precipitation can cause serious erosion even on well-vegetated areas. Sampson and Weyl (1918) reported on the relation between range preservation and erosion in central Utah and concluded that much of the damage to western ranges was due to erratic runoff and erosion following loss of vegetative cover. They further noted that while topography, climate, and soil are primary factors determining erosion, erosion is slight where the native groundcover remained, and the severity of erosion is largely determined by the extent to which groundcover is maintained. Our understanding of the relationship between native plant cover and erosion potential is limited, but the above conclusions are as relevant today as they were then (Gifford 1984). In short, vigorous groundcover minimizes erosion and maintains soil.

Nutrient cycling is receiving increased attention as a controlling factor in rangeland productivity (West 1991). Because most rangeland soils of dry environments are not subject to through-leaching, most elements other than nitrogen are in adequate supply for plants (Charley 1977), except where parent material is deficient in an essential element. More than 30 percent of the available nitrogen and organic matter in typical rangeland soils is found in the top 10 cm of the soil (Charley 1977), the layer most subject to erosion. Furthermore, even slight soil erosion can result in large reductions (35 to 75 percent) in relative growth of plants, apparently due to loss of nitrogen.

Disturbance Regimes

Rangelands are subject to a wide variety of disturbances, including fire and trampling by animals. Fire often keeps woody shrub and tree species from invading grasslands. On the other hand, some brushlands and grasslands with a moderate component of grasses do not burn well if they have been severely overgrazed. In the West, fire suppression has allowed woody species such as big sagebrush, pinon pine, and juniper to invade millions of acres of grasslands or mixed grassland/shrublands (Branson 1985). Thus, disturbances can interact in complex ways, few of which are well understood. Although we noted earlier that species diversity may be highest at some intermediate frequency and intensity of disturbance, disturbances may leave communities vulnerable to invasion by exotic species (Orians 1986, Hobbs and Huenneke 1992), with concurrent changes in ecosystem structure, function, and composition (Vitousek 1986).

Herbivory—A Process or a Disturbance?

Herbivory is a natural process on rangelands as in all ecosystems and involves a wide variety of insects and other invertebrates as well as vertebrates such as rabbits, prairie dogs, and ungulates. Depending on the scale at which it is observed, herbivory may be considered as a continuous process of nutrient cycling or as a disturbance.

Many rangelands evolved in the presence of large herding ungulates, which play such a dominant role that they must be considered keystone species. Two examples in North America are the bison on the Great Plains and the caribou on the arctic tundra. Under natural conditions, these animals move in large herds and by grazing an area for a few hours or days may remove vast quantities of vegetation and set succession back to a lower seral stage.

The question thus arises: "If rangeland plants evolved with large grazers, why are rangeland plant communities so susceptible to excessive removal of vegetation?" Or put another way, "Aren't domestic livestock simply performing the ecological functions (grazing, removal of vegetation) previously performed by native ungulates or other herbivores?" To address these questions, some ecological and historical factors must be considered. Milchunas *et al.* (1988) proposed a model to explain differing responses of vegetation to grazing depending on both the moisture regime and the evolutionary history of grazing (Fig. 7.1). These factors can help us predict not only how native vegetation may respond to livestock grazing, but how susceptible a given rangeland may be to invasion by exotics (Mack 1986).

Grazing will be discussed in more detail later under "threats," but two aspects of the coevolution of grazers and plants should be pointed out here. First, many plant species have evolved in the presence of herbivores and are well adapted to withstand grazing. Populations of some species even have a competitive advantage under conditions of grazing (McNaughton 1993). Plants have many life-history strategies to survive grazing. Among rangeland plants, herbivore defense strategies are common and often well developed. Such adaptations include physical features such as thorns on shrubs or awns on grass seeds, growth forms that limit vulnerability of plants, and production of chemicals that make plant material unpalatable or indigestible.

A controversy has developed in ecology over the response of plants to grazing, particularly regarding whether grazing can enhance growth and productivity of grasses ("compensatory growth"), and the implications of this question for livestock management (Belsky 1986, Levin 1993, Painter and Belsky 1993). The controversy is well displayed in the ecological literature (see Forum on Grazing Theory and Rangeland Management, *Ecological Applications* 3(1), 1993). We will not review the controversy here, except to

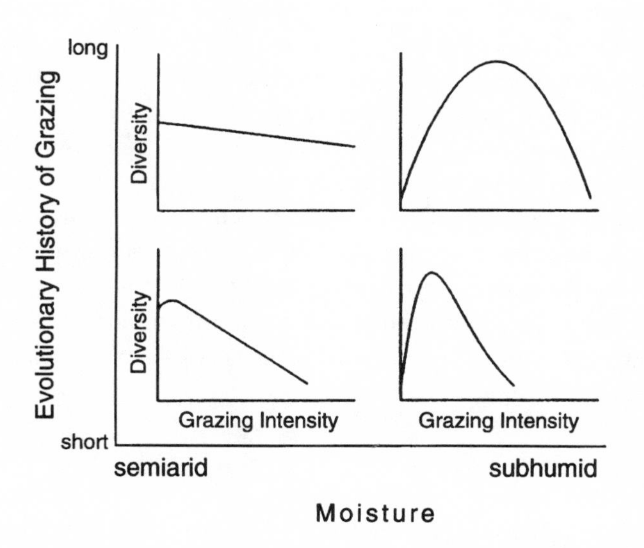

FIGURE 7.1 Plant diversity of grassland communities in relation to grazing intensity along gradients of moisture and evolutionary history of grazing (from Milchunas *et al.* 1988).

note that much of it is traceable to confusion over temporal and spatial scales, and to attempts to generalize across vastly different rangeland systems and to equate naturally evolved herbivore grazing systems with current livestock management practices. We question, as others have (McNaughton 1993), how much real influence science has had on rangeland management. We further concur with Imanuel Noy-Meir's (1993) response to the issue:

> There are several well-known mechanisms by which grazing can reduce the subsequent growth rate of plants, and several other well-documented mechanisms by which grazing can enhance plant growth rate . . . The question of compensatory growth is of relevance to management of Western rangelands for livestock production, but of relatively little relevance to conservation goals. No region-wide answers can be expected.

A second aspect of herbivory that is critical to understand is that on most North American rangelands, evolutionary pressure to develop resistance to grazing has not been strong for the last 10,000 years because populations of large herbivores were limited by factors other than year-round availability of forage. A well-established principle of game management is that many ungulates on northern ranges are limited by quantity and quality of forage on winter ranges. For example, mountain goats are to some degree limited by the number of windblown snowfree patches on which they forage during the winter; these patches typically make less than 5 percent of their range. Since these areas can support only so many mountain goats, the population never gets large enough to severely deplete the vegetation on the remaining 95 percent of the range that is snow-covered for half the year. In fact, most wild ungulates of North American rangelands concentrate seasonally on either (1) open, low-elevation winter ranges (in the colder, snow-covered northern rangelands), or (2) near sources of free water (springs, seeps, etc.) during the dry season (in the hot, arid southern rangelands). These concentration areas are usually a small percentage of the year-round range of the species, but most limiting to the population in terms of forage. Thus on large portions of rangelands, plant species evolved with minimal grazing pressure from large herbivores.

Succession

Succession on rangelands is a continuing source of confusion. Much of the theory of succession in North America was influenced by observing forests where, within limits, succession appeared to be linear and predictable in the sense that cutover forests would revert to brushlands and eventually forests. Unfortunately, rangeland succession is neither so linear nor so obvious.

Retrogression is the term range managers use to describe changes in species composition of a plant community away from the idealized pristine condition. Grazing or changes in environmental conditions can cause retrogression, which may lead to a loss of diversity, net primary production, and groundcover. A typical retrogression on grasslands involves a plant community moving from perennial grasses to annual grasses to annual herbs. Within this process, complex shifts in species abundance may occur even within a larger category such as perennial grasses. For example, with slight grazing, a grassland plant community may shift from primarily rough fescue to primarily bluebunch wheatgrass.

Two different models of vegetation dynamics on rangelands have been used to explain the changes in vegetation communities as a result of (over)grazing and other disturbances. The traditional model was the Clementsian succession model named after the pioneer plant ecologist F.E. Clements, who developed a theory of rangeland succession that looked very much like models developed for forest systems. This model assumes that there is only one stable vegetation state—the climax—and that movement away from this state from overgrazing (retrogression) is a continuous, reversible phenomena. In other words, retrogression and secondary succession after disturbance will be more or less mirror images of each other (Fig. 7.2). Thus this model predicts that if ranges are retrogressing from overgrazing, removal of the grazing should set them back on the pathway to the climax stable state.

An alternative model, the "state-and-transition" model, is based on a much more complex and realistic view of plant dynamics. The state-and-transition model, as proposed by Friedel (1991), Westoby *et al.* (1989), Laycock (1991), and others, assumes that more than one stable or steady state condition is possible, as well as several less stable states. The state-and-transition model is supported by much empirical evidence and we believe that it represents a more realistic understanding of rangeland vegetation dynamics.

An important prediction of the state-and-transition model is the existence of transition "thresholds." When a transition threshold is exceeded, the plant community will move in a positive feedback mode to a new vegetation state. An example would be moving from a grassland community to a shrub dominated community with heavy grazing. The model predicts that when some density of shrubs is reached, the vegetation will tend to move rapidly toward a shrub dominated community even with no further grazing

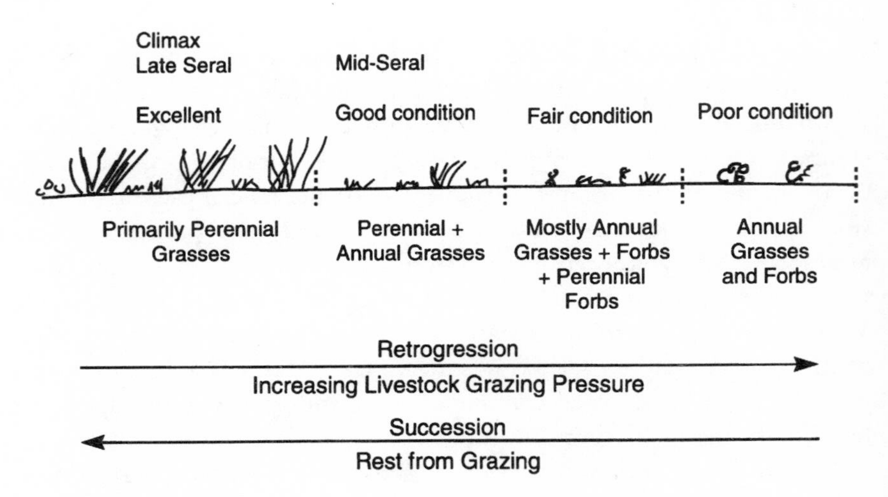

FIGURE 7.2 The classic succession and retrogression model for rangelands.

pressure. Retrogression is seen as step-wise rather than continuous; that is, a plant community may be rather resilient up to a certain threshold, after which it can no longer return to its previous stable state (Archer and Smeins 1992, Schlesinger *et al.* 1990).

As an example, rangelands invaded by pinyon–juniper can revert to grassland if grazing pressure is reduced and fire is not suppressed, but only up to a point. After that point, the juniper trees outcompete grasses for soil moisture and sunlight, and grasses can no longer become established. Fire regimes cannot easily be restored when the biomass of grasses is too low to carry fires that would burn out pinyon and juniper trees.

A case history of grazing pushing a plant community beyond a threshold is described by Archer (1989) for southern Texas savannas. In that region, grazing pressure alters herbaceous composition while decreasing fire frequency. If grazing pressure is relaxed before a critical threshold, then succession toward a higher quality grassland can occur. If, however, enough woody plants have become established, then new successional and positive feedback processes may move the system toward a new stable state—a woodland. Once in the shrub or woodland state, the site may not return to grassland even if grazing ends (Fig. 7.3).

After a critical threshold has been passed, a site may degenerate further. Schlesinger *et al.* (1990) describe a mechanism by which long-term grazing on rangelands can lead to site deterioration. They state:

> Studies of ecosystem processes on the Jornada Experimental Range in southern New Mexico suggest that long-term grazing of

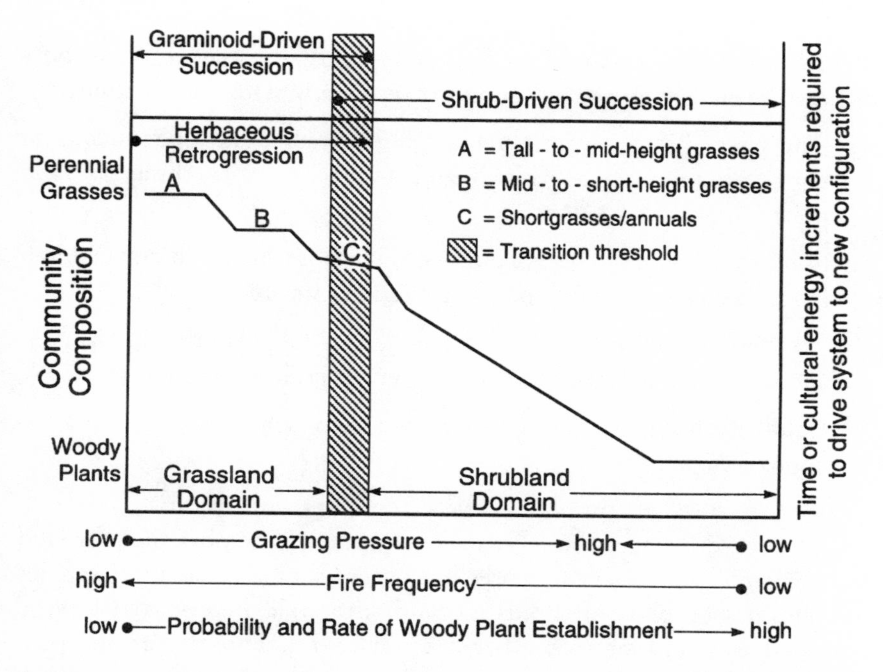

FIGURE 7.3 Conceptual diagram of changes in community structure as a function of grazing pressure pushing a community past a threshold. Within the grassland domain, grazing alters the composition and productivity of herbaceous species while decreasing fire frequency and intensity, thereby increasing the probability of woody-plant establishment. If grazing pressure is reduced before some critical threshold, succession toward higher-condition grasslands may occur. However, if sufficient numbers of woody plants become established, shrub-driven successional processes begin to predominate and the site moves toward a new steady-state configuration. Once in the shrubland or woodland domain, the site will not revert to grassland after grazing has ceased, expecially if the displaced grasses had originally established under a different climatic regime. Human manipulation can alter grass–shrub mixtures, but subsequent succession may result in a rapid return to a community dominated by woody plants (from Archer 1989).

semi-arid grasslands leads to an increase in the spatial and temporal heterogeneity of water, nitrogen, and other soil resources. Heterogeneity of soil resources promotes invasion by desert shrubs which leads to further localization of soil resources under shrub canopies. In the barren area between shrubs, soil fertility is lost by erosion and gaseous emissions. This positive feedback leads to the desertification of formerly productive land in southern New Mexico and in other regions . . .

Archer and Smeins (1992) list four examples of apparent regional shifts of North American vegetation to alternative steady states in recent history:

1. The Great Basin—from perennial bunchgrasses and open stands of sagebrush to dense sagebrush and annuals such as cheatgrass and medusahead.
2. Southwestern Desert Grasslands—from tobosa and black grama grasslands to creosotebush, tarbush, or mesquite shrublands.
3. Southern Grasslands and Savannas—from tallgrass prairies and savannas to oak, juniper, mesquite, or thorn scrub woodlands.
4. California Mediterranean Grasslands—from perennial bunchgrasses to annual grasses.

The presence of threshold levels together with positive feedback processes beyond those thresholds suggests a few basic objectives for land management on rangelands. First, management should avoid altering plant communities beyond thresholds. Second, areas that have moved beyond thresholds but not yet into another steady state (i.e., those that are in a positive feedback mode and are continuing to degrade) should receive a high priority for restoration to a stable state.

Threats to Rangeland Biodiversity

The biodiversity of rangelands is threatened by many factors, including mining, suburban encroachment, roads, and water diversions. Our discussion here will focus on threats from livestock production because, as suggested earlier, it is most responsible for the deterioration of rangelands in western North America. Direct impacts of grazing are only one facet of the problem, since livestock production includes predator control, irrigation, fencing, and much more. For example, the amount of water that is diverted to grow hay to feed cattle in winter has impacts that can rival the effects of direct removal of vegetation by cattle. Nevada, the most arid state in the nation with an average of less than 9 inches of annual precipitation, allots 90 percent of its water to fields of hay (Wuerthner 1992). We will focus here on two aspects of livestock production—livestock grazing itself and livestock management practices. We will also touch briefly upon two other major threats, exotics and recreation.

Livestock Grazing

Livestock grazing has proven to be the most insidious and pervasive threat to biodiversity on rangelands (Ferguson and Ferguson 1983, Jacobs 1991,

Voigt 1976, Wagner 1978, Wuerthner 1992). In the western United States, few areas have not had cows, sheep, or horses on them for at least some period in the last 200 years. We have already alluded to the vast changes in plant communities resulting from livestock grazing, especially in the Intermountain West. While the cowboy may have enriched our culture with movies, stories, and song, his general lack of understanding of how to manage cows and rangelands has caused serious damage to the West.

Livestock grazing has impacts on individual plants or species, plant communities, soils and watersheds, and native animals. In addition each of these impacts have secondary or ecosystem-level effects. In this section we first describe how animals graze, then consider each of these impacts. Finally, we revisit in more depth the issue of livestock grazing compared with grazing by native ungulates.

Animal grazing. For the purpose of discussion, we use the term *grazing* in the generic sense to refer to herbivores eating the annual production (including vegetative parts—not just fruits or seeds) of any plant regardless of whether it is a grass, shrub, or herb. Some biologists distinguish grazing (the eating of grasses and forbs) from browsing (the eating of parts of shrubs and trees). Our focus here will be on large mammalian grazers, particularly ungulates, but many of the principles apply to other grazers, such as rabbits, that survive on the vegetative parts of plants.

The key to understanding grazing is to remember that all grazing ungulates are opportunistic to some degree. They can eat quite a variety of species, but given a choice they will be selective. For example, the mule deer has been reported to eat over 700 plant species (Kufeld *et al.* 1973) but prefers certain species and even certain plant parts.

As a consequence of selective preferences, grazers continuously highgrade the range, eating the most palatable and digestible plants first. Plants vary in their palatability not only between species but among individuals of the same species, from season to season for the same plant, and even from one part of the plant to another (e.g., leaves versus stem on a grass). Grazers constantly try to maximize their nutritional input per unit of effort. Different species of grazers have different foraging strategies and forage preferences (Cooperrider 1986). Nevertheless, at least among North American ungulates, food habits and preferences overlap considerably; when two or more ungulates occupy a common range they often are selecting for the same plant species. The result of such selectivity is that with continuous grazing on a rangeland, some plants/plant species will be grazed much more severely than others due to their higher palatability at some time of year for one or more species of ungulate.

Impacts of grazing on the individual plant. When an individual plant is grazed, it may respond in several ways. If it is a grass and is grazed above the intercalary and apical meristems (the points from which cell growth is initiated), it may regrow if there is still enough soil moisture. The growth form of a plant is thus an important adaptation to grazing. Tufted bunchgrasses (termed *caespitose*) with a meristem well above the ground, and grasses with stems growing horizontally below the ground (termed *rhizomatous*) represent two extremes of adaptation to grazing (Mack and Thompson 1982). Caespitose grasses are poorly adapted to heavy grazing and are thought to have evolved in the absence of significant grazing, whereas rhizomatous grasses are well adapted and in many cases have evolved with large herding herbivores. There are exceptions to this generalization, however, as plants have many other mechanisms for responding to grazing pressure. In any case, grazed plants lose a certain amount of leaf area and this loss inhibits the ability to capture energy through photosynthesis and ultimately to store energy back into roots (for grasses and herbs) or roots and stems (for shrubs).

Impacts of grazing on plant communities. If individual plants are continuously grazed so that they cannot store enough energy for their reserves to last through the dormant season and regrow during the next season, they eventually die out. In less severe cases such plants merely become more and more stunted and do not reproduce, whether by seed or vegetatively.

The selective nature of grazing combined with the limited tolerance of some plant species to grazing can result in substantial shifts in species composition of rangeland plant communities. With heavy and continuous grazing the most palatable and nutritious plants will first become weakened, will not reproduce, and will eventually die. Ultimately these plants may disappear from the community. At the same time, other plant species that are less palatable or more tolerant of grazing, such as annual grasses, annual forbs, and particularly exotic species, may increase in the area. Exotic weeds, which include what many range managers consider noxious pests such as leafy spurge or cheatgrass, are usually well adapted to grazing and to invading overgrazed sites. Some scientists, such as Bartolome (1993), have questioned the relative importance of selectivity as opposed to overall grazing intensity, but most range managers agree that moderate or heavy grazing over several years will usually change plant community composition.

The shift in plant community composition with grazing is the retrogression referred to earlier. The retrogression typically involves a reduction in perennial grasses accompanied by increases in less palatable annual grasses and herbs. In many cases, it may involve loss of palatable woody species (such as willows or cottonwoods) or a shift from herbaceous to woody species.

Loss of palatable woody species is often a serious and obvious impact in riparian zones on western rangelands. Overgrazing by cattle in riparian zones can eliminate the willows and sedges that stabilize stream banks and can eventually lead to a lowering of the water table and a drying of the site (Fig. 7.4) (Kovalchik and Elmore 1992). The same situation can result from a local overabundance of native ungulates such as elk, but it is far less common, particularly when natural predators are present to keep ungulates on the move.

Secondary impacts of changes in plant community composition. Changes in plant communities have many secondary effects on the animals that inhabit them. For example, many birds depend on woody riparian vegetation for foraging and nesting. Eliminating willows or cottonwoods along streams may reduce bird species richness in the landscape. Similarly, ungulates such as antelope require shrubs of a certain height and density on their fawning grounds. If livestock grazing alters the structure (height, density, or cover) of this vegetation, antelope fawns are much more susceptible to predation by coyotes and other species, and the antelope population may decline.

Secondary impacts of grazing on soils and watersheds. The secondary impact of grazing on soils and watershed function is probably even more serious than the immediate effect on plants. By reducing the amount of cover of both live plants and dead ungrazed leaves, livestock grazing can both increase the amount of overland erosion and decrease the amount of infiltration of water into the soil (Blackburn 1986). Changing the plant species composition can also substantially affect both erosion and infiltration (Fig. 7.5). Erosion not only results in decreased soil fertility, but can exacerbate siltation of streams with concomitant decreases in aquatic productivity, fish populations, and species diversity. Loss of woody vegetation around streams is particularly harmful to stream health. Woody vegetation is critical in maintaining the stability of banks and reducing bank erosion. This vegetation also shades the streams, resulting in cooler temperatures and increased oxygen availability, needed by fish such as trout.

Decreased infiltration results in less recharge of aquifers and can reduce seasonal flows in springs, streams, and rivers. In the West, some once-perennial streams have dried up, apparently due to continuous livestock grazing

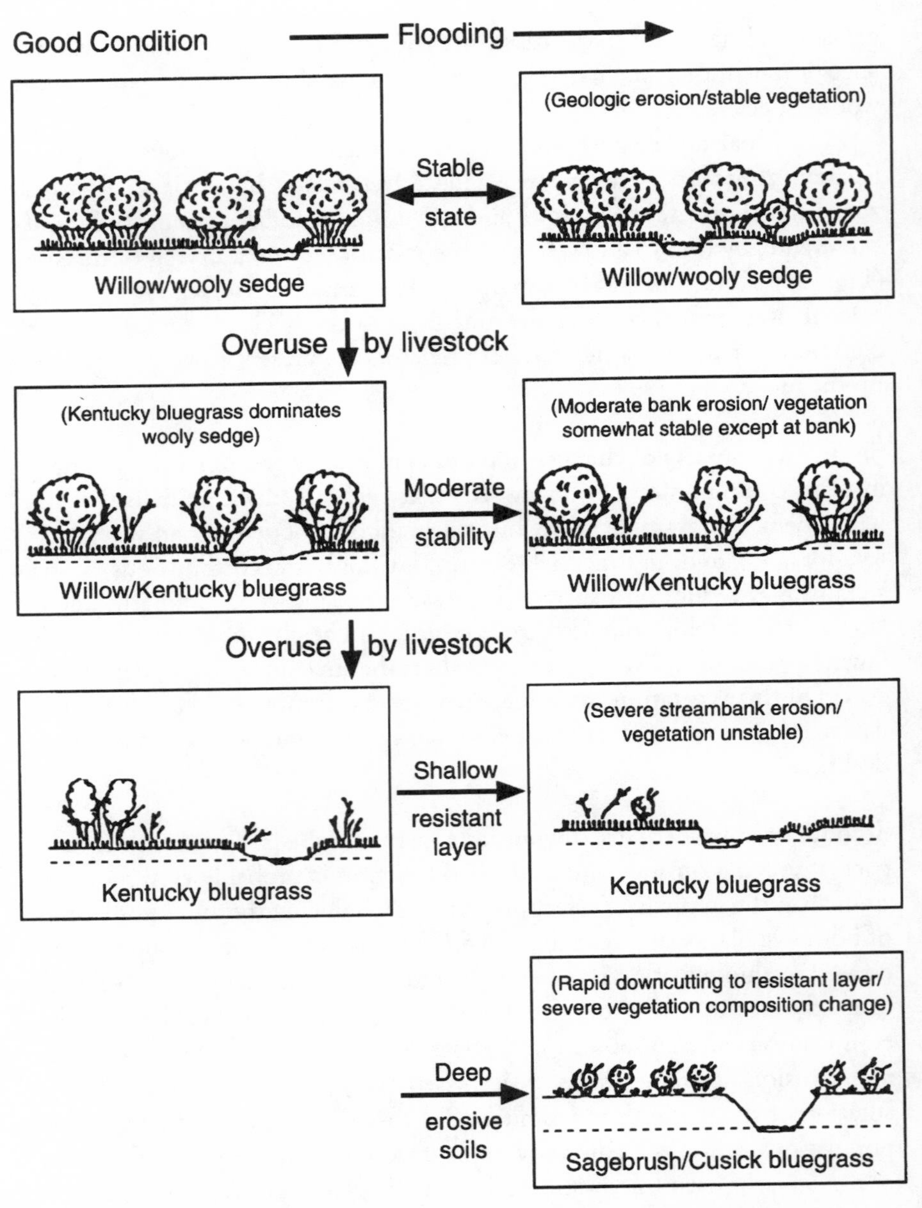

FIGURE 7.4 Deterioration of sites supporting the willow/wooly sedge plant association with flooding and improper use by livestock (from Kovalchik and Elmore 1992).

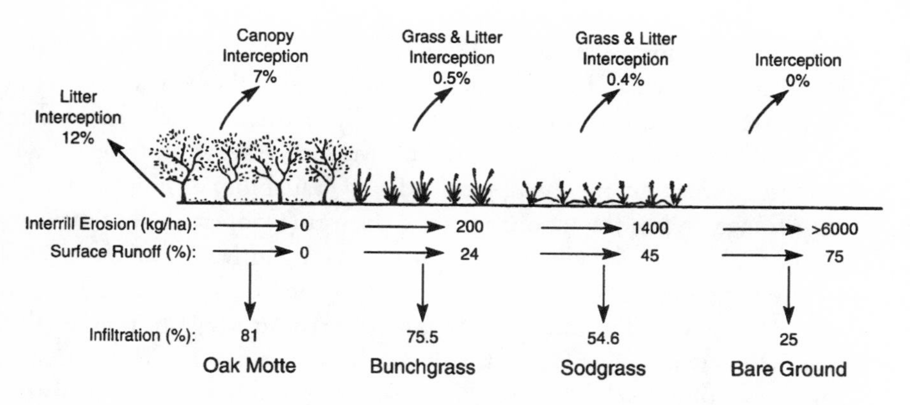

FIGURE 7.5 Effect of plant community composition on interception, infiltration, and runoff (from Blackburn 1986).

to the point of lowering water tables. In a few cases, where livestock grazing has been eliminated or drastically reduced, year-round flows have been restored to these streams and trout fisheries have been reestablished.

A controversy exists about the damage or alleged benefits caused by soil compaction and other direct effects of livestock (Savory 1988). We agree with Allan Savory that the effects on soils and vegetation of a short burst of trampling by many animals differ greatly from continuous trampling by fewer animals, as can be observed around water sources. Savory (1988), however, claims that animal impact, which he describes as "all things animals do besides eat" is necessary in arid lands ("brittle environments" as he terms them) to advance succession. Little evidence supports this contention. Animal impact, as Savory defines it, is hard to separate from impact of livestock eating. On the Appleton-Whittell Research Ranch Sanctuary (a relatively brittle environment) grazing was eliminated in 1968. Savory's theory predicted an initial improvement (following the elimination of the stress of grazing), followed by deterioration as the residual beneficial effect of animal impact wears off (Savory 1986). The deterioration he predicts would eventually lead to a loss of diversity and relative instability of the ecosystem. Sixteen years later, with neither grazing nor animal impact, both plant species diversity and diversity of several animal groups studied (birds, small mammals, grasshoppers) had all increased (Bock *et al.* 1984). Brady *et al.* (1989) concluded that the data on vegetation and wildlife changes after 16 years did not support the hypothesis that continued animal impact is needed to prevent ecosystem deterioration. Rather, cessation of grazing allowed recovery.

Competition for forage, water, and space. Livestock compete with wild animals for food, water, and space. Competition is one of the best studied impacts of livestock grazing on wildlife, especially ungulates (Peek and Dalke 1982, Van Dyne *et al.* 1984). The impact of livestock grazing on forage supplies for wildlife is a complex subject involving availability of forage, forage quality, forage preferences, and many other factors (Cooperrider and Bailey 1982, 1984). Historical failures to appreciate competitive effects have resulted from simplified notions of what was adequate forage. A vast body of evidence from wildlife biology and animal nutrition suggests that small changes in nutritional quality of forage can have significant effects on ungulate population levels. Because livestock are typically turned out onto rangeland during the period of peak growth and nutritional value of forage, they have an opportunity to graze the best and most nutritious forages first—that is, to high-grade the range. Livestock thus graze areas used later in the year by wild animals, forcing the wild animals to forage and survive in a habitat that has already been degraded nutritionally.

Although impacts of forage usurpation on ungulates are relatively well documented, impacts on other animal species such as insects are virtually unknown. Where evidence is available, the same patterns seem to be true: herbivores forced to survive in areas high-graded by livestock will have reduced survival and their populations will eventually decrease. Affected herbivores include species like the desert tortoise, which feeds on similar vegetation as cattle and sheep, yet must compete with these more mobile species (Berry 1978, Campbell 1988). Carnivores may also decrease if their food supply declines. Animals adapted to earlier successional stages of vegetation may increase with grazing, depending on the structure and composition of the plant community.

Livestock and wildlife may also compete for water, particularly in desert areas where water is scarce. Such competition has often been mentioned but most of the information is anecdotal rather than based on scientific research.

Finally, evidence suggests that wildlife species compete with livestock for space; that is, some native species seem to avoid domestic animals. Again, most of the information is anecdotal. One of us (Cooperrider) has often observed elk departing sites in spring within hours of cattle arriving, a phenomenon observed through radiotelemetry studies in other areas. The reasons for departure are not well known, although we suspect that elk prefer not to forage in areas high-graded by cattle.

Impacts of disease transmission. Livestock can serve as disease vectors to wildlife. This topic is poorly understood, particularly since, until the last 20 years, most studies of diseases carried by both wild and domestic animals have focused on the effects on domestics.

In extreme cases, livestock carry diseases that can extirpate populations of native animals, as has been observed many times with domestic sheep and bighorn sheep. Entire populations of bighorns have died off within a year or two of coming in contact with domestic sheep (Goodson 1982). In other cases, livestock carry diseases that cause chronic health problems in wild species. Probably only the most obvious cases of disease transmission have been noted, much less studied. Any time exotic animals (including livestock) are brought into an area, there is a risk of disease transmission to the detriment of the native species. Because native species may not have been exposed to these diseases, they may lack genetic or acquired resistance to the new pathogens or parasites.

Miscellaneous impacts. We have described some of the more obvious impacts of livestock on biodiversity. Many other effects go largely unnoticed. For example, livestock can change the way other organisms move through the landscape. Cattle are believed to be responsible for spreading mesquite into grassland areas by eating the mesquite beans and then moving out onto grassland where they are deposited, undigested and viable, with water and fertilizer to get them established. Some evidence shows that livestock accelerate invasion of sites by exotic weeds by helping them disperse around the landscape. Another important impact involves the relationship between the brown-headed cowbird and cows. Cowbirds, as evident from their name, are closely associated with cows; when cows are moved into a new area, cowbirds often invade shortly thereafter. Cowbirds are brood parasites on other bird species, often causing population declines of the species with which they come in contact. The cowbird and the cow have been implicated in the decline of the federally endangered least Bell's vireo (Lowe *et al.* 1990).

Summary of livestock grazing effects on biodiversity. The effects of livestock grazing on biodiversity on rangelands are many and complex. Some problems are only temporary but many are long term. Some effects apply in certain areas and not in others. Intensity and timing of grazing are both important considerations, and some impacts may not show up for many years. Do not assume that, because there are no obvious impacts of livestock grazing, that no impacts exist. Biotic impoverishment is often subtle. We have summarized some impacts in Table 7.1 to help those analyzing threats from livestock to particular areas. Keep in mind, however, that because several impacts often occur concurrently and that overall effects may be synergistic rather than additive, ecological impacts are difficult to

Table 7.1 Impacts of Livestock Grazing on Biodiversity

Grazing activity	Primary effect	Secondary and tertiary effects
Selective grazing/overgrazing of individual plants	Decreased viability of individual plants	Plant decadence and death over time; loss of species from plant community; extirpation; extinction
Selective grazing of most nutritious forage (high-grading)	Loss of forage for other animal species	Decreased density of native herbivores; decreased density of carnivores and omnivores
	Overgrazing of riparian vegetation	Loss of vegetation in riparian areas; loss of riparian habitat structure and riparian dependent species; erosion of banks along streams; increased erosion; degraded hydrological cycle
Excessive removal of primary production	Retrogression of plant community	Shift in animal community; lowered stability and vegetative cover of plant community; decreased interception and infiltration of precipitation; increased runoff; increased erosion and degraded hydrological cycle
	Change in disturbance cycles—especially decreased frequency and intensity of fire	Change in plant community composition, especially increase in woody species; decrease in other species; extirpation; extinction; impairment of nutrient cycling; impoverished production for all organisms

Trampling of ground—especially around concentration areas (waterholes; shaded areas)	Compaction of ground	Decreased infiltration; increased runoff leading to increased erosion
	Erosion of banks	Decreased bank stability of streams; increased sedimentation and erosion; degraded hydrological cycle
Livestock watering	Fouling and contamination of waters	Disease problems in native animals
	Water usurpation	Decreased water available for native wildlife in arid regions
Importation of disease	Disease transmission to native wildlife	Decreased viability of native wildlife; extirpation; extinction
Presence of livestock on range	Avoidance of areas by native wildlife	Decreased effective habitat available for native wildlife; depressed abundance of native wildlife; extirpation; extinction

study or analyze with traditional reductionist methodologies. For example, cattle grazing may simultaneously reduce cover, change species composition, increase erosion, and decrease infiltration. The collective impact of all these processes on the landscape and on biodiversity may be far more severe than any impact in isolation.

Comparison of Livestock Grazing with Grazing by Native Herbivores

Some advocates of livestock grazing in North America have alleged that livestock grazing is merely substituting a nonnative animal for native ones that have been removed from the system. The implication is that these animals are replacing the ecological functions that were previously provided by the extirpated native species. In its mildest form, the assertion is that cattle are merely being substituted for the herds of bison that once existed in North America only 150–200 years ago. In another form, some argue that livestock grazing is similar to the effects of not just bison but other native ungulates such as elk and deer.

If the objective is to conserve biodiversity then both arguments are untenable, untrue, or irrelevant. Cattle do have similar (not identical) forage preferences to bison, but their foraging habits differ (Van Vuren 1982). Areas that were historically grazed by large herds of bison, namely the Great Plains, have grass species well adapted to grazing. But, bison grazed in a much different way than free-ranging cattle. During the late Pleistocene and early Holocene, the wild ox or auroch, the now-extinct wild ancestor of cattle, was widespread in Europe and Asia, where it inhabited forests and scrublands (Clutton-Brock 1981). This was a time when climates were wetter and warmer. Modern-day cattle, in spite of thousands of years of domestication, still retain a tendency to stay in the riparian zones—those remnants of their ancestral Pleistocene habitat. Cattle are thus a riparian species, and their natural tendency on North American rangelands is to concentrate in small scattered groups near water, where they can do considerable damage. The practice of confinement of cattle with fences may exacerbate this damage.

Bison, on the other hand, moved in large herds across vast areas, and did not concentrate in riparian areas for long periods of time. Furthermore, the bison has many adaptations to more northern climates such as the ability to get moisture from eating snow during the winter when free water is not available. The impacts of these two disparate foraging patterns on the vegetation and landscape are predictably quite distinct.

Finally, we note as many others have that bison were found in only token numbers in both the Great Basin and the southwestern deserts (Hall 1981,

MacDonald 1981), the two areas where livestock grazing is most contentious and, in the opinion of many ecologists, most damaging. At least since the Pleistocene, large herding herbivores were very rare or nonexistent over most of these regions.

Extending the argument of substitution to other native wild animals is even more suspect. Native North American ungulates have different forage preferences from livestock. More importantly, native ungulates move about (especially away from riparian zones) much more freely than livestock. Most wild ungulates on western ranges are migratory and thus only graze an area for one season (Patten 1993). Moreover, because wild ungulate populations are generally limited by something other than total year-round forage supply, most North American species (with the possible exception of bison) existed in much lower densities than livestock currently do.

Livestock Management Practices

Just as consequential as the direct and indirect impacts of livestock are the impacts of livestock management practices. We noted that, in harvesting forests, associated activities such as skidding logs and building roads go on concurrently with falling and removing trees and cause more damage to the ecosystem. On rangelands, livestock grazing is accompanied by fencing, water development, predator control, vegetation manipulation, and other practices that can affect biodiversity. The effects of livestock production in its totality must be considered.

Fencing. Fences, by design, are barriers for animals. Although most North American ungulates can move through or over traditional three-wire barbed-wire fences, some problems exist. First, not all fences are as loosely constructed, and tighter fences such as woven wire can severely impede movement of native wildlife. Pronghorn antelope are particularly limited in their ability to cross fences, and woven wire fences can effectively fragment their habitat and ultimately cause population decreases or extirpation (Yoakum 1978).

Second, even though species such as mule deer can and do easily jump fences, a certain number (especially juveniles) get tangled in them and die every year. In times and areas where these animals abound, these losses may not limit population densities. However, as populations decline, such losses become more significant.

Fences cause other primary and secondary problems. For example, fences often are accompanied by roads or cleared rights-of-way, which provide corridors for invasion of exotic plant or animal species. Many of these effects have been described by Jacobs (1991).

Water development. Since water is scarce on rangelands, water development is often used to distribute livestock grazing across the landscape. Unfortunately, developing water sources for livestock often involves taking water from streams, springs, or seeps, where it was used by native plants and animals, and moving it somewhere else for livestock. In other cases, springs have been drilled, resulting in overuse of aquifers and eventual drying up of water sources historically used by native species. Areas around natural water or water developments tend to become sacrifice areas when livestock are present. Livestock typically denude these areas of vegetation and compact the ground. Of little value to any native species, sacrifice areas do provide nodes for establishment of exotic plants and diseases.

A variety of other primary and secondary effects of water development were reviewed by Jacobs (1991). We have described here development of free water for livestock to drink. Such water is usually developed on a small scale. On a larger scale, water development for livestock includes the diversion of vast amounts of water to irrigate crops (such as alfalfa) used for feed.

Predator control. Historically, and still in many regions today, livestock grazing has been accompanied by attempts to remove native predators. In many cases these attempts have been successful with the result that wolves, mountain lions, and bears have been extirpated from large portions of their historic ranges. In other cases, populations of more resilient species such as coyotes have been reduced in numbers, if not in distribution, by efforts that have killed hundreds of smaller predators exposed to the same traps or poisons.

Most controlled species, such as wolves, coyotes, and mountain lions, are at or near the top of the food chain and may often influence the structure and function of the entire ecosystem. For example, predators may not only limit abundance of prey at times, but they may strongly influence their distribution and movement patterns, thereby influencing impacts of herbivores on vegetation. One can make a good case that all large predators are likely keystone species.

Thus, where wholesale predator control (i.e., attempts to extirpate or drastically reduce densities of predators over a large area) is a component of livestock management, then livestock grazing must be considered a serious threat to biodiversity even if direct effects on vegetation are minimal. It is important to distinguish between wholesale predator control and the occasional control of a problem individual. The latter may be called for when, say, an individual cougar suddenly switches food habits and starts attacking domestic sheep (or humans) instead of deer. If control of individual predators can be carried out with no reduction in year-to-year density or health of the predator population, then it is arguably not a serious threat to biodiversity.

Vegetation manipulation. Livestock grazing is often accompanied by massive vegetation manipulation, including herbicide spraying, plowing and seeding, mechanical control such as chaining, and controlled fire. These projects are often extremely expensive and cannot be justified by cost/benefit ratios. However, many are carried out at taxpayers' expense on public lands.

Vegetation control typically reduces plant species diversity. For many years, BLM plowed and seeded large areas of the Intermountain West with a monoculture of crested wheatgrass, an exotic grass from Eurasia that was considered good forage for livestock. The impact on biodiversity (in hindsight) was predictable—reduced species richness, often with increased erosion and decreased water infiltration as a result of the plowing and seeding. Yet, interpretive signs glorifying the benefits of crested wheatgrass remain on BLM lands.

Similar problems exist with other practices such as chaining (dragging an anchor chain between two tractors to mow down brush) and herbicide application. Prescribed burning, however, can both improve forage and increase biodiversity if it is designed to restore or mimic natural fire cycles and burn patterns. Other practices such as herbicide or mechanical treatments may sometimes be required to return an area to a condition where it can be maintained by fire.

Unfortunately, few if any vegetation manipulation projects have been carried out to restore biodiversity or even species diversity. Rather, these projects have been conducted with the single-minded purpose of increasing livestock forage, and they have been evaluated by their success in achieving that limited objective over a short time period. Vegetation manipulation to restore or increase livestock forage can severely affect biodiversity and should not be confused with the more difficult task of restoring biodiversity on rangelands. Vegetation manipulation can play a role in restoration, when applied cautiously.

Summary of impacts of livestock management practices on biodiversity. Livestock grazing is accompanied by an entire complex of management practices that may be as damaging to biodiversity as the grazing itself. Four of them were described briefly above—fencing, water development, predator control, and vegetation manipulation. But there are others. These practices and some potential impacts on biodiversity are summarized in Table 7.2. Some of these practices, such as wholesale predator control, are

TABLE 7.2 Impacts of Livestock Management Practices on Biodiversity

Practice	Primary effect	Secondary and tertiary effects
Fencing	Barriers to native wildlife	Decreased habitat for native wildlife; decreased abundance; extirpation
	Injury to individual animals	Individual animal death
Water development	Usurpation of water used by native wildlife	Decreased abundance of native wildlife populations where water is limiting; extirpation; extinction
	Creation of sacrifice areas of heavy trampling and annihilation of vegetation	Decreased habitat available for native wildlife; decreased abundance of native species
	Spreading of livestock impact into new areas	Increase in area of livestock grazing impact; primary, secondary, and tertiary effects (as shown in Table 7.1)
Salting	Spreading of livestock impact into new areas	Increase in area of livestock grazing impact; primary, secondary, and tertiary effects (as shown in Table 7.1)
Roads	Increase in human access to sites	Increased human disturbance; decreased abundance of vulnerable species; extirpation; extinction
	Soil disturbance	Increase in invasion of exotic species, especially plants; decrease in native species

Predator control	Decreased density of native carnivores	Extirpation or extinction of keystone predators; disruption of biotic community with unpredictable effects on species and ecosystem function
Vegetation manipulation—chaining	Removal of vertical structure	Decrease in animal species richness and diversity
Vegetation manipulation—plowing and seeding	Soil exposure and disturbance	Increased invasion by exotic plant species; decreased abundance of native species; disruption of soil community; loss of native plant and animal species richness
	Introduction of exotic (seeded) plant species	Decreased density of native species
Vegetation manipulation—burning	Removal of vertical structure	Decrease in animal species richness and diversity
	Loss of nutrients from system	Decreased productivity of biotic community
	Soil exposure and disturbance	Increased invasion by exotic plant species; decreased abundance of native species; disruption of soil community; loss of native plant and animal species richness

not a necessary part of a viable livestock operation, but the linkages between such practices and grazing management need to be clarified and recognized. If practices harmful to biodiversity are considered essential by the livestock industry, then decisions on whether to allow grazing should evaluate the whole operation (i.e., livestock grazing together with water development, predator control, water diversion, and all the other operations that accompany grazing).

Introduction of Exotics

No discussion of rangeland threats would be complete without mention of exotics. Biodiversity on rangelands is threatened by intentional and unintentional introduction of nonnative (exotic) species. However, the problems posed by exotic plants as opposed to animals are sufficiently different to warrant separate discussion.

Animals. Many nonnative wild ungulates have been introduced into the rangelands of Texas and New Mexico. Fortunately, only a few of them have spread to the rest of the West. On the other hand, feral horses and burros are well-established on many western ranges and (unfortunately) are protected by federal law. These animals and other exotics compete with native species for forage, water, and space in much the same way as livestock. For example, feral burros are one of the major threats to the desert bighorn sheep of the Southwest (Seegmiller and Ohmart 1981). Strong evidence suggests that, if uncontrolled, exotic invaders such as barbary sheep and feral burros would outcompete and eventually extirpate native species such as desert bighorn sheep (Cooperrider 1985).

Mammalian exotics are not the only problem. The range extension of the brown-headed cowbird in conjunction with cattle grazing has already been mentioned. The impact of exotic fishes on endemic fishes of western rangelands will be discussed in Chapter 8.

Plants. Rangelands are also subject to invasions by exotic plants. For example, the Intermountain West has been invaded by cheatgrass, medusahead, leafy spurge, and saltcedar, all of which are closely associated with the arrival of European settlers (Mack 1986). The grasslands of California with a Mediterranean climate have been heavily invaded by exotic annual grasses and forbs from the Mediterranean. This latter invasion was also closely linked with human disturbances of the native grassland (Mooney *et al.* 1986).

In contrast to large herbivores such as burros, which appear to be competitively superior to native ungulates under certain conditions, many plant species invade apparently because livestock grazing has changed the environment rather than because they are inherently better competitors. However, once established, they can inhibit or prevent restoration of native biodiversity.

Some authors have downplayed the distinction between a native and an exotic. For example, Johnson and Mayeux (1992) state: "An unbiased appraisal of ecological plant performance suggests that no special significance should be attributed to the label 'native'. Clearly, the status of nativity for plant species making up today's natural vegetation is relative." They further write: "We must now be bold enough to accept the challenge of shaping and synthesizing new ecosystems, even in the 'natural' environment." This suggestion should not go unchallenged. We do not dispute the idea that species composition of vegetation varies continuously in time and space, but this natural variation should not give humans carte blanche to rearrange ecosystems. We are doing enough rearranging by default and should not encourage more. We have a poor understanding of how ecosystems work, yet Johnson and Mayeaux selectively ignore a vast array of evidence that exotics can not only displace native species but also disrupt ecosystem functions (Vitousek 1986).

Recreational Impacts

Highways, mining, urban and suburban development, and recreation all can cause problems, although their effects on rangeland biodiversity do not differ much from similar practices in nonrangeland areas. We briefly note two recreational impacts—camping and off-road vehicles—because of their unique role in rangelands.

Camping. We fully recognize the spiritual renewal and unique experiences afforded by hiking, backpacking, and camping in wild rangelands, including deserts and alpine tundra. The problem is that many rangeland areas cannot sustain a great deal of this activity without damage. The central problem is that people, be they alpine hikers or weekend car-campers, are attracted to precisely the areas that are rare within rangelands—areas with trees and areas around water. Furthermore, these are typically the areas with the greatest species richness and are most susceptible to practices such as burning wood, diverting water, and simply disturbing animals.

Human camping needs to be carefully controlled, especially in the most sensitive rangeland reserves. However, the demand for this activity should be accommodated where compatible with biodiversity objectives, even at the expense of other uses such as grazing or mining. On most rangelands the damage being done by livestock to waters, riparian areas, and forest patches far exceeds any done by humans. In some cases camping and related recreational activities can be accommodated by simply moving such activities away from the most sensitive areas.

Off-road vehicles. In contrast to nonmotorized recreation, use of off-road vehicles is extremely damaging. In the past 20 years pressure has increased to open rangeland areas, particularly desert lands managed by BLM, to recreational use of off-road vehicles. In addition to disturbing wildlife and damaging vegetation and soils, off-road vehicles also substantially increase water and wind erosion (Sheridan 1979, Wilshire and Nakata 1976). At a time when our society needs to reduce its use of petroleum, we see no reason for government policies that encourage a form of recreation that damages natural ecosystems while squandering gasoline.

Management of Rangelands

On rangelands as elsewhere, planning should begin at the regional level, as outlined in Chapters 4 and 5, with inventory and then landscape zoning. We focus here on landscape-level issues that are particularly pertinent or problematic on rangelands. As with other types of ecosystems, we need to establish regional priorities. In addition, all players (landowners, government agencies, citizen groups, businesses, and others) must understand their roles and responsibilities.

Once the status and role of each area within a landscape has been designated for a particular purpose (i.e., core reserve, class I buffer, class II buffer, corridor, corridor buffer, multiple-use matrix), then needed modifications in management and status designations must be made. For example, an area of BLM land recognized as a hot spot of biodiversity should then be designated as an Area of Critical Environmental Concern (ACEC), wilderness area, research natural area, or whatever protective designation is feasible. Management should then be adjusted to meet landscape-level conservation goals.

Some changes in land designation may involve purchase, trade, or other acquisition of lands. BLM, the major federal rangelands management agency, has authority to trade lands, and as manager of roughly 325 million acres of public lands, has lots of stock in trade. Some of these lands have already been designated for trade, and others can be identified rather easily. BLM has many small parcels near cities like Las Vegas and Phoenix, which are difficult to manage and of limited value for biodiversity; yet they are valuable stock. Trade of parcels such as these has been used to acquire areas of high biodiversity, such as the San Pedro Riparian National Conservation Area in southern Arizona, with no cost to the taxpayers for purchase.

Another land-use tool that will likely be used more and more in the future is the conservation easement. Under this arrangement, a landowner agrees to manage land to maintain conservation/biodiversity values. Stipulations under such an arrangement will vary and may require not grazing cattle, not grazing portions of a ranch such as the riparian areas, or other more subtle or sophisticated covenants. Conservation easements are entered into by the consent of the landowner and often involve payment in compensation for uses or revenue precluded by the easement. For a discussion of such arrangements, see Barrett and Livermore (1983).

Management of the two basic types of areas—reserves (nodes) and multiple-use lands (matrix)—will be discussed separately. Corridors and buffers will be discussed in conjunction with both types. Before turning to these topics, we consider briefly some ways in which rangeland goals are determined and needs evaluated.

Determining Rangeland Goals

A major issue for rangelands has to do with goals and how they are described and measured. Traditionally, public rangelands were rated by condition classes which were supposed to reflect successional stages. Each range site was evaluated by the percentage of climax plants found in the overall composition. Range resembling the purported potential natural (climax) vegetation was rated as "excellent" and earlier seral (successional) ranges were rated as "poor," "fair," or "good" depending on the similarity in species composition to the described climax community. This rating system was even formalized in the way that agencies like BLM reported on range condition to Congress.

But this system has some inherent flaws. First, the use of percentages rather than some measure of production (pounds per acre or kilograms per hectare) or cover can lead to very misleading results, since percentages tend to fluctuate greatly depending upon growth of annual forage.

Second, in such a system all plants are treated as equal, even though some are probably more important as indicators of a natural state or as keystone species. For example, a major perennial grass species might be absent yet the range could still be rated as excellent.

Finally, the system treats rangeland succession as if it operates linearly with a gradual increase in percent composition of climax species. Strong evidence suggests that succession (or more appropriately, plant community dynamics) on rangelands is much more complicated than the linear succession model, as discussed earlier (e.g., Archer and Smeins 1991, 1992; Friedel 1991, Laycock 1991, Niering 1987, West 1992).

These three objections are all valid ecologically and highlight problems that need to be remedied. Another objection raised by ranchers and range managers is that attaching values (excellent, good, fair, poor) to successional stages is misleading since a range manager might want to keep a range site at a lower successional stage for a variety of reasons. For example, a manager might desire a range with 25 percent cover of sagebrush and other shrubs found in mid-successional stages, rather than in the climax stage that consists primarily of perennial grasses. A mid-successional range might be most valuable, for example, for fall sheep grazing or winter deer and elk range.

In response to these concerns, the federal agencies and the Society for Range Management (SRM) are adopting the concept of "desired plant communities" (BLM and SRM) and "desired future condition" (Forest Service). With this approach, the range manager or management agency would decide what plant community is most desirable for the site, and the range would be evaluated in terms of how it approximates this "desired plant community."

This subjective approach has some significant risks. One risk is that the classification system might be used to manipulate data and misrepresent what is happening. This concern is based on a cynical but largely well-deserved mistrust of the agencies by many people. As recently as 1990, BLM reported that its ranges were "in better shape today than ever before in this century" (USDI–BLM 1990) even though they have virtually no data to back up this assertion (Keystone Center 1991). In fact, BLM's own statistics show that only 3 percent of its rangelands are in Potential Natural (Excellent) condition.

A legitimate fear is that an agency like BLM may simply modify its goals (desired plant communities) to resemble more closely the successional stages already present. The agency could then blithely report that 99 percent (or whatever) of rangelands were meeting agency objectives (i.e., they resemble the "desired plant communities").

A second objection relates to the static nature of such a designation. In a natural landscape, disturbances generally maintain a shifting mosaic of seral stages. Few areas in a natural landscape remain permanently in lower seral stages; rather each proceeds through succession, though not necessarily to climax. Thus, 90 percent of a natural landscape might at any time resemble the climax vegetation with the rest consisting of some combination of seral or less stable states recovering from fire or other disturbances. From 50 to 100 years later, the landscape would have similar proportions of seral stages but not in the same locations. Late seral stages would have moved toward mature conditions (potential natural), whereas other areas would be in lower seral stages due to more recent disturbances. By designating a lower successional stage as the desired plant community, managers may try to freeze succession or convert a naturally dynamic system into a static one. This problem raises the question of whether early successional stages can be truly sustainable, that is, whether grazing can be used to maintain lower seral conditions without long-term degradation. This question is unanswered for most rangeland types, although the generally poor condition of rangelands in the western United States suggests that rangelands cannot be maintained in early stages without long-term degradation.

The issue of how rangeland goals should be determined can be partially solved by defining goals at a proper scale, which generally is the landscape or regional scale. At this scale, a reasonable goal (at least for maintaining biodiversity) would be to maintain a shifting mosaic of seral stages that resembles that which would occur over large areas in the absence of human interference. If the natural (pregrazing) mix was, say, 70–10–10–10 of excellent, good, fair, poor (or 70 percent of sustainable stable states), then landscape management would attempt to approximate that dynamic mix or "desired landscape condition." Similarly, seminatural or multiple-use landscapes could be judged by the degree to which they deviate from this desired landscape condition. We recognize that the situation may be more complex due to the presence of multiple steady states, some or all of which could be considered climax or potential natural vegetation. Also, with changing climate, the proportions and even the composition of communities can be expected to change, adding further complexity. However, the principle of measuring the landscape against some defined mix of vegetation communities would still apply over a planning period of decades.

Use of a landscape measure of range condition, or indeed of any other condition rating, requires a great deal of understanding about range ecology. Unfortunately, the pre-European landscape condition of few, if any, rangeland areas has been described in detail. Similarly, the successional pathways of many range sites are poorly known.

The Society for Range Management (SRM) has adopted new rangeland assessment guidelines based on ecological sites (Society for Range Management 1992). These guidelines incorporate the concepts of nonlinear succession and of thresholds (termed "site conservation thresholds"). In short, they are based on the "state and transition" model of plant dynamics, which is much more realistic than the traditional linear succession model.

The SRM has also adopted the concept of desired plant communities and of a site conservation rating (SCR) that would be either sustainable or unsustainable depending on the ability of the vegetation to protect the site from accelerated erosion. The guidelines suggest that only vegetation communities that have an SCR of "sustainable" should be selected as a desired plant community, thus alleviating some but not all of the concerns about misuse of the "desired" concept.

We support the intent of these guidelines and believe that they will help make rangeland evaluation more congruent with current ecological thinking. But we don't think these guidelines go far enough in recognizing the limitations of using individual sites as a basis for management and evaluation. A site's capability is not independent of its context in the landscape. To take an extreme case, a one-acre site of remnant grassland in the middle of Chicago does not have the same potential vegetation dynamics as one acre in the middle of 10,000 acres of native tallgrass prairie because of differences in seed sources, disturbance, propagation, and other landscape level effects. Ecology in recent years has emphasized the importance of landscape processes. The SRM guidelines should be expanded to incorporate landscape-level thinking. In particular, the concept of "desired plant community" needs to be expanded to the landscape level as well as to incorporate the ideas of a desired landscape condition and desired landscape processes. Desired landscape condition would include the entire dynamic biotic community, not just plants. Specific knowledge of sites will, of course, be used in developing any desired landscape condition. But determining goals site-by-site, as has been done in the past, should be avoided.

Applying these new concepts on the ground will require much more knowledge of basic rangeland ecology than is presently available. So much of the research and thinking on rangelands has focused on livestock production, that in many cases the fundamental ecological questions have not even been posed, much less studied or answered.

Managing Rangeland Reserves

Recall that two purposes of reserves are commonly emphasized: (1) to serve as core areas for protecting sensitive elements of biodiversity, and (2) to serve as benchmark or control areas with minimal human manipulation so that the effects of multiple-use management can be compared. Most rangeland landscapes suffer from a lack of reserves large enough to serve either purpose adequately. The possible exception is alpine tundra, which as we have noted is well represented in national parks and wilderness areas because of its recreational attraction and lack of commercial value. But not all alpine wilderness areas are true Class I reserves since many are grazed by livestock, particularly sheep, and because recreational impact is often highest there (Fleishner 1992). Grasslands and deserts are poorly represented in United States reserves, particularly where they do not exist in landscapes with scenic geological features.

Designating a comprehensive network of reserves should be a high priority in rangelands. Even areas well represented in nominal reserves, such as alpine tundra, need to be reexamined to determine if the designation is serving its intended purposes and providing adequate protection. Three types of areas that need careful examination are: (1) units of the National Park Service, (2) federal and state wildlife refuges, and (3) designated BLM and Forest Service wilderness areas that remain open to livestock grazing.

The national parks and monuments are often cited as biodiversity reserves, but large portions of these units cannot legitimately be considered as such because their naturalness has been compromised by facilities and highways. To take an extreme example, Yosemite Valley within Yosemite National Park is as much a biodiversity reserve as Coney Island is a marine sanctuary. A regional landscape plan must consider which parts of parks are or could be functioning as biodiversity reserves.

Federal and state wildlife refuges have similar limitations. In many cases, federal refuges are being managed for a few species (particularly waterfowl) with active manipulation of water levels, planting of food crops, burning, and other activities. Such activities can and often do serve a legitimate function in wildlife management or species-level conservation. These areas, however, are not true core reserves—they are not examples of natural marsh ecosystems.

Finally, the role of designated wilderness areas needs to be examined. The often negative impact of recreationists on biodiversity has already been mentioned, although this harm is less of a concern where motorized recreation is prohibited. The main problem is the provisions of the Wilderness Act that allow livestock grazing to continue in designated wilderness areas. Over one-third of the wilderness areas in the United States allow livestock grazing (Reed *et al.* 1989). These grazed areas cannot possibly serve as reserves or as benchmark areas for evaluating effects of grazing.

We are not advocating that management on all these areas be changed radically. In some cases, reasons may be compelling for retaining current management practices. However, in zoning the regional landscape, areas should not be considered reserves if they are not managed as such. Units such as national parks should be subdivided into core areas, corridors, buffers, and multiple-use lands, just like other lands, rather than pretending that the entire park serves purposes of natural areas. This is not an argument for adding new uses in parks or failing to control uses that are detrimental. We simply recommend that certain long-standing uses such as recreation be recognized and adequately accounted for in regional planning for biodiversity. How much land should be in reserves is a difficult question (see Chapter 5) and implementing such a decision can be even more arduous. Often little scientific basis exists for saying that a given percentage of a landscape should or should not be in reserves to maintain biodiversity, although as noted earlier estimates made by scientists are usually quite high. In many areas, options for reserve establishment are already limited due to well-established land uses for urban/suburban development and agriculture. Restoration may be centuries away, but not having a scientific answer does not alleviate the need for some working goals.

A reasonable short-term goal may be to have at least 25 percent of any given rangeland landscape in Class I and Class II reserves. In these reserves human extractive activities would be prohibited, but active management may be needed to mimic or restore ecological processes. Management would include activities such as prescribed burning and control of exotics.

The following are recommendations for management of rangeland reserves. They apply to some extent to any type of reserve but emphasize problems particularly relevant on rangelands.

1. **Make management and legal/administrative designation congruent.** The first responsibility for management of a reserve is to make sure that the legal and administrative designation is clearly established. This is true whether it is a unit of a national park, research natural area in a national forest, BLM area of critical environmental concern (ACEC), or Fish and Wildlife Service wildlife refuge. Without clear management direction, these areas will be forever muddled in controversy over their roles. A great many problems in conservation have come from unclear or contradictory objectives.

2. **Remove or control incompatible uses.** For a Class I reserve, certain uses should be removed or controlled to the greatest extent possible. These include:

   - Any form of livestock grazing, unless it can be clearly established that grazing is necessary to substitute for a natural process (e.g., native ungulate grazing) that for some compelling reason cannot be reestablished;

   - Any form of mining;

   - Most recreation except that which is of minimal impact such as day hiking, birdwatching, and in some cases backpacking;

   - Most motorized activity, with the possible exception of that necessary for administration/enforcement;

   - Any usurpation or diversion of water; and

- Any human settlement except for the minimum necessary for administration or enforcement.

The exceptions will clearly require careful thought and planning based on local ecological conditions and trends. The above are meant to be guidelines, not absolute proscriptions.

3. **Control exotics.** A high-priority goal should be the control or removal of exotic plants or animals wherever possible. Some persistent plants such as cheatgrass may not feasibly be extirpated, but appropriate goals such as reducing them to a minor component of the flora should be formulated and pursued. The cost and difficulty of control, as well as other biological issues (Westman 1990), combined with limited public understanding make control programs difficult and often contentious. But it must be done. Set priorities, plan, educate the public, and get on with the job.

4. **Distinguish internal and external problems and implement appropriate programs.** Managers of reserves should distinguish between problems that can be solved internally and those that are regional in nature, then develop strategies for each. For example, problems with wide-ranging exotic animals such as burros must be dealt with regionally. On the other hand, problems such as recreational impacts at prime fishing spots are usually manageable on a more local scale. Issues that cannot be dealt with locally should be addressed at the appropriate scale, which would likely involve working with many other agencies and interests.

5. **Establish connectivity between reserves.** In some cases, identification and protection of natural corridors that provide connectivity will be straightforward. For example, riparian areas in rangelands are not only zones with high species diversity and high concentrations of endangered species (Johnson 1989) but they are naturally connected. Where rangelands are intermixed with forests, connectivity may be provided by ensuring that open areas remain. The bighorn sheep of the Northern Rockies often live and survive on patches of grassland on south-facing slopes within large areas of forest. Bighorns are reluctant to move through heavily forested areas or across large expanses of flat land. Therefore, providing connectivity for bighorns will require ensuring that open areas remain between patches of grassland and that traditional travel routes between mountain ranges are maintained in a relatively open condition.

6. **Determine key ecological processes and manage to maintain or restore them.** Management must pay attention not only to structure and composition of reserves, but also to function. As a start, the ecological processes reviewed in Chapter 2 should be considered and the question should be asked "are these processes functioning as they did before large-scale human intervention?" Paradoxically, in some cases active intervention may be required to maintain or restore a natural process such as fire.

7. **Identify opportunities or needs for expansion of reserves and take appropriate action.** Since rangelands have traditionally had so few reserves, years may be needed to establish an adequate network. And, as new biological information is acquired, new needs will be recognized. Managers of reserves should keep current lists of priorities for expansion so they can take advantage of opportunities that arise for funding or for land exchanges. Expanding a reserve should not become an end in itself, if expansion is not justified by biological considerations. Similarly, expansion of a reserve should not be viewed as requiring a change of agency jurisdiction or land ownership. If a reserve or reserve system can be expanded by changing management or management designation or through a conservation easement, then one of these may be the easiest and fastest way to expand the functional reserve. For example, many rangelands contain national parks and monuments surrounded by BLM lands. In many cases a reserve portion of a park could be expanded by designating a BLM reserve (ACEC or other designation) and a change in management. If BLM or another landowner is reluctant, unable, or lacking in authority to change management, then other options such as park expansion must be considered.

8. **Identify needs for buffering and take appropriate action.** Deciding how large or wide buffer zones should be is also a complex subject involving many factors and poorly understood ecological processes. For instance, how might we buffer an area from livestock grazing? In theory, livestock can simply be fenced out of reserves. But fences can cause other problems, as previously mentioned; therefore, fencing every reserve may not be desirable. Furthermore, few fences are leak proof—they are blown down, wires are cut, gates are left open, as any rancher can attest. Having a few cows grazing for a few days may not be a serious threat in terms of competition for forage, but other factors such as disease transmission must be assessed.

   Consider bighorn sheep. In Lava Beds National Monument an introduced population of bighorn sheep within a large fenced enclosure all died after making contact through the fence with a band of domestic sheep moving through an adjacent Forest Service grazing allotment. Diseases, like many other impacts, can travel through fences. In fact, many of the diseases that affect bighorn sheep are carried by insect vectors. The flight range of these insects is poorly known, although disease transmission beyond 10 miles is considered unlikely. Thus, if domestic sheep are kept 10 miles from bighorn sheep range, disease transmission is minimized. Unfortunately, 10 miles is not always enough, as occasional domestic sheep stray from flocks, and an occasional bighorn ram will travel long distances. A single bighorn ram can pick up a disease and carry it back to the entire herd. Considering all this information, one may question whether a reserve needs to be better buffered, or if the appropriate action is to remove domestic sheep from the entire landscape.

Multiple-Use Lands

Management of the semi-natural matrix (multiple-use lands) will be the greatest challenge. These lands will continue to be a large portion of the regional land base for a long time. Yet, to sustain their health and ability to support human uses, these lands must also maintain their biodiversity, albeit not in the same manner or degree as in reserves. The challenge is to learn how to use these lands while preventing long-term deterioration (Brussard 1991).

Management of multiple-use rangelands, particularly BLM lands, has been hampered by lack of understanding and lack of willpower to control damaging activities. Until recently BLM had no credible research program in spite of many well-known and publicized problems with degraded ranges. Therefore, the first need is for a much better understanding of rangeland ecosystems and how they work. While developing that understanding, BLM and other range managers should proceed cautiously when authorizing uses, trying always to avoid irreparable damage.

Any resource management activity will affect the structure and function of an ecosystem to some degree. In the past, this fact has often been ignored, particularly for grazing. The relevant question is not "Is the system going to be altered?" but rather "How much is it going to be altered?" and "For how long will it be affected?" and "What is the probability that it can be returned to its natural state?"

Recall the threshold effects described earlier in this chapter. Management should be designed to avoid moving plant communities beyond those irreversible thresholds that lead to unstable or impoverished ecosystems. Practices are not justifiable if they result in loss of species or in soil erosion in excess of soil formation. Some practices such as strip or heap leach mining inherently set back succession to something comparable to primary succession (from bare rock or mineral soil). Only with massive inputs of energy and money can these areas be restored to their original condition if it can be done at all. Practices that result in such destruction must be avoided until (if ever) there is evidence that true restoration is possible.

As is clear from our earlier discussion, the most widespread impact on multiple-use rangelands is livestock grazing. Grazing is also one of the most complex impacts ecologically. Therefore, guidelines for grazing management cannot be rigid. Rather, general criteria must be used and adapted in an experimental fashion to the particular landscape where grazing is taking place, keeping in mind evolutionary history, physical environment, and other factors.

In light of the many detrimental effects of livestock and the difficult challenge of grazing an area sustainably, livestock will need to be removed from many areas where they are now grazed, particularly those areas of the West that receive less than, say, 10 to 15 inches of annual precipitation. A policy such as this may require removal of livestock from over 50 percent of the West. There are many considerations in livestock grazing other than just numbers, such as timing as emphasized by Savory (1988). Nevertheless, it is difficult to see how grazing in these areas, which supported few large herbivores in pre-Columbian times, can be made sustainable. As a society we need to ask if it is wise to subsidize destructive grazing practices on arid and semiarid lands in order to export meat to areas with greater precipitation (and thus productivity), particularly since the latter produce the vast majority of the nation's meat anyway.

Some suggested criteria for livestock grazing potentially compatible with conserving biodiversity are listed below.

1. **Scale.** Grazing management must be planned first at the landscape level. Many ecological processes such as disturbance and succession operate at this scale, and planning pasture by pasture or allotment by allotment without a larger regional perspective will be fragmented and counterproductive.

2. **Wholesale predator control.** Livestock grazing enterprises should not require predator control by widespread trapping, poisoning, or shooting to be economically viable. The need for wholesale predator control (as opposed to control of an occasional problem animal) may be alleviated by change in class of livestock, by guard dogs, by herding, or by other techniques.

3. **High-grading of forage.** Grazing practices must recognize and deflect the inherent tendency of grazing animals to high-grade the forage. Three methods to minimize the tendency of high-grading and resulting range retrogression are: (1) no year-round grazing of the same area; (2) no grazing of the same area every year; and (3) no grazing beyond an intensity that allows the most palatable species to restore carbohydrate reserves following grazing.
4. **Sensitive areas.** Grazing must not degrade sensitive or highly preferred areas such as riparian zones. In many cases the only feasible way to avoid such degradation is exclusionary fencing. Alternatively, removal of livestock from the landscape is often preferable.
5. **Natural grazing patterns.** Where an area historically had a large herding herbivore such as bison, grazing should try to mimic the natural pattern. Ideally, instead of using livestock the native herbivore should be reintroduced.
6. **Natural disturbance regimes.** Grazing should be planned and carried out with an awareness of the area's natural disturbance regimes and should try to keep such processes operating. For example, if fires followed by succession were a regular occurrence in a landscape, then such processes (not just the fires but also the successional pathways) should be re-created. Grazing that chronically maintains a site in a lower successional stage, unstable condition, or denuded state should be prohibited. Such areas are vulnerable to invasion by exotics and are also at risk of excessive erosion.
7. **Thresholds.** Grazing practices should minimize the risk that plant communities will move past irreversible thresholds and retrogress into conditions where restoration will be difficult or impossible.
8. **Averages and extremes.** Management must be designed to recognize the inherent variability of processes such as precipitation and plant growth on rangelands. Managers should avoid designing grazing using the average year since the average year rarely occurs. Variability is the norm. Grazing practices must recognize that 50 percent of the time precipitation will be below normal. As Westoby *et al.* (1989) have suggested, removing all livestock for situations that arise once every 10 years on the average may be better than reducing stocking by 10 percent on a continuing basis.
9. **Flexibility.** Grazing practices should be based on recognition of the importance of rare events. Whether opportunities or threats, rare events

dictate much greater flexibility in management than is now the norm (Westoby *et al.* 1989). For example, conditions of optimal temperature and soil moisture for establishing certain range plants may occur only once a decade. No grazing or deferred grazing of such areas may be the best management to allow establishment. Unfortunately, flexibility in the past has been mainly thought of as using more forage during good years.

10. **Maintenance of wildlife/biodiversity.** Grazing should not result in the loss of any native plant or animal species from the regional landscape, or impoverishment of any other measure of native biodiversity (genetic integrity, natural relative abundance patterns, etc.). This criterion should be the bottom line whether the potential loss is from direct impacts of grazing or from associated livestock management practices such as predator control.

Recovery/Restoration

So little effort in rangeland management has focused on restoring biodiversity that there is not much to report on this topic. Major efforts have been made to rehabilitate strip-mined lands over the last 20 years, and some efforts have been made to restore rangelands degraded by livestock grazing. But 20 years is not long enough to evaluate success. However, some innovative and promising attempts at restoration have begun (Betz 1992, Steuter 1992, DeLoria 1992, Scott 1992, Rieger 1992, Allen and Jackson 1992). Much of this work is described or summarized in Allen (1988), Jordan *et al.* (1987), and Berger (1990), and we will not review it here.

Some efforts at large-scale rehabilitation of ranges damaged by overgrazing have been made. But these projects were designed to restore livestock productivity, not biodiversity. One of the largest and best publicized rehabilitation efforts on public lands is the Vail rangeland rehabilitation effort in eastern Oregon (Heady 1988, Heady and Bartolome 1977). These and similar efforts on western ranges may have aided the livestock industry, but the benefits to biodiversity have been marginal at best and in many cases quite destructive. Rehabilitation often consisted of plowing and seeding a monoculture of nonnative grass. Indeed, in the Vail project, where 267,000 acres were seeded, the standard seeding was only crested wheatgrass (Heady and Bartolome 1977). Although marginal benefits to some species of wildlife are touted for these rehabilitation projects, little evidence exists to support such claims. In fact, some of the benefits attributed to these programs, such as increased populations of ungulates including elk, deer, and antelope, represent longer term, region-wide trends traceable to better regulation of harvest by state agencies.

Finally, many areas that were overgrazed and then rehabilitated are now subject to severe invasions of noxious weeds (exotic, undesirable, unpalatable plants) such as medusahead. How a massive program of herbicide use, mechanical brush control, plowing, and seeding of monocultures may have contributed to region-wide invasions of exotics is not known, although we suspect that these artificial disturbances increased the spread of invaders. In any case, because rangeland problems have developed over many decades, full rehabilitation may also take a long time. The urge to provide a quick-fix, such as by plowing and seeding, may do more damage in the long run.

General Guidelines for Rangeland Management

To summarize, we provide here a set of general recommendations to guide management of rangelands for biodiversity. They have been developed and modified from some guidelines put together in outline form by Fred Smeins (1992). We acknowledge his considerable influence but warn that he may disagree with some of our modifications or interpretations of his original recommendations.

1. **Understand the organisms and systems being managed.** We can no longer afford deliberate ignorance in our management of rangelands. We have enough ignorance without trying. We need new and innovative research on the ecology of rangelands; major rangeland management agencies such as BLM should provide leadership in such endeavors. Of central importance in such research is the recognition that rangelands are more than cows and grass. True ecological research needs to look holistically at the entire ecosystem—at soils, soil microbes, plants, insects, and all vertebrates, and the myriad ecological processes such as nutrient and hydrologic cycles.

2. **Recognize that changes will occur with or without management.** As with other landscapes, changes are occurring and will continue to occur on rangelands with or without humans or their impacts.

3. **Distinguish between changes driven by past and current events.** Many ranges are suffering more from past damages than from current practices. To attribute poor condition to current practices may be not only unproductive and invite conflict, but may hinder restoration.

4. **Recognize the role of episodic and catastrophic events (including disturbance cycles).** Fires, floods, extreme weather, and other rare or catastrophic events play a role in rangelands over the long term. Accept this fact and plan management to account for and take advantage of these events.

5. **Think long-term; act short-term.** Many rangeland processes occur slowly, and management must recognize this. Rangeland recovery is certainly in this category. Nevertheless, actions may need to be taken now to influence change in the future. Rangeland monitoring must be designed to clearly distinguish between short-term changes and long-term trajectories of change.

6. **Manage at the appropriate spatial scale.** Much management has focused at too low a spatial scale, such as the site or pasture, ignoring the fact that many ecological processes such as disturbance cycles, wild animal movements, and even succession are processes that can only be understood or managed at a higher level such as the landscape or region. Management decisions at the site, pasture, or allotment level need to be made in the context of larger scale objectives and plans.

7. **Recognize the difficulty of rangeland restoration.** Recovery processes are slow and will involve random or unpredictable events. Some changes may not be reversible, at least without massive human intervention (at a great cost in time and energy) and will still require long periods of time for full recovery.

8. **Manage livestock grazing creatively and sustainably.** Discussions of issues such as public lands grazing, red meat consumption, and vegetarianism tend to ignore the fact that livestock grazing on rangelands has been in existence for several thousand years globally and several hundred on this continent, and is likely to continue. This is not to say that it will or should continue in all the places where it has been in the past. We need to take a hard look at the desirability of producing beef on arid or semiarid public lands. Creative livestock grazing on rangelands will require both temporal and spatial exclusion. Similarly, much of the controversy over grazing has focused only on numbers, whereas class and distribution of livestock and timing of grazing may be equally important. Sustainability, including maintenance of native biodiversity, should be the fundamental criterion for determining if livestock grazing should be allowed. The threshold ecological question should be "Can livestock be grazed without loss of biodiversity at the landscape level?" Livestock grazing that mimics grazing patterns of native large herbivores is most likely to be sustainable.

9. **Maintain or restore ecological processes.** Many rangeland ecosystems have passed beyond threshold levels of recovery, more due to loss of ecological processes than from changes in species composition, although the two are related. As long as processes continue, species re-

covery may often be relatively easy to attain. Conversely, loss of key processes may set off positive feedbacks that will cause further degradation. Once thresholds have been passed, rehabilitation is likely to be slow and expensive.

10. **Treat rangelands gently**. Many rangelands are fragile, existing in areas with thin soils and harsh and unpredictable climate. Furthermore, our knowledge of such areas is very limited. As with human health, prevention is cheaper and less painful than treatment. By preserving biodiversity we preserve our options for the future.

CHAPTER EIGHT

MANAGING AQUATIC ECOSYSTEMS

> Mechanized man, having rebuilt the landscape, is now rebuilding the waters. The sober citizen who would never submit his watch or his motor to amateur tamperings freely submits his lakes to drainings, fillings, dredgings, pollutions, stabilizations, mosquito control, algae control, swimmer's itch control, and the planting of any fish able to swim. So also with rivers. We constrict them with levees and dams, and then flush them with dredgings, channelizations, and the floods and silt of bad farming.
>
> Aldo Leopold (1941), *Lakes in Relation to Terrestrial Life Patterns*

Nowhere is the fragmentation of our thinking and institutions and the arrogance of our management more pronounced than in our stewardship of aquatic ecosystems. Water flows through ecosystems in a continuing hydrological cycle that most elementary school students would have little trouble grasping. Yet we have partitioned the study and management of water into a multitude of relatively independent academic disciplines and subdisciplines. Not surprisingly, our governmental agencies have carved up responsibility for "water management" so that even on a short stream segment, over ten federal, state, or local agencies may have mandated responsibilities.

The European settlers of North America treated water resources carelessly from when they first arrived, and we continue that tradition today. We have acidified entire watersheds with strip-mine drainage in the Appalachians, where virtually the only stream life is now the orange scum of iron-fixing bacteria. We have allowed acid rain to kill all the fish in nearly half the small lakes in the Adirondacks. We have dammed some of the most beautiful rivers in the world, such as the Colorado River through Glen Canyon. We have virtually destroyed salmon fisheries with dams and diversions and poor land-use practices in the Northwest and tried to replace them with fish factories (hatcheries). We have converted streams through cities like Los Angeles and Albuquerque into cement ditches. And, of course, we have used practically every running stream in the country as a garbage disposal system.

Not content with converting the water system, we have tried to improve upon the fisheries it contains, deliberately and successfully introducing more than 70 species into North America (Courtney and Moyle 1982a) and translocating at least another 158 North American species into drainages to which they were not native. These introductions have had disastrous effects on native biodiversity.

Furthermore, because water moves throughout landscapes, aquatic systems tend to integrate and reflect all that is being degraded at a regional scale. Excessive erosion of topsoil, salinization of water from irrigation, runoff of pesticides from fields and oil from roads, and influx of inadequately treated wastes and sewage all accumulate downstream where they are compounded and often have synergistic effects.

Considering our brutal, or at best, unthinking treatment of our waters, we should not be surprised that most aquatic ecosystems are degraded and that such a high percentage of aquatic species are endangered, threatened, or candidates for listing (Table 8.1). Over 25 percent of the fish and amphibians of North America are classified as rare, imperiled, or some more severe form of endangerment; for crayfish and unionid mussels, the figure is over 60 percent (Master 1990). This compares with percentages of less than 15 for mammals, birds, and reptiles (Table 8.2).

To further compound the problem, because of their inherent interconnectedness with other waters and with land, aquatic ecosystems are some of the most complex ecologically. These systems cannot be studied or managed effectively without considering both upstream and downstream effects, including the effects of activities on land and the pollutants we pump into our atmosphere. Furthermore, there are few truly pristine or natural rivers or lakes, so studies must always contend with a high degree of "noise" from human perturbations. We have eliminated our benchmarks.

In this chapter, we can only briefly outline the ecology of aquatic ecosystems and their management. Our intent is to provide an overview to help managers and conservationists pose the proper questions when dealing with aquatic resources. Answers to such questions will of necessity come from a variety of sources and kinds of expertise. However, the overriding need seems to be for citizens and agencies to view these systems holistically and organize and implement integrated approaches to conserve them.

We review some principles of ecology as they apply to aquatic systems,

TABLE 8.1 Federally Endangered, Threatened, and Candidate Species in Freshwater Ecosystems of the United States[a]

Taxon	Endangered	Threatened	Candidate	Total
Reptiles	3	4	17	24
Amphibians	6	5	60	71
Fishes	53	34	173	260
Mollusks	46	8	271	325
Insects	13	9	201	223
Crustaceans	8	2	124	134
TOTAL				1,037

[a] From Williams and Neve 1992.

discuss the basic interconnectedness of aquatic and terrestrial ecosystems, and introduce some new approaches to conservation of aquatic systems and their biodiversity. At the same time, we point the reader to some of the more accessible and pertinent papers and reference books. We pay particular attention to physical properties of aquatic systems since they have been implicated in so much loss of biodiversity and because conservation biologists often have such a weak background in the subject, considering it the domain of engineers and hydrologists. We believe that aquatic ecosystems have been left to the engineers for too long.

TABLE 8.2 Global Status of Selected Aquatic Animal Groups of North America[a]

Number of U.S. species ranked	Mammals	Birds	Reptiles	Amphibians	Fishes	Crayfishes	Unionid mussels
GX (extinct)	1	20	0	3	18	1	12
GH (historical; possibly extinct)	0	2	0	1	1	3	17
G1 (critically imperiled)	8	25	6	23	78	62	88
G2 (imperiled)	23	9	10	17	72	49	49
G3 (rare, not imperiled)	19	23	25	26	110	84	35
G4–G5 (widespread and abundant)	330	628	251	153	549	106	73
G? (not yet ranked)	62	55	9	3	24	9	26
TOTAL	443	762	301	226	852	313	300
% of Total + GX–G3	13	11	14	28	34	65	73
% of Total = E or T[b]			5	6	3	7	111

[a] From Master, 1990.
[b] E, federally listed as endangered; T, federally listed as threatened.

Ecological Principles

To try to summarize in a few pages the ecology of the multitude of aquatic systems in North America would be pretentious. Literally thousands of volumes of symposia, treatises, and textbooks have addressed issues and aspects of aquatic ecology and conservation. Each discipline has developed its own terminology and classifications to the extent that a glossary of aquatic terms would fill an entire volume. Thus, even seasoned biologists faced with problems in aquatic systems tend to be overwhelmed by the complexity of the task and the literature, and fall back to a narrow focus on immediate problems, such as sedimentation from a particular clearcut. Not surprisingly, managers, faced with daily crises and little time to gather information, also seek refuge in piecemeal and overly expedient responses to degradation of aquatic systems.

In 1971 Barry Commoner elucidated his four laws of ecology:

1. Everything is connected to everything else;
2. Everything must go somewhere;
3. Nature knows best; and
4. There is no such thing as a free lunch (Commoner 1971).

These seemingly simplistic "laws" are clearly exemplified in aquatic systems, yet humans seem most tenacious in denying the applicability of such laws to water issues. The first two laws provide a useful framework around which to organize the following discussion of aquatic ecology; the last two are fundamental to understanding threats to aquatic systems.

Everything Is Connected to Everything Else

The central importance of the hydrologic cycle in ecosystems has been noted in Chapter 2. The water molecule is inherently connected to all living things and a mover of many abiotic components of ecosystems. This importance derives from two characteristics of water. First, water is required by all organisms during at least part of their life cycle. An axiom of biology is that "the fires of life burn only in water," meaning that metabolism can only occur in an aqueous medium. Second, water is the great transporter of everything from blood and sap to nutrients, fish, and ships. And, of course, water is a carrier of energy. Thus water essentially unites the world, circulating through it in a continuous cycle and carrying life's most basic needs—food, nutrients, and energy.

The essential interrelatedness of entire river systems has been recognized and elucidated by aquatic ecologists, most notably in the "river continuum concept," which suggests not only that biotic communities along the length of a stream are interconnected but that they tend to maximize the capture and use of energy and nutrients (Vannote *et al.* 1980). What has been less well recognized in government and management agencies is the interrelatedness of the terrestrial, wetland, and aquatic realms. However, many biologists have noted the central importance of the riparian zone, the link between the terrestrial and aquatic systems (e.g., Gregory *et al.* 1991, Schlosser 1991).

Although we focus on aquatic systems in this chapter, we want to emphasize the need to understand and manage whole landscapes rather than separating the aquatic from the terrestrial. Similarly, we make no major distinction between river systems and lakes. Most lakes are connected to some river system (i.e., they have water flowing into them) and most have both inlets and outlets. They are thus not closed systems and should not be treated as such. Even relatively isolated water bodies such as desert sinks (lakes with no outlet) like Great Salt Lake have intimate connections with the surrounding landscape. Lakes have many unique biological properties that differ from rivers, but this topic is beyond the scope of this chapter and is well treated elsewhere (Hutchinson 1957, LaBounty 1986, Wetzel 1975).

Everything Must Go Somewhere

Understanding what is carried downstream and upstream in an aquatic system and how these flows are altered by human activities is central to enlightened management of these systems. The hydrologic cycle of evaporation, precipitation, and infiltration/runoff best illustrates this second principle. Water captures the energy of the sun by warming and evaporating (changing from a liquid to a gas) and rising into the atmosphere. In a gaseous state, water may move inland, where the influence of landforms such as mountains may cause condensation and thus precipitation as rain or snow. At this point water has much potential energy, and that energy eventually leads the water back to the sea unless the water evaporates again or is incorporated into a living organism. Freshwater is a great determinant of the density of life (biomass) in terrestrial systems and is more usable by most life forms than is saltwater. Yet freshwater is an exceedingly small percentage of the water in the world, and the turnover is rapid compared with other sources (Table 8.3). Keeping freshwater within the terrestrial portion of its cycle is critical to maintaining the overall biomass and diversity of life in terrestrial systems.

Water flows downhill and it takes many things with it—nutrients, silt, pollutants, dead wood, and many other living and nonliving components of the ecosystem. Of prime importance is its ability to carry nutrients downstream. Working against this tendency are the plants and animals of the

TABLE 8.3 World Water Resources and Rate of Renewal[a]

Location	Percentage of world supply	Average rate or renewal
Oceans	97.134	3,100 years (37,000 years for deep ocean water)
Atmosphere	0.001	9–12 days
On land		
Ice caps	2.225	16,000 years
Glaciers	0.015	16,000 years
Saline lakes	0.007	10 to 100 years (depending on depth)
Freshwater lakes	0.009	10 to 100 years (depending on depth)
Rivers	0.0001	12 to 20 days
In land		
Soil moisture	0.003	280 days
Groundwater		
To a depth of 1,000 meters (1.6 miles)	0.303	300 years
1,000 to 2,000 meters (1.6 to 3.2 miles)	0.303	4,600 years
TOTAL	100.0	

[a] From Miller 1982.

aquatic ecosystem that capture such nutrients. Aldo Leopold (1941) summarized this process nicely:

> All land represents a downhill flow of nutrients from the hills to the sea. This flow has a rolling motion. Plants and animals suck nutrients out of the soil and air and pump them upward through the food chains; the gravity of death spills them back into the soil and air. Mineral nutrients, between their successive trips through this circuit, tend to be washed downhill. Lakes retard this downhill wash, and so do soils. Without the impounding action of soils and lakes, plants and animals would have to follow their salts to the coast lines.

Both terrestrial and aquatic systems have developed mechanisms that slow down or retain water, nutrients, and energy in its flow back to the sea. Forests and trees intercept moisture, thus lessening the energy of impact on the ground and increasing infiltration into the soil, and reducing runoff. The multilayered canopies of ancient forests in the Northwest are able to intercept fog, which condenses and adds water to the forest during the summer dry season. These forests also can absorb up to 12 inches of rainfall per hour without any perceptible overland flow (runoff) except in the streams (F. Euphrat, personal communication). Similarly, forest vegetation in snowy regions can prolong melt and runoff, thus dampening flushes of water, soil, and nutrients.

Herbaceous vegetation and litter provide the same function on rangelands—increased infiltration and decreased runoff. However, most rangeland soils are less capable of absorbing rain. Even though precipitation in rangelands is typically much less than in forests, there is often more runoff and thus greater sedimentation of streams. For this reason, some scientists have suggested that rangelands contribute more to sedimentation than do forestlands. Branson *et al.* (1981) concluded that whereas streamflow in the western United States originates mostly in the forested mountains, sediment loads are derived mostly from the rangelands. They cite, as an example, a report of Dortignac (1956) that the Rio Puerco, a major rangeland area that represents less than 20 percent of the Upper Rio Grande Basin, contributes less than 8 percent of the water yield but almost half the measured sediment loss.

Riparian and wetland vegetation serve similarly to trap and hold sediment and nutrients and keep them in the system. This process is closely related to the way in which riparian systems reduce or buffer the energy of high flows (VanHaveren and Jackson 1986). Riparian and wetland vegetation slows down the flow of water and allows sediment and organic matter to settle out and be deposited. At the same time, alluvial riparian zones often function as shallow aquifers that recharge at high flows and drain at low flows, again slowing water on its downhill journey.

In a certain sense, clean water is not just a byproduct of healthy ecosystems, something desired by humans but of limited ecological significance. Rather, clean water shows that the "wealth" of the ecosystem—soils and nutrients—is not being flushed away.

A certain amount of sediment and nutrient loss is natural and serves necessary ecological functions. For example, places for attachment are often limiting for many small plants and invertebrates in streams. Small rocks in cobble and gravel beds provide sites of attachment for insect eggs and larvae and other macroinvertebrates. Clean gravel beds support a high algal and invertebrate biomass that can in turn support other organisms higher on the food chain such as fish. If these beds are filled with sediment, they cannot provide this habitat. However, periodic high flows in rivers, termed flushing flows, tend to clean out gravel beds of accumulated silt. Gravel beds are important not only for small invertebrates, but also as attachment sites for the eggs of fish such as trout and salmon.

Near the ocean, estuaries form the final collecting ground for most silt, organic material, and nutrients. Estuaries often have high animal species richness and among the highest animal biomass of any habitat on earth. As with upstream riparian/wetland areas, estuaries enable birds, fish, and other animals to capture nutrients and move them back upstream or onto land. For example, the oysters of Chesapeake Bay could once filter all the water in the bay in three days. Today, the population has been reduced to less than 1 percent of its previous abundance, and it takes three years for oysters to filter the water.

The role of animals in moving nutrients uphill has received little study. But one may logically assume that animals must be a primary means for returning rare nutrients to terrestrial systems, since other mechanisms such as wind are so limited. Williams and Davis (1992) pointed out that salmon and other anadromous fishes provide major movements and releases of biomass and minerals upstream. These authors also noted the potential importance of mayflies and other emergent aquatic insects in moving energy and nutrients upstream. Thus, riverine systems provide corridors for movement of energy and nutrients upstream as well as downstream. Gravity helps move energy and nutrients away from the upland areas. Plants and animals counter this trend by capturing or moving these elements upstream and inland. Of course, many rivers today also carry human wastes and pollutants.

Threats to Aquatic Biodiversity

The threats to aquatic biodiversity are as numerous and complex as the waters and ecosystems. In their attempt to modify aquatic systems, people have ignored Commoner's Third Law of Ecology, "Nature knows best." Intricate river and lake ecosystems that have evolved over thousands of years have been remade or drastically modified in a few years or decades. But, of course, we have paid the price, for as Commoner's Fourth Law states "There is no such thing as a free lunch."

Three basic categories of human threats to biodiversity have been recognized: (1) resource misuse; (2) pollution; and (3) exotic species (Coblentz 1990). All three are extremely important in aquatic systems.

Resource Misuse

Resource misuse includes a variety of activities ranging from dams to livestock grazing. Some of the major misuses are described here.

Dams and diversions. Some of the impacts of damming and diverting waters are obvious and predictable—loss of stream habitat, blockage of fish runs, and loss of downstream nutrients. But we are only beginning to see many of the more subtle and longer term effects. For example, in 1983 selenium poisoning from water diverted for agriculture in California's San Joaquin Valley was found to be killing waterfowl. Selenium, a trace element, is common in western soils including those of the San Joaquin and is toxic at quite low concentrations. When these soils were irrigated, they carried selenium into Kesterson National Wildlife Refuge, managed by the U.S. Fish and Wildlife Service for waterfowl. Over several years selenium accumulated in the reservoir and then caused heavy die-offs of waterfowl (Harris 1991).

Thus, not only does diversion of water cause problems in the system from which it is diverted, but it can create damage in its new path. There is indeed no such thing as a free lunch, but the bill for lunch may not come for many years. The selenium poisoning at Kesterson was at first thought to be an isolated and rare problem. Later investigation showed it to be not only a widespread and insidious problem in California (Jensen *et al.* 1990), but throughout the West. Moreover, the problem had been identified by government scientists (and ignored by government) over 30 years earlier (Harris 1991). A remedy to this disaster and similar problems will require profound changes in powerful institutions—including agriculture, water marketing, and water law, among others (Jensen *et al.* 1990).

Diversion of water for irrigation also has damaging effects, including salinization and waterlogging of soils. Since most farmlands have already been converted from their natural vegetation, their value for conserving biodiversity today is limited. But if their agricultural potential is destroyed by waterlogging and salinization, then the demand may increase for more farmland, which means further pressure on remaining natural areas. Furthermore, salinity problems are not necessarily confined to the farmland itself, as shown by the Kesterson example. One solution to salinity problems is to flush the accumulated salts out of the soil. This flushing water must go somewhere, and it usually ends up adding a lot more salt to the same river from which the water originally was taken. The Colorado River, for example, collects 4.7 million tons of salt in its trip to the Gulf of California, thus having at its mouth a salt concentration of about 900 parts per million—enough to stunt crops and damage plumbing and industrial fixtures (Marston 1987).

Dams cause a profusion of problems, from destroying riverine habitat to obstructing movement of aquatic organisms (Table 8.4). All can have severe impacts on biodiversity both directly and immediately, but also more subtly due to secondary, tertiary, or cumulative effects. Many of the latter may take years to be noticed—for instance the effects of dams on the big fishes of the Colorado River such as the humpback chub and Colorado squawfish.

TABLE 8.4 Impacts of Dams and Diversions on Biodiversity

Dam activity	Primary effect	Secondary and tertiary effects
Flooding of backwaters	Destruction of riverine and riparian habitat	Loss of species dependent on such habitat
	Creation of a barrier for movement along riverine and riparian corridor and across it	Loss of riverine and riparian connectivity; depressed species populations; extirpation; extinction
River obstruction by the dam	Interception of downstream flow of nutrients, silt, etc.	Decreased nutrient supply to downstream habitats; biotic impoverishment of downstream communities
	Blocking of upstream movement of migratory fishes	Depressed population abundance; genetic deterioration; extirpation; extinction
Regulation of water flow	Decrease in peak flows (flood control; hydroelectric dams)	Decrease in natural disturbance cycles; decrease in functions such as flushing flows; degradation of downstream habitat; biotic impoverishment of downstream communities
	Irregular flow (hydroelectric power)	Unnatural fluctuation in downstream habitat; depressed populations; extirpation; extinction
	Decrease in winter flows (irrigation)	Decrease in available habitat in winter; depressed populations; extirpation; extinction
	Temperature increase in water	Degradation of downstream aquatic habitat; depressed populations; extirpation; extinction
Water diversion	Decrease in downstream water flows	Decrease in available habitat downstream; depressed populations; extirpation; extinction
	Diversion of water to develop agricultural lands	Loss of wildlands to agriculture; decreased available habitat; depressed populations
	Irrigation of fields with water return to river with increased salinity	Increased salinization of river system; degraded quality of habitat; depressed populations

Most of the major river systems in the United States have not just a single dam or diversion, but a whole system of them. Thus, the impact is cumulative and complex. Consider the Tennessee River and its tributaries, which drain large areas of Tennessee, Alabama, and Kentucky and portions of other states. Beginning in 1936, this watershed, which contains 32 endemic taxa (mostly mollusks and fish) and 224 native fish, has had 36 multipurpose dams built on it. Nearly 40 percent of the large waterways have been affected by dams (Neves and Angermeier 1990). Or consider the Colorado River, which drains portions of 7 western states and in which 74 percent of the fish are endemic. The dams and diversions on this river allow virtually no water (less than 1 percent of the flow) to reach its mouth. Not surprisingly, both systems have high numbers of extinct and endangered taxa.

Channelization. Many streams and rivers have been channelized (i.e., straightened out, widened or deepened, and typically lined with concrete, boulders, or other retaining materials). Streams are typically channelized to prevent seasonal overbank flooding and to provide drainage to water-saturated soils of a floodplain. These practices essentially destroy functional rivers and associated wetland or riparian vegetation with their high species diversity. In extreme cases, channelization essentially turns natural riverine ecosystems into cement plumbing systems.

Most channelization is done by governmental agencies with taxpayers' money. For example, over about 15 years beginning in the mid-1950s, the U.S. Soil Conservation Service altered more than 8000 miles of channelways of smaller streams not affected by flood regulation activities of the Army Corps of Engineers.

Logging. Some impacts of logging were described in Chapter 6. By removing a substantial portion of the vegetation in a watershed, logging decreases interception and infiltration and increases runoff. With increased surface runoff, erosion increases and more sediment is dumped into stream systems. Furthermore, activities associated with logging, such as road building and log skidding, greatly increase the sedimentation of streams. Besides contributing sediments from their surfaces, logging roads destabilize slopes, causing landslides and massive deposition of sediments into streams.

Many other effects of logging on watersheds are less clearly defined. One concern is the effect of removing woody debris from river systems. In the natural forest, many trees would fall into streams and decay or be deposited for varying times along the way to the sea. These logs and other woody debris provide important habitat and nutrients for many aquatic and riparian species and influence other riverine processes (Maser *et al.* 1988). For example, stream segments with large woody debris in an Alaskan stream were shown to support 5 to over 50 times the densities of juvenile salmon as logged segments with no large woody debris (Sedell *et al.* 1988).

Livestock grazing. As with logging, overgrazing by livestock can affect streams, wetlands, and lakes. A serious impact on watersheds is the effect of cattle on riparian vegetation. Cattle prefer to stay and graze within the riparian zone and can drastically change the species composition of these areas (see Chapter 7). Over time, with continual grazing, all woody vegetation along the stream may be removed, with consequent lessening of bank stability, increased sedimentation, decreased stream shading, and many other primary and secondary effects such as increased water temperature and overfertilization.

Roads and urbanization. Another upland activity that can seriously affect aquatic systems is road building and general urban development. Dirt roads, as are built in forestry operations, contribute sediment to streams. Paved roads, on the other hand, may not expose as much dirt, but they essentially reduce infiltration to zero, forcing water to be drained off elsewhere, often with automobile-related pollutants such as lead, oil, and gasoline byproducts.

With its massive paving of streets and parking lots, urbanization has similar but greater effects. When periodic heavy rains hit urban areas, houses, pavement, and concrete storm sewers replace the natural systems for deflecting the energy of rainfall (vegetation), for allowing infiltration (soil covered with vegetation), and slowing down runoff (small ephemeral streams). Thus, with little to slow it down, rainwater is likely to cause much more severe flooding than would have occurred in the natural system.

Because humans are attracted to water, and rivers have played such important roles in transportation and commerce, human settlements are usually concentrated in and close to riparian zones. This development has degraded one of nature's most sensitive and species-rich ecosystems.

Instream disturbances/extraction. Finally, resource misuse often occurs right in the stream. In the past, and continuing in some areas today, many streams have been severely damaged by hydraulic mining and other instream extraction. Similarly, logging operations such as log skidding often took place in streambeds. Some great log runs scoured rivers and riparian zones for dozens of miles. Many of these activities have been curtailed or are better regulated, but some continue. Streambeds are often good sources for sand and gravel, which are still extracted directly from streams, causing the loss of these substrates from the stream and substantial disturbance and siltation of water in the process.

Pollution

The effects of pollution are often clearly observed—dead fish, sulfurous smelling water, beverage containers, and the extreme case of the Cuyahoga River in Cleveland, Ohio, catching on fire. Indeed, much of the awareness and publicity that energized and emerged from the first Earth Day in 1970 dealt with obvious pollution of this sort. However, the causes and effects of pollution are complex and difficult to sort out. And, even though the case for stopping pollution is compelling for reasons of both human and ecosystem health, efforts to date in this country have been inadequate.

The types of pollutants that end up in our rivers are myriad—everything from sewage and garbage to chemical pesticide residues. This complexity makes control and regulation immensely difficult. To compound the problem, it is not always easy to trace the origin of certain pollutants. The easiest sources of pollution to deal with are point sources—sources that can be traced to a single place such as a factory or a sewage drain entering a river.

Point sources. Historically, European settlers in North America treated rivers as natural sewers, dumping raw sewage and most industrial wastes directly into the water. Indeed, a major reason factories were located next to rivers was the ease with which wastes could be discharged directly into the water. The river provided a means to ship raw materials to the factory and manufactured goods away, after which the byproducts could be dumped in the stream. This system worked well, it could be argued, if you lived upstream or if the amount of pollution was small. Dilution, however, if it really worked at all, worked only when the amounts of pollution were small. Few citizens still believe that the solution to pollution is dilution, but government still lacks the backbone to stop such practices.

With the creation of the Environmental Protection Agency in 1970 and passage of many federal laws, most notably the Clean Water Act, standards for "acceptable" levels of effluents have been developed. Enforcement of such standards has been slow, but some progress is being made. However, many waters are still severely degraded, and point-source effluents such as fecal coliforms from municipal discharge are still a leading source of degradation of rivers in the United States (Table 8.5).

Table 8.5 Assessment of Water Quality in the United States[a]

Type of waterbody (percentage assessed)	Beneficial uses (percentages)	Leading causes of degradation (percentage)	Leading sources of degradation (percentages)
Rivers (45)	Full (70), partial (20), none (10)	Siltation (42), nutrients (27), fecal coliforms (19)	Agriculture (55), municipal discharge (16), habitat modification (13)
Lakes (73)	Full (74), partial (17), none (9)	Nutrients (49), siltation (25), organics/low DO (25)	Agriculture (58), habitat modification (32), storm sewers/runoff (28)
Estuaries	Full (72), partial (22), none (6)	Nutrients (50), pathogens (48), organics/low DO (29)	Municipal discharge (53), resource extraction (34), storm sewers/runoff (28)

[a] From U.S. Environmental Protection Agency 1990a.

Non-point sources. Non-point pollution sources, particularly agriculture, continue to be the leading cause of degradation of this country's rivers and lakes (Table 8.5) (U.S. Environmental Protection Agency 1990a). The major sources of water-quality problems for aquatic fauna in streams are non-point pollution other than from agriculture (38 percent) and non-point pollution from agriculture (30 percent) (Judy *et al.* 1984).

Garbage/landfills. Of increasing concern are the effects of landfills or garbage dumps on aquatic systems. The traditional method of disposing of garbage was, and still is in many areas, to dump it in unlined, uncapped landfills. Rainwater flowing through such landfills picks up pollutants and toxic chemicals and eventually carries them into aquifers and streams. Years may pass before the effects are noticeable in downstream waters, but by then the problems of clean-up are massive since so much of the underground aquifer has been contaminated. Problems of aquifer pollution occur with other point-source pollutants such as gasoline leaking from underground storage tanks.

Mining. Mining threatens biodiversity in many ways. Virtually all mining requires road building, and surface mining destroys the surface flora and fauna. Strip mining may destroy large acreages, and reclamation of mined land is still experimental. Nevertheless, the greatest impact of mining is probably on water resources, not land. Mining requires bringing large amounts of underground material to the earth's surface where it is exposed to rain. This material or ore usually includes high concentrations of carbon and sulfur (coal) or high concentrations of metal ions. When exposed to rainwater, these materials form runoff that is highly acidic or has high concentrations of metal ions. Both are toxic to aquatic organisms.

Furthermore, processing this ore may require washing with water or smelting. Ore processing leaves highly toxic water that must go somewhere, and smelting typically puts pollutants into the air to be deposited downwind.

Consider, for example, the Silver Bow Creek/Clark Fork River area in Montana, which is the largest of the Environmental Protection Agency's "Superfund" clean-up sites. The Clark Fork Basin was subjected to mining and smelting for more than 100 years and included what was at one time the largest open pit in the world, the Berkeley Pit copper mine. More than 130 miles of the Silver Bow Creek and Clark Fork River have been contaminated with arsenic, lead, zinc, cadmium, and other metals, and this contamination has spread to nearby aquifers used for drinking water. Soils throughout the local valley are contaminated with smelter emissions (Young 1992). The Berkeley Pit and a network of underground mines contain more than 11 billion gallons of acid mine water that rises a little higher every year, further threatening local aquifers and the already contaminated rivers. The estimated cost of cleaning up the site is over 1 billion dollars (Young 1992).

Exotics

The introduction of exotic (nonnative) aquatic organisms has been so widespread in North America that few natural communities are not affected by them (Williams and Neves 1992). Introductions have been of two types: (1) introducing fish or other aquatic organisms from other continents (e.g., introducing carp into midwestern waters); and (2) transplanting fish or other aquatic organisms native to North America from their native region to an area outside that of their historical distribution (e.g., transplanting eastern brook trout into western streams). For convenience, we will refer to the former as "introductions" and the latter as "transplants."

In many cases, the introductions and transplants were done deliberately in a naive effort to somehow "improve" a natural fishery. In other cases, aquatic organisms were transported accidentally in bait buckets, ballast tanks, or other water containers. Whether deliberate or accidental, introductions and transplants have devastated native biodiversity. While practices such as dumping of bait buckets continue to be carried out by uninformed citizens, it is shocking to note that agencies responsible for fish management still spend huge sums of money releasing exotic sport fishes into natural water bodies (Wilcove *et al.* 1992; Courtenay and Moyle 1992, 1994).

Although the best data on introduced aquatic species concern fish, other introduced organisms have also created major problems. For example, the opossum shrimp was widely introduced into lakes and reservoirs in the western United States and Canada to improve coldwater fisheries. Unfortunately, this shrimp was able to outcompete gamefish for cladocerans, a group of small aquatic invertebrates, extirpating native *Daphnia* populations and contributing to the collapse of other populations of large cladocerans. When opossum shrimp were introduced into Flathead Lake in Montana, they not only caused the collapse of the fishery but had secondary effects on bald eagles, grizzly bears, and even tourists (Spencer *et al.* 1991). In about 1985 the zebra mussel, a filter-feeding Eurasian bivalve, became established in the North American Great Lakes system, probably having been carried there in ballast water. Since then, zebra mussels have spread into all five Great Lakes and many rivers and lakes of the eastern United States, competing with native species for food and space. Zebra mussels are predicted to inhabit eventually most lakes of temperate North America with unknown long-term impacts on community structure and ecosystem function. Similar problems have occurred with aquatic plants, though they have not received as much attention. For example, a major exotic aquatic weed, *Hydrilla verticillata,* has become widely established in U.S. waters, largely from discarded aquarium plants. Finally, introductions and transplants of aquatic or semiaquatic vertebrates other than fish also harm biodiversity. For example, the bullfrog, introduced to much of the West, has proven to be a serious predator on (and sometimes competitor with) native frogs, fish, turtles, and other organisms.

Scope of introductions and transplants. By World War II, 14 species of nonnative fish had been established in North America. Today there are at least 70. In addition, at least 158 and possibly over 200 transplanted fish species have become established (Courtenay and Moyle 1994).

A significant proportion of the fish of most states now consists of nonnatives. Of 108 species of fish now known to exist in Nevada, 63 (58 percent) were introduced (Deacon and Williams 1984). Other states have percentages that range down to 2 percent, with Alaska being the only state lacking established exotic species. On a more local basis, a survey of Cataract Canyon, a portion of the Colorado River running through Canyonlands National Park, found 28 species, of which only 8 were native (Deacon and Williams 1984).

Effects of introductions and transplants. The effects of aquatic introductions and transplants on native species are much the same as with terrestrial organisms and include competition/displacement of natives, predation on native species, disease transmission, and genetic mixing or "swamping."

Displacement of native aquatic species through competition or predation can be quite pronounced. Consider Clear Lake in northern California, one of the oldest large natural lakes in North America. Its original fish fauna consisted of eleven species, of which at least three were endemic (Moyle 1976). Over the years, sixteen fish species were successfully introduced into the lake. Twenty-one fish species now inhabit Clear Lake but six native species were extirpated, including two that are now globally extinct (Courtenay and Moyle 1992). While stream diversions have contributed to some of Clear Lake's native species displacement, Courtenay and Moyle (1992) concluded that the introduction of large predatory centrarchid fish is the most likely cause of the extinctions.

Diseases from exotic fishes have been implicated in other losses. The exotic red shiner, for example, was introduced from bait buckets into the Virgin River of Arizona, Nevada, and Utah. The red shiner carries an Asiatic tapeworm that has infested the endangered native woundfin and contributed to its decline (Deacon 1988). The tapeworm originated from another introduced species, the grass carp, which was first introduced into Arkansas.

Hybridization, the interbreeding of closely related subspecies, is a major problem in many areas, causing loss of genetic purity and, in some cases, decreased fitness. Courtenay and Moyle (1994) suggested that hybridization is rare with introductions but fairly common with transplants. Hybridization may be very gradual and the effects hard to detect, but it can sometimes be rapid. Importantly, evidence suggests that hybridization can result in reduced fitness of native species. In the 1980s, the sheepshead minnow was released into the Pecos River, where it began to hybridize with the endemic Pecos pupfish. Five years later, hybrids could be found along more than 250 miles of the stream. The pure Pecos pupfish essentially no longer exists.

Finally, introduced or transplanted fish can have a substantial impact on other taxa. For example, Wilcove *et al.* (1992) describe growing concern and evidence that introduced game fishes are reducing populations of native amphibian populations in many areas.

Role of fishery management. Wilcove *et al.* (1992) summarized the conflict between native species and species introduced to enhance sportfishing. As of 1991, the U.S. Fish and Wildlife Service listed 86 species, subspecies, and populations of fish as threatened or endangered. Of these, 44 were threatened to some degree by introduced fishes, and 29 were threatened by species introduced for sport fisheries. These introductions include both accidental or deliberate release of bait fish by anglers, but also the deliberate introduction of game fish by fisheries managers. At least one native species, the Miller Lake lamprey, was deliberately poisoned to extinction by the state of Oregon because it preyed on introduced trout (Miller *et al.* 1989).

Trout have been moved around so extensively in the West that it is hard to find a stream with only native species. Rainbow trout from the Pacific region have been introduced to the Great Basin and Rocky Mountain region, which had native cutthroat trout. And some cutthroat were moved to the Pacific states. Many local fishes were translocated with little thought to genetic origin or purity. On top of this, brown trout from Europe and brook trout from eastern North America have been transplanted into many western lakes and streams. The result has been many local extirpations and frequent loss of pure strains of trout. Thus five trout populations—Apache trout, Gila trout, Greenback cutthroat, Little Kern golden trout, and Paiute cutthroat trout—are now protected under the Endangered Species Act. The Alvord cutthroat trout, native to streams of the Great Basin, now is extinct due to introductions of nonnative trout (Miller *et al.* 1989).

Fisheries managers have been most adamant about the need for introductions in the case of reservoirs. They insist that such waters are not natural to begin with, thus the argument for avoiding introductions does not apply, and that there is little threat to natives. To some extent the record supports this contention for parts of the country. Many reservoirs have been built in the Southeast, a region with high rainfall and high fish diversity. Riverine species that declined in abundance were replaced by introduced predatory species to enhance sportfishing. Courtenay (1990) and Courtenay and Moyle (1994) concluded that these introductions into reservoirs in southeastern states apparently affected few native species.

The same practice in the Southwest has had quite disastrous results for native fish. The Southwest is a region of low rainfall and naturally low species diversity (Courtenay and Robins 1989, Courtenay and Moyle 1994). Reservoir construction on the Colorado and Rio Grande rivers drastically disrupted the life cycles of native fishes. But Minckley (1991) suggested that most species would have persisted had exotic predatory species not been introduced. The combination of dam building and introductions has caused nearly all native fishes in the Colorado River and a rapidly increasing number in the Rio Grande drainage to be threatened or endangered; many are close to extinction (Minckley and Deacon 1991).

At least three major arguments can be made for banning exotic introductions and transplants altogether. First, the presence of exotics in a seemingly innocuous situation, such as a reservoir, increases the chance that they will be accidentally or intentionally spread to other areas—the so-called "bait bucket release."

Second, stocking a reservoir with predatory fish further strengthens the fragmentation of upriver and downriver stream segments. The dam itself blocks travel, and the reservoir creates a stretch of unfavorable habitat for stream fishes. Moreover, the presence of exotic predatory fish in the reservoir can virtually eliminate interchange between upstream and downstream fishes, thus fragmenting populations (Sheldon 1988).

Third, in most situations the problems caused by disease transmission, hybridization, displacement of natives, and other harmful effects of introductions and transplants far outweigh any purported benefits. We hope that the era of intentionally introducing and transplanting fishes outside of their native ranges is coming rapidly to an end.

Conserving Biodiversity in Aquatic Systems

Biodiversity conservation in aquatic systems has not received as much attention as in terrestrial systems until recently. For example, one of the first and most influential books in modern conservation biology, *The Fragmented Forest* by Larry Harris (1984), virtually ignores the aquatic portion of forest landscapes. We (the authors) are terrestrial ecologists by training, and have also been guilty of shortchanging aquatic ecosystems in our writings. Because aquatic systems have received less attention, biodiversity strategies for aquatic systems are just beginning to emerge. We will first look at some of the historical approaches to conservation in aquatic systems and then outline a more comprehensive approach to conserving aquatic biodiversity.

Conserving biodiversity in aquatic systems at the ecosystem or watershed level poses some unique problems. First, these systems are linear and branched, so that the flow of water forms a continuum from headwater to sea (or sink, in the case of landlocked systems). Thus, upstream events such as pulses of pollution can have effects far downstream. Second, few reserves have been designed or designated for aquatic resources, although early designation of federal lands as wilderness areas, national parks, and national monuments often provided fortuitous protection for native fishes (Williams 1991). Furthermore, except for a few small coastal watersheds, no river systems exist that have not been severely modified by humans and that might serve as controls or benchmark aquatic systems. Finally, since aquatic systems are inherently connected, it is difficult to establish downstream reserves or refugia that are reasonably protected or buffered from both upstream and downstream influences, much less atmospheric influences.

Upon careful scrutiny, problems of conserving biodiversity in aquatic systems parallel in many ways concerns of terrestrial systems. Sheldon (1988) (and others he cites) have shown that species richness of fishes is correlated with drainage area and discharge, both of which can be thought of as surrogates for available stream habitat. Thus, river systems can be thought of as islands, with large river systems having high species richness and small systems having low species richness. This idea suggests that fragmentation of drainage networks will lead to extinctions (Sheldon 1988), and implies that water diversion should also increase aquatic extinction rates by decreasing available stream habitat. Circumstantial evidence supports this assertion, although it is difficult to separate the effects of water diversion (habitat loss) from the physical effects of the dams. The Colorado River, where essentially all water has been diverted and where the existing habitat has been fragmented by a series of dams, has a high proportion of endangered fish taxa. Many other factors, however, have affected the Colorado River fish.

The problems of aquatic reserves and their surroundings are thus similar to those of terrestrial systems. As noted earlier, most terrestrial reserves were designated and managed for purposes other than conservation of biodiversity, are not large enough to contain viable populations of all native species, and are inadequately buffered from outside influences.

Given the similarity of aquatic and terrestrial conservation problems at the landscape/watershed level, we conclude that many of the principles of landscape design reviewed in Chapter 4 apply to aquatic systems. In general, these principles require zoning the landscape (watershed) into nodes/reserves (areas of concentrated biodiversity), buffers of increasing human uses around such nodes, a well-managed matrix, and corridors between nodes. The main divergence from terrestrial reserve design would be that the buffers and corridors in an aquatic reserve network would essentially be the same and would generally be linear. More progressively, however, the aquatic network must be viewed as inherently imbedded in a matrix of land with which it interacts in complex ways. Thus, aquatic and terrestrial zoning need to be accomplished in an integrated and coordinated way. Before exploring how these principles can be applied, it is useful to consider the conservation approaches used to date.

Historical and Current Approaches

Conservation of aquatic ecosystems has been piecemeal and small-scale, focusing on a few species (usually game fish), striving to "improve" systems by introducing new species, relying on high-technology approaches such as hatcheries or structures for enhancement or mitigation, and evaluating success by limited criteria.

The effort has been fragmented largely because of the diversity of disciplines and agencies that deal with aquatic systems. Furthermore, many efforts to restore streams have focused narrowly on the water itself or the riparian corridor, even though the activities on the uplands were having the greater impact. Thus, for many years fisheries biologists in state agencies noted severe damage to riparian areas and streams from livestock grazing but were unable or unwilling to control such activity because their agency lacked authority or willpower to change these practices.

Management has often focused on small portions of a watershed such as a single lake or a stream segment of a few miles, with no overall plan for the watershed. For example, a General Accounting Office investigation of BLM and Forest Service management concluded that both agencies had successfully restored riparian vegetation on a few isolated areas but that this restoration represented only a small fraction of the degraded riparian areas managed by these agencies. Since neither agency had comprehensive inventories of riparian areas, the percentage that had been rehabilitated could not even be determined (U.S. General Accounting Office 1988).

The focus of aquatic management on a few species is exemplified by the fact that, until the last two decades, few if any state fish and wildlife agencies had any biologists assigned to work with nongame fish, much less with amphibians, mussels, or other elements of the aquatic system. The pathological emphasis on trying to "improve" waters by introducing exotics and transplanting natives has already been mentioned. In spite of the well-documented consequences of such activities, they continue to be common in many agencies.

Hatcheries. The well-entrenched faith of our culture in technology to solve problems has affected aquatic management—most notably in the development of fish hatcheries. Hatcheries were originally built to augment stocks of game fishes and allow higher harvest without depletion. Later they were viewed as a way to augment populations in decline due to habitat deterioration. They thus represent an attempt to address symptoms rather than causes (overharvest, habitat deterioration)—what Lewis Thomas called a "halfway technology."

The term *halfway technology* was first used to describe medical practices that treat symptoms rather than eliminate causes of disease. In natural resources, they are practices that attempt to postpone the inevitable damage being done to ecosystems by human activities (Frazer 1992, Meffe 1992). Halfway technologies do little to address the root causes of problems or to alleviate damage except in the short term. But they often provide cosmetic relief (or in some cases outright deception) by making the damage less obvious or visible to the public.

For many reasons halfway technologies not only fail to cure the problem but instead do considerable damage. The plight of Pacific salmon is a classic case. Salmon and steelhead in California, Oregon, Idaho, and Washington once numbered in the tens of millions and provided a stable source of food for indigenous peoples. In the 1880s, the Columbia River Basin was estimated to have an annual return of 12 to 16 million fish, compared with less than 3 million fish now. The riverine habitat first began to degenerate with the building of dams, increased logging and agriculture, and other developments, accompanied by overharvesting of some stocks. Nehlsen *et al.* (1991) considered 214 stocks of Pacific salmonids to be at high or moderate risk of extinction. The attempted solution to declining productivity has been technological—the building of hundreds of hatcheries to spawn thousands of fish and produce millions of eggs to stock back in the environment. However, the fundamental problem remains—the natural environment remains largely unsuitable for salmonid survival, and millions of fish are being placed into a degraded habitat where they have little chance of survival to maturity and reproduction. Furthermore, hatcheries are expensive, they cause genetic problems when hatchery fish breed with wild fish adapted to local conditions, they encourage overfishing of mixed stock fisheries, and most importantly, they hide or mask the real problem of habitat degradation.

Meffe (1992) summarized six reasons why the hatchery approach is counter-productive and should be reconsidered:

1. Data show that hatcheries are not solving the problem—salmon continue to decline despite decades of hatchery production;
2. Hatcheries are costly to run and divert resources from other efforts, such as habitat restoration;
3. Hatcheries are not sustainable in the long term, requiring continual input of money and energy;
4. Hatcheries are a genetically unsound approach to management that can harm wild populations by introducing maladaptive genes;
5. Hatchery production leads to increased harvest of declining wild populations of salmon; and
6. Hatcheries conceal from the public the truth of real salmon decline.

Unfortunately, hatcheries are not the only examples of halfway technologies. A great deal of effort has been put into instream structures either to control erosion or to restore stream health or productivity for fish. Usually termed *stream improvement practices,* these structures include such things as "gabion deflectors" and "fish locks and elevators" (Payne and Copes 1986). Like hatcheries, however, these efforts are doomed to fail if not accompanied by reform of the original land uses, such as logging or livestock grazing, that have created the problems.

Habitat Evaluation Procedures. A final aspect of historical and current approaches to aquatic systems that deserves attention is the way in which conservation efforts have been planned and evaluated—particularly those designed to mitigate impacts from such practices as damming streams, draining wetlands, or dredging waters. In particular, the Habitat Evaluation Procedures (HEP) and similar procedures deserve scrutiny.

HEP was developed by the U.S. Fish and Wildlife Service and is widely used to evaluate impacts and design equitable mitigation for water projects. Under provisions of the Fish and Wildlife Coordination Act, whenever agencies such as the Army Corps of Engineers or Bureau of Reclamation conduct projects that affect wildlife habitat, they are required to mitigate for the losses. The Fish and Wildlife Service has responsibility to ensure that the mitigation is equitable and that it is carried out. HEP was developed to determine equitable mitigation.

HEP is a structured and quantitative way of evaluating habitat before and after a project and determining how much mitigation is needed to compensate for damage. It has been applied to both terrestrial and aquatic ecosystems. We describe HEP briefly here; for a more in-depth review we recommend one of the lengthier descriptions, such as Schamberger and Farmer (1978). In its most basic form, HEP consists of a sequence of three steps. In the first step, a set of indicator species—usually more than one but fewer than ten—are selected. In the second step, a habitat relationship model [one of the so-called Habitat Suitability Index (HSI) Models] is used to rate the habitat for each species before and after the project on a scale from 0 (= totally unsuitable habitat) to 1.0 (= optimal habitat). Finally, through a complex set of bookkeeping procedures, these values are used to produce a number for habitat units lost (the sum of acreages multiplied by HSI ratings) for before and after each project. The difference between the two then becomes the basis for the amount of mitigation that must be done. Consider, for example, a project that will involve draining 70 acres of prime wetlands. In step 1, biologists could determine that mallards and pintails are the best indicator species; then they would use the HSI models for both pintails and mallards to determine how many habitat units were being destroyed by the project. This result would be used to estimate that a mitigation proposal to create a new wetland by flooding, say, 50 acres 10 miles to the west where water was available, was equitable.

The HEP type of approach has many limitations for conserving biodiversity. Through a biased selection of species and models, one could replace a pristine marsh with a city duck pond and show on paper that equitable mitigation had been achieved. Examples of such abuse are numerous. Similar problems exist with instream flow methodologies such as the U.S. Fish and Wildlife Service's "instream flow incremental methodology" (IFIM) which attempt to predict the amount of river flow necessary to sustain fisheries. Deacon (1988) described how attempts to define flow requirements quantitatively for the endangered woundfin in the Virgin River of Utah, Arizona, and Nevada turned out to be overly simplistic. Even though an initial water allocation was made, flows were not maintained, and other ecosystem functions were altered. The exotic red shiner became dominant in the system. Deacon concluded that management of endangered species in the Virgin River requires a continuous influx of information on populations, habitats, and biotic interactions that must be integrated into the numerous water management decisions in the system.

Even in the hands of a conscientious user, systems such as HEP and IFIM embody many of the limitations, biases, and arrogance of our society's approach to aquatic resources and nature in general. For instance, HEP assumes that the complexity of ecosystems can be captured (and quantified) through a few habitat requirements of a few species. Corollary assumptions are that those people charged with selecting indicator species have the wisdom to choose appropriate ones, and that the available HSI models are accurate and dependable. Second, the technique ignores the spatial context and its importance to the ecosystem being considered. For instance, if the original marsh was a critical migratory stop for waterfowl, the procedure implicitly assumes that the mitigation area will be just as useful even if it is 100 miles away. Finally, at the heart of the HEP procedure is the assumption that anything can be mitigated, or said another way, that no piece of land or habitat is unique. Good biologists know better.

Evidence suggests that human ability to reconstruct habitat is imperfect at best. In summarizing mitigation work on saltwater wetlands, Zedler (1986) stated "it is premature to conclude that an artificial tidal wetland can develop and replace functions of a natural one. Furthermore, there is no evidence that restoration of degraded wetland habitat can compensate for lost habitat area." The same statement could be made for freshwater wetlands and in fact for all riverine ecosystems.

Thus, in summary, historical and current approaches to aquatic system management and evaluation have been narrowly focused on a few species and small in scale, have relied heavily on high-tech but halfway approaches, and have arrogantly assumed that humans can improve the aquatic system or at least adequately mitigate for any damage.

New Principles for Conserving Aquatic Biodiversity

A holistic program for conserving biodiversity in aquatic systems must be based on new and ecologically defensible principles. The following principles and the strategy that derives from them are based on ideas developed by Frissell (1992), Leopold (1990), Moyle and Sato (1991), Reeves and Sedell (1992), Rinne (1990, 1992), Williams and Williams (1992), and Williams and Davis (1992), among others. Although some of the papers just cited (particularly Frissell, Moyle and Sato, and Rinne) deal nominally with conservation strategies for fish, their approach seems to be more broadly based, focusing on maintaining ecosystem integrity. Such a strategy is likely to conserve a much broader array of taxa than traditional approaches.

We propose twelve principles to guide the development of strategies for conserving biodiversity in riverine systems.

1. **Scale.** The proper unit for management is the watershed, and because watersheds are hierarchically ordered, for watersheds within watersheds. Ultimately, strategies must be developed for entire river systems (e.g., the Mississippi River or the Columbia River drainages). These strategies must be general but detailed enough to determine goals and objectives for components such as water quality and discharge from tributaries. To complement these system-wide strategies, lower level strategies must be developed for tributary watersheds in the context of the larger goals and objectives.

2. **Baseline.** The baseline for management and evaluation of effects on hydrology should be historical flow patterns, including the variance in these flows. Historical flow patterns are the aquatic counterpart of pre-European vegetation dynamics in terrestrial landscapes. To the extent that they can be reconstructed, historical flow patterns provide a useful baseline from which to measure how humans have modified the abiotic portion of the system.

3. **Integrated management of land and water.** Instream conditions largely result from human activities on land and in riparian and wetland areas. Coordinated planning for land and water resources—or better yet, for interlinked aquatic and terrestrial ecosystems—is an absolute necessity.

4. **Reserves and refugia.** To conserve biodiversity in riverine systems, a well-dispersed network of reserves or habitat refugia—including headwater watersheds and relatively intact lower-river reaches—should be maintained and restored. That few pristine watersheds remain should

not be used as an excuse for not designating areas to serve this function. Naturalness is relative, and many watersheds can be restored.

5. **Priorities for reserves.** Priority in selecting, designating, and restoring refugia should be given first to areas of high native species diversity or endemism, or that are of critical importance to the aquatic system, and secondarily to habitat next to such areas. The guidelines for identifying biodiversity hot spots given in Chapter 4 also apply to aquatic ecosystems.

6. **Restoration goals.** Restoration should focus on underlying processes, not on cosmetic improvements or halfway technologies. Halfway technologies (including most hatcheries) should be discontinued so that scarce resources can be applied to the root causes of environmental problems.

7. **Restoration and "time bombs."** A principal function of restoration should be to proactively defuse existing "time bombs," areas that, from human activity or neglect, could cause extreme damage to watersheds in the future. Time bombs exist at all scales. At the landscape scale they include watersheds destabilized by logging and contaminated areas such as mined waste areas and Superfund sites. At a smaller scale they include road systems with undersized culverts that will fail with the next large flood.

8. **Restoration priorities.** Priorities should be based on doing the most good for the least investment, with comparisons of cost and benefit made on the appropriate scale. We should not condone setting priorities based simply on cost/benefit ratios of individual projects.

9. **Dams and diversions.** Dams and diversions are among the major causes of biotic impoverishment in aquatic systems. Conserving biodiversity cannot be effective if new dams are being built. Restoring biodiversity will require removal of existing dams (G.W. Hayduke, personal communication).

10. **Exotics.** Exotics, including both introductions and transplants, as previously defined, greatly threaten biodiversity in aquatic systems. No new introductions should be allowed in any waters. Established exotics should be controlled or eliminated where possible.

11. **Gene and species level programs.** Recovery programs dealing with gene and species level problems will be needed to complement watershed level efforts.

12. **Information needs.** Better information from inventory, monitoring, and research will be needed to restore and conserve aquatic biodiversity.

Strategy and Guidelines for Conserving Aquatic Biodiversity

The principles listed above provide the basis for the strategy and guidelines described here. We recognize at the outset that we must plan for and manage aquatic systems in an integrated way, guided by top-down watershed level guidance. We also recognize that several promising strategies are already developed or under development. The National Research Council has convened a Committee on Restoration of Aquatic Ecosystems to analyze current efforts and make recommendations (National Research Council 1992). Another effort called the National River Public Land Development Project is drafting a proposal for the U.S. Congress, outlining measures needed to protect the health of the nation's riverine ecosystems (Dewberry 1992). A comprehensive strategy for managing habitat of at-risk fish species and stocks in national forests within the range of the northern spotted owl has recently been published (Thomas *et al.* 1993). We incorporate elements from these proposals but focus on biodiversity conservation, as more broadly defined.

The guidelines we propose are designed to provide some guidance for what needs to be done to conserve biodiversity of aquatic habitats. We do not attempt to provide guidance on how to get such activities funded or accepted politically. Similarly, we do not address other goals or uses for aquatic habitat such as clean water for drinking or recreational fisheries. We do suggest that a systematic program to conserve aquatic ecosystems and their biodiversity will have many spin-off benefits for human society, but it will also require some sacrifices. Finally, we note that most of the elements of the program described here can be applied at various watershed levels; thus, they provide a structure for organizing a conservation plan for a major riverine system such as the Colorado River as well as for a 20,000-acre watershed.

Coordinated Planning and Management

To counteract the fragmented approach to managing riverine systems, our planning and management must be integrated. The need for integration has been recognized by groups such as The National Research Council (1992) and the Oregon National Rivers Policy Task Force (Dewberry 1992) in addition to many scientists and managers. Taking a landscape or watershed perspective will require not only that different agencies work together for common goals, but that scientists and managers from different disciplines interact cooperatively. Goals for major river systems will have to be developed at a national or international level. These goals can in turn provide a framework for developing plans and programs for smaller watersheds.

Classification and Zoning of the Landscape/Watershed

A key element of any workable conservation plan is the zoning of the landscape or watershed according to biological criteria and needs. Frissell (1992) proposed a classification of watershed habitats for use in restoration and conservation projects. The classification was developed for anadromous salmonids but has some general applicability, particularly when dealing with large and mobile species that use a variety of habitats in their life cycle. Based on Frissell's ideas we propose a classification of waters and associated watersheds as follows:

1. Focal or refuge habitats (hereafter called refuges): Areas of high-quality habitat that support a high diversity or high productivity of native species. They are usually relatively undisturbed.
2. Adjunct habitats: Areas adjacent to refugia that have been degraded by human or natural disturbances.
3. Nodal habitats: Areas distant from refuges (upstream or downstream) that serve critical life history functions for aquatic organisms originating in refuges throughout the watershed.
4. Source areas: Watersheds that do not support a high diversity of aquatic life, but are, or could be, stable watersheds in terms of sedimentation. These watersheds provide important sources of high-quality water for downstream refuges or nodal habitats. Source areas will include headwater areas with intermittent streams.
5. Degraded habitats: Heavily disturbed habitats that now support few natives, even though some of them were once the most productive habitats in the riverine system. These habitats include areas with high human population density such as towns and other settlements along rivers. Many of these areas are so degraded that significant recovery within the span of several human generations is unlikely. This category includes both the "lost cause" and "grubstake" habitats of Frissell (1992).

Once a river system has been mapped and classified in such a manner, this information can provide a basis for land-use zoning and restoration priorities. There are many other classification systems for waters and watersheds and more will be developed in the future. The system used is not too important as long as it is biologically defensible, truly reflects functional use and interrelationships of the larger watershed, and provides a sound basis for zoning watersheds and setting priorities. If the watershed of concern is large, then classification and zoning can provide the basis for more localized planning and management. Management is described below under three categories: reserves, multiple-use areas, and restoration.

Reserve Management

Reserves would be centered on areas where the biotic communities are largely natural and are managed mostly to protect their natural features. Ideally all refuges as described above would be designated and managed as reserves. In addition, portions of adjunct habitat, source areas, or nodal habitat would be designated and protected as reserves. The criteria and considerations outlined in Chapter 5 for reserve design should be applied here. These considerations include the need for areas large enough to be protected from edge effects and other external influences, the need for replication, and other factors described there and in publications such as Moyle and Sato (1991).

Moyle and Sato (1991) suggest a three-step process for identification of aquatic preserves:

1. Identify geographic regions for which an aquatic preserve system is desirable.
2. Within each region, identify potential preserves (waters with the highest percentage of native fishes or other taxa). If this list does not include all native species, then add waters where those taxa occur as potential preserves.
3. Develop a priority list for acquisition and management based on: (a) class of water; (b) presence of intact, native biotic communities; (c) amount of drainage included and other indicators of size; (d) protection against external, edge, and boundary effects; (e) ability to support minimum viable populations of large or otherwise important species; (f) redundancy as a positive feature; (g) difficulty of management; (h) presence of rare or endangered species; and (i) economic considerations.

Various categories of reserves and buffer zones can be designated that allow increasing human uses moving away from the center of core reserves. One of the key management issues in need of resolution is the degree to which fishing should be restricted. While limitation of hunting has been a common practice in terrestrial reserves, fishing has rarely been excluded from any aquatic system, freshwater or marine. Even national parks, which have a long history of excluding hunting, have traditionally allowed fishing. Yet if an aquatic reserve system is to conserve biodiversity and provide benchmarks for comparison with exploited areas, core areas will need to be closed to fishing, for the same reasons that core areas of terrestrial reserves are closed to hunting. Moving away from core areas, flexible and innovative ways of limiting fishing impact such as catch-and-release buffer areas may be used.

Ideally reserves would be large enough to include year-round habitat for all species. This objective can be achieved in aquatic systems in which most of the biota can obtain all life history functions within reasonably small geographic areas. For example, isolated populations of desert pupfish and endemic headwater trout populations can be encompassed in single reserves.

However, the single-reserve approach is of limited use in dealing with wide-ranging species such as salmon or the big river fishes of the Colorado River system. Reserves in this context must be for particular life history functions of one or more species, such as spawning areas, and a network such as Frissell (1992) proposed will be more suitable. Two-tiered management of this sort is analogous to management in terrestrial systems. Terrestrial managers often rely on preservation of blocks of habitat, carefully chosen and large enough to conserve nonmigratory species with narrow niches or small home ranges. This approach works for most small mammals, many birds, and a variety of plants and invertebrates. However, for larger or migratory species such as bears, elk, or whooping cranes, management must select and protect reserves for several critical life history functions (nesting areas, wintering areas, seasonal foraging areas, migratory staging areas, etc.).

Management of Multiple-Use Waters and Watersheds

Multiple-use waters and watersheds will continue to be all those not designated as reserves or buffer areas. Although most of these waters and watersheds are degraded to some degree, they still have important ecological functions. For migratory or anadromous species, these areas must provide habitat for passage without excessive mortality. Furthermore, most activities in these areas will affect other portions of the aquatic system, particularly downriver. Since these areas include such a wide variety of water and watershed types, and so many kinds of land and water ownership, we cannot make specific or comprehensive recommendations here. But, we list a few guidelines:

1. **Zoning.** Source areas and nodal habitats should be zoned to exclude incompatible human activities. For example, if a source area is a relatively intact forested watershed that contributes clean water to the system, then uses such as logging, grazing, and off-road vehicles should be excluded while camping, hunting, fishing, and hiking could be allowed. The same sort of limited human uses would apply to nodal habitats.

2. **Historic flows.** Historic flows should be maintained or mimicked to the greatest extent possible. Multiple-use areas are normally where watersheds have deteriorated and dams have been built. Thus, a return to historic flow patterns may be unattainable in the short term. Nevertheless, in many cases, with modest efforts or purchase of minimal water rights, flows that approach historic patterns can be approximated. One of us observed water management on the Cache la Poudre River in Colorado for over 10 years. As is typical in Rocky Mountain streams, water managers for the Cache la Poudre virtually shut off downstream flows in fall once the irrigation season is over. This is a period when water flow is normally low and little water is collected in reservoirs. It is also a stressful period for aquatic organisms because the habitat has contracted. Thus, shutting off flow at this time worsens the stress but gathers little water for people. Most of the runoff comes in spring with snowmelt, and water managers, having spent all winter trying to slowly fill reservoirs, are often forced to release large amounts during peak runoff. With more enlightened management a mid-winter flow that approximated historic levels could be released at relatively minimal cost but with great benefit to the biotic community.

3. **Pollution.** Excessive pollution or developing "time bombs" for future disaster should be proscribed. In the United States, the federal antipollution laws address this issue fairly well. But enforcement needs to be expanded and penalties stiffened.

4. **Floodplain development.** New floodplain development should be prohibited and existing structures removed following catastrophic floods. Development of subdivisions and other human structures in floodplains is a longstanding problem for human health and safety and the biotic integrity of the system.

5. **Dams and diversions.** No new dams or diversions should be built, and existing dams should be removed over time. Evidence is adequate to conclude that in the long term most dams seriously harm river ecosys-

tems (e.g., Stanford and Ward 1979). Although, in general, small diversions, carefully located, designed, and constructed, may pose little threat to biodiversity, dams generally were a mistake. We must face this reality and begin to develop a program for phasing out most existing dams and restoring natural flow patterns. The National River Public Land Development Project is preparing guidelines for meeting such a goal.

6. **Exotics.** No exotics (introductions or transplants) should be introduced into any waters and existing nonnative species should be eliminated where possible. Our only caveat is that exotics should be eliminated with great caution. The historic way to eliminate nonnative fish is by poisoning lakes or stream segments with rotenone. This kind of action should not be taken unless biologists can confirm that native fauna such as mollusks or other gill-bearing invertebrates will not be harmed.

Restoration

Restoration of aquatic systems is still highly experimental and can be extremely costly. Priorities for restoration must be carefully set and we must learn as much as possible from our experiments. Chapter 9 discusses adaptive management—the approach of combining management or human intervention (restoration in this case) with research and monitoring.

The record clearly shows that past restoration has been piecemeal and cosmetic rather than coordinated and designed to curtail underlying causes. The National Research Council (1992) has comprehensively reviewed restoration projects; one should refer to that publication to learn what has worked in lakes, rivers and streams, and wetlands.

In planning restoration projects, we suggest six guidelines, most of which have been discussed or follow directly from the principles offered earlier. But they are so critical and so often ignored, that we reiterate them here.

1. **Priorities.** Priority for restoration should go to areas of high species richness or high native species productivity. In the past much effort has gone into areas that Frissell (1992) described as lost causes, resulting in what he terms the "rathole strategy of habitat restoration." Projects on lost cause areas may seem to have high benefit–cost ratios if the economic analysis is limited in scale and looks at local projects in isolation from the larger aquatic ecosystem. From a broader perspective, however, these projects are often not worthwhile.
2. **Natural functions.** Restoration should be designed to restore natural functions of aquatic systems. If used at all, structural or other high-technology approaches should be only temporary measures to speed

rehabilitation while underlying causes are being remedied. Thus, the Dexter National Fish Hatchery in New Mexico, where endangered desert fishes are reared and maintained, serves a critical function while native habitat is being identified and restored (Johnson and Jensen 1991). On the other hand, we cannot justify substituting salmon hatcheries for native stocks whose habitat has been and is still being destroyed by human activity.

3. **Time bombs.** A high priority should be given to restoring time-bomb areas that may cause disasters in the future. Toxic waste dumps and destabilized watersheds can potentially cause major problems. Restoration of such areas is often delayed until it is too late because they are perceived as posing little immediate threat.

4. **Determining costs and benefits.** Priorities for restoration should be based on cost/benefit or other economic and ecological analyses at the appropriate scale. We recognize that social and political forces will affect priorities. However, priorities are often determined largely by administrative boundaries, public relations considerations, or other factors unrelated to biology. In addition, analyzing benefits and costs at too small a scale (the project rather than the watershed, for example) may result in less effective programs. The effectiveness (benefits) of several projects may be linked, so that doing one project without another might be of limited value.

5. **Work from headwaters down.** Riverine systems are inherently connected, and unrestored headwaters may prevent restoration farther downstream.

6. **Caution.** Large-scale restoration without knowledge of what works best should be avoided. Little is known about the effects of active restoration techniques for aquatic or any other systems; we may do more harm than good when intervening in natural recovery processes. Allowing natural processes to accomplish restoration is often safer than drastic human intervention. If great uncertainty exists, smaller pilot efforts should be preferred over large-scale experiments.

The emphasis in this book and chapter on landscape/watershed/ecosystem level management does not mean that species-oriented programs should be abandoned. They will continue to be an important part of aquatic system management. And of course, biologists need to be concerned with genetic conservation to provide for long-term adaptability of populations and evolution. The larger scales or levels of organization simply provide context for decisions at lower levels.

Reintroductions of endangered fishes back into native habitats are important programs. Yet, if not done properly, reintroductions can have negative effects. In at least one instance, a reintroduced fish eliminated a population of another rare native organism (Williams *et al.* 1988). The American Fisheries Society has developed and adopted guidelines for reintroductions, including genetic considerations, and these should be followed in conducting such operations.

Among aquatic species, genetic conservation has received the most attention with fish (e.g., Meffe 1987, 1990), but consideration of genetics needs to be expanded to other taxa. Using desert fishes as an example, Meffe (1990) described six ways that information on the extent and distribution of genetic variation in rare fishes can be applied in recovery programs:

1. Describing the amount and geographic distribution of genetic variation in species;
2. Estimating historical levels of natural isolation and gene flow among populations;
3. Selecting unique gene pools for special protection;
4. Contributing to taxonomic clarification;
5. Choosing stock to release into the wild; and
6. Monitoring hatchery populations.

Though this sort of information is critical, our understanding of conservation genetics is quite limited, even for fish. Our understanding of genetics for most other aquatic organisms is far more primitive. Nevertheless, the applications cited above for fish should have relevance for recovery programs for other aquatic taxa. And indeed, recovery is sorely needed.

CHAPTER NINE

MONITORING

You live and learn. Or you don't live long.

Robert Heinlein (1973)

If we change the way we manage our forests, rangelands, and aquatic systems, how will we know if our new approaches are actually conserving biodiversity? The only possible answer lies in monitoring. We know little about how ecosystems work, yet we tamper with them extensively. We certainly do not know if we can repair the damage already done. Yet we continue to tinker—adding species here, eliminating species there, moving waters around, digging up the earth, releasing poisons into the air and water, and generally exploiting resources we little understand. Even our efforts at conservation or restoration are largely untested. The only way we can begin to understand what we are doing is through long-term, systematic monitoring of biodiversity at many levels of organization and spatiotemporal scales.

Monitoring must be an integral part of management. By monitoring our management experiments, we can determine which practices are compatible with protecting biodiversity. Conversely, by learning which practices harm biodiversity, such management can be curtailed. This process of linking management with monitoring within a research framework is termed *adaptive management* (Holling 1978, Walters 1986).

Although few activities in conservation are more important than monitoring, it is rarely done well. Some biologists and managers perceive monitoring as blind or mindless data gathering, and sometimes it is. Monitoring is often done poorly or incompletely, partly due to budget constraints. Funding sources, from Congress on down through agencies, rarely give monitoring a high priority; monitoring is usually the first thing to be dropped when budgets are reduced or when competing activities need more money. Yet in spite of this spotty track record, agencies and land managers often use the promise of monitoring to justify questionable projects. How many times have we heard the statement: "We will go ahead with this (timber sale, management plan, dam, grazing program, etc.), but we will monitor the results"? A survey of monitoring in BLM's wildlife and endangered species program identified 3043 objectives that required monitoring, of which only 1855, or 60 percent, were actually being monitored (Keystone Center 1991). Finally, the reward system for monitoring is typically unrelated to how well the monitoring has been done. Rather the reward (or punishment) is related to whether the news is good or bad, not to whether the job was done in an efficient, systematic, thorough, honest, or scientific manner.

The good news about monitoring biodiversity is that the process is similar to programs that have already been developed for everything from business (the so-called quality control and quality assurance, as well as auditing) to deer or duck management. Thus, the basic procedures have already been worked out and there is no reason to reinvent much of anything. The bad news is that we have not been doing a good job of monitoring natural resources, even when goals and objectives are straightforward. Monitoring biodiversity is more complex and challenging than monitoring goose productivity. In particular the task of selecting "indicators" is a difficult one, for which we have little experience.

We should design monitoring programs just as carefully as we conduct any scientific study, using principles of sampling theory and experimental design. This chapter will not attempt to deal with statistical procedures but instead refers the reader to the many texts and courses on the subject, or better yet to a real live statistician.

In this chapter, we introduce the concept of adaptive management for biodiversity, discuss problems with monitoring, describe some types of monitoring, and outline the steps in a monitoring program. We then describe a hypothetical biodiversity monitoring program and finally suggest some guidelines for monitoring biodiversity. We will focus on monitoring at the regional landscape level and lower.

The Monitoring/Adaptive Management Cycle

We need to monitor biodiversity because we don't know much about how humans are affecting the many components of ecosystems. If we clearly understood the effects of our activities we would not need to monitor. We could simply take the needed measurements and design programs and then implement them, much like architects, engineers, and builders construct a building. We would not need to monitor because we would be confident about the outcome. This approach to resource management has been termed *linear comprehensive* (Bailey 1982) or more recently *linear management.*

Regrettably much resource management is carried on with just such an attitude, another reflection of the arrogance of our culture toward nature.

The alternative approach of adaptive management considers human activities with a degree of humility and recognizes how ignorant we are about biodiversity and how to maintain it. The underlying assumptions of adaptive management as applied to biodiversity can be summarized as follows:

1. Maintaining optimally functioning ecosystems with all their components (i.e., biodiversity) is an overriding goal.
2. Ecosystems are extremely complex, and human understanding of them is rudimentary.
3. Human activities may have severe and largely unpredictable effects on ecosystems, and these effects can be irreversible or require centuries for restoration.
4. Management should therefore be conservative, erring on the side of minimal risk to ecosystems.
5. Careful, systematic monitoring of ecosystems and how we affect them can help us learn how to avoid causing further harm to them.

Monitoring is thus the cornerstone of adaptive management; without monitoring we cannot learn and cannot adapt.

Consider, for example, how this philosophy could be applied to an issue such as salvage logging after a large fire or insect outbreak. Assume that there was economic demand for salvaged timber, but also a concern about the effects of such logging on the integrity of watersheds. One approach would be simply to forge ahead with massive salvage logging in many watersheds, confident that there would be little risk. An alternative approach would be to conduct salvage logging in only one small watershed and monitor it carefully along with appropriate control watersheds to determine effects. If results of monitoring showed that there were no serious impacts from salvage operations, then managers could feel more secure in authorizing such activities in the future. On the other hand, if damage was detected, only a small area would have been affected rather than entire watersheds. In summary, the risk to the ecosystem would have been minimized, and we would have learned from the experiment regardless of the outcome. These contrasting approaches are summarized in Table 9.1.

Adaptive management can be applied to a problem of any scale, and is not unique to biology. Note that the process is cyclic (Fig. 9.1); it is not a one-time experiment but rather an incremental approach for continually gaining insight while minimizing risk to ecosystems.

TABLE 9.1 Comparison of Adaptive Management and Linear Comprehensive Management in the Context of Ecosystem (Biodiversity) Management

Parameter	Linear comprehensive management	Adaptive management
Concern with ecosystem	Minimal concern with ecosystem due either to belief in human ability to manipulate or restore ecosystem or to lack of concern with ecosystem degradation	Recognition of overriding value of ecosystem and necessity of conserving a properly functioning ecosystem for many reasons
Knowledge of ecosystem	Assumes that ecosystems, ecosystem processes, and effects of humans on them can be easily understood and predicted by traditional reductionist science	Recognizes that ecosystems and ecosystem processes are beyond human ability to understand except in the most rudimentary way and that effects of human actions on them are to a large extent unpredictable
Method of predicting effects of human actions	Emphasizes traditional reductionist science aided by modern high-tech tools such as computer models	Emphasizes using experience to learn incrementally, starting with small-scale experiments and slowly and cautiously gathering new knowledge
Risk	Assumes that human actions pose little threat to ecosystem or that such risks are not a concern	Emphasizes minimizing risk to ecosystem
Scale—spatial	Assumes that knowledge about ecosystems and effects of humans on them can be extrapolated across large regions; bases management on assumptions that effects are local	Recognizes that local ecosystems are unique and that extrapolating across large regions is risky; recognizes that all ecosystems are connected and that local actions can have major effects on other or larger regions up to the global level
Scale—temporal	Assumes that effects of human activities on ecosystems are generally short-term and reversible	Recognizes that effects of human activities may be long-term and/or have time lags before effects are observed
Learning/monitoring	Assumes that learning from management actions is not necessary; monitoring not necessary since outcomes are predictable	Recognizes that careful and systematic monitoring is essential in order to learn how to manage ecosystems sustainably

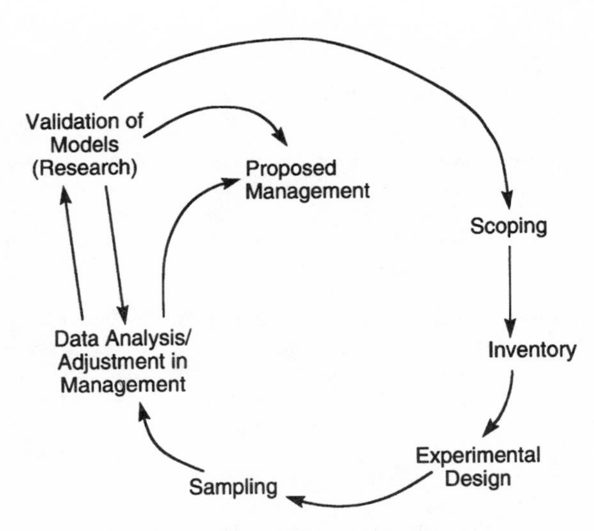

FIGURE 9.1 The adaptive management and monitoring cycle.

Types of Monitoring Programs

Monitoring here refers to the periodic measurement or observation of a process or object. Monitoring can be conducted to determine if restoration is effective or to assess the effects of human activities on an ecosystem. The procedures are similar in both cases. Monitoring is usually done in conjunction with some sort of human activity. The management action may be large scale, such as implementing a regional biodiversity strategy, or it may be a small project such as restoring tree cover to a cutover forest block or a stretch of degraded riparian vegetation. An activity in this context can be an "action," such as restoring a stream, or a "nonaction," such as designating an area as wilderness and excluding some uses.

Three types of monitoring are often distinguished: implementation monitoring, effectiveness monitoring, and validation monitoring.

Implementation monitoring is the process of determining if a planned activity was accomplished. Thus it asks the question "Did we do what we said we were going to do?" If the planned activity was a habitat enhancement plan for a rare species, implementation monitoring would be done to determine if all the plan's provisions had been implemented on schedule. As the proverbial bean counting, implementation monitoring is typically the role of Congress, budget offices and accounting offices, and other bureaucracies. It may also be done by citizen watchdog groups, whose goal is to make sure that the agencies do what they have promised.

Effectiveness monitoring is the process of determining if an activity achieved the stated goal or objectives. It asks the basic question "Did it work?" For example, if a stream restoration project was conducted to restore populations of a rare stock of coho salmon that had been limited by lack of good spawning habitat, then effectiveness monitoring would be done to see if the salmon population had become more abundant after the project was completed.

Validation monitoring determines if assumptions and models used in developing the plan are correct. For example, if the stream restoration for coho involved putting logs and large rocks in the stream to create more pools and hiding cover, then concurrent validation monitoring might be needed to determine if this is the element of spawning habitat that is limiting. Many biologists consider validation monitoring to be synonymous with research.

Implementation monitoring and effectiveness monitoring are often confused. This is the age-old problem of confusing means and ends—tactics and strategy. In the previous example, agencies typically will report that they have improved so many miles of stream habitat and thus achieved their goal. But what if the salmon population does not increase? Other factors such as offshore overharvest may also be limiting salmon or maybe the project was implemented sloppily. When confronted with this situation, agency employees will often respond "that is not our responsibility" or "we don't have the authority" (for a more comprehensive effort). We interpret such responses to mean that the problem is not being dealt with at the proper scale, geographically or administratively. In this example, the proper geographical scale would be the entire marine and terrestrial ecosystem that supports salmon at all stages of their life cycle, and the proper administrative scale would be a cooperative effort among all agencies and jurisdictions that have any influence on salmon productivity. If goals and objectives are not set at the proper scale, then implementation and effectiveness monitoring are easily confused and both are ineffective.

A fourth type of monitoring is often discussed. This is the so-called "baseline monitoring." Baseline monitoring is directed at some element or process that is not expected to change. For example, baseline monitoring is often done on water by measuring a set of chemical and biological variables such as nitrogen content, pH, and coliform count to determine if the water is clean and is no threat to humans or other animals. In most cases, this is simply another form of effectiveness monitoring, the objective being to ensure that human activities are not degrading water quality. The planned action is a little more nebulous, but the desired goal (clean water) is implicit if not explicitly stated. Without a goal, baseline monitoring would qualify as the mindless data gathering so often criticized.

In other cases, baseline monitoring simply refers to an initial inventory or set of measurements taken at the beginning of a monitoring program. For example, before a rehabilitation program for a watershed, water quality might be monitored. This might involve a set of measurements being taken of the water in the watershed to be treated as well as on an adjacent control area. In this case, the measurements are merely the first stage in a program of effectiveness monitoring.

Most of the remainder of this chapter will be concerned with effectiveness monitoring as described above. If we do not make the distinction, the term *monitoring* will refer to effectiveness monitoring.

Problems with Monitoring Programs

Before describing the basics of a monitoring program, it is instructive to look at some ways that monitoring programs can go awry. Few programs for monitoring biodiversity are being conducted anywhere. But, resource management agencies in the United States and elsewhere have had years of experience in trying to monitor particular elements of biodiversity. Since many of the problems are similar, we should take advantage of that experience to guide broader programs.

From a scientific point of view, monitoring can go wrong because either the wrong indicators were chosen or the wrong measurements or observations were made: in short, because of poor experimental design. Poor experimental design results in the wrong variables being measured, in the wrong place (or geographic scale), at the wrong time (or frequency), or with poor precision or reliability.

The result of poor experimental design is that the effect of a human activity may be missed or detected too late, the geographic scale of the effect may not be determined, or artifacts not related to human activity or mere randomness or noise in the system may be detected and attributed to humans. The detection of noise instead of real effects is the so-called "Type 1 error" of statistics, the type of error that scientists have been most concerned about avoiding. But in conservation a "Type 2 error," the failure to detect a real effect, may be more worrisome. Damage to an ecosystem may occur without us perceiving it.

A common problem is that results of monitoring are inconclusive or that the monitoring program is perceived as costing too much money. Piecemeal or poorly designed monitoring with inconsistent or inadequate funding and inconsistent data collection has been extremely common in past monitoring of natural resource problems. Not surprisingly, results of such monitoring are typically inconclusive or the information gained is never used. Thus, failures of monitoring programs are caused as much by administrative or institutional problems as by purely scientific ones.

The root cause of most of these problems is that those doing the monitoring have not had a clear vision of what they wanted to achieve and why they were monitoring. Furthermore, they often had neither a real commitment to monitoring nor to using the information in decision making. We discuss some ways to overcome these problems later in the chapter.

Steps in a Monitoring Program

Monitoring within the context of adaptive management is described below in terms of six overlapping steps that can be applied at any spatial scale (regional landscape, watershed, project), or any biological scale (global, continental, bioregion, regional landscape/ecosystem, community, species, population, gene pool). Note that the organization of the steps and the terminology is our own as tailored to biodiversity issues; it does not necessarily conform to other descriptions of adaptive management such as those of Holling (1978) or Walters (1986).

Scoping

The first step in monitoring is scoping or problem definition. This is the process of identifying and refining problems and issues, determining data needed to address these issues, and ranking issues and data needs (Jones 1986). The process begins with the preparation of proposed management and conservation plans as outlined in Chapters 4–8. In adaptive management the goals and objectives of management will determine the goals and objectives of monitoring. Thus the goals and objectives of the monitoring program may be tiered just like management goals and objectives (Table 9.2).

Management goals and objectives need not be quantified, but they must be defined in a way that people can clearly understand. In the minds of many people, monitoring is virtually synonymous with measurement or quantification. Agencies have virtual dictums that "objectives must be measurable!" This often results in objectives being chosen based on their measurability, rather than on more relevant ecological criteria. Or it causes objectives to be chosen and quantified arbitrarily (e.g., we will increase the cover of western wheatgrass from 10 to 25 percent within 5 years). More important, it often results in crucial goals or objectives being avoided because we do not know how to quantify them. Yet if we examine our lives, we realize that we monitor many things, big and small, every day without measurement. We do not,

TABLE 9.2 Examples of Tiered Goals for Management and Monitoring

Management (goal/subgoal)	Monitoring (goal/subgoal)
Goal: Maintain viable populations of all native species in natural patterns of abundance and distribution	Goal: Detect any drop in abundance, distribution, or health of any endangered, threatened, or vulnerable species before it is too late to reverse the trend
Subgoal: Maintain viable populations of pine marten	Subgoal: Measure abundance of pine marten to determine if there is any significant change in abundance or distribution
Goal: Maintain ecological processes	Goal: Detect any significant change in major ecological processes before it is too late
Subgoal: Ensure that nutrient and sediment losses do not exceed natural rates of loss	Subgoal: Measure sediment and nutrient content of second-order streams to determine if human activities are causing accelerated loss of these components from the system.
Goal: Maintain evolutionary potential of the biota	Goal: Detect any chronic loss in genetic diversity of species most likely to be losing it
Subgoal: Ensure that there is no loss of genetic diversity due to forest cutting practices	Subgoal: Measure changes in genetic composition and fitness in key commercial tree species to determine if selection cutting is causing loss of genetic diversity and fitness

for example, go out with calipers or ruler to decide when the lawn needs mowing. For a great number of things in life we use "visual data" or other quite reliable sensory perceptions.

Measurement does have a useful purpose in management and monitoring. We need quantification because we disagree with each other or do not trust each other. After all, scales were developed in the market so that people would not get cheated. If we could all agree on what is a "healthy" riparian area by looking at a picture, we might not have to take so many tedious measurements of vegetation height, pool–riffle ratios, and so on. Numbers should be tools, to be used when helpful but not as an end in themselves. If they help clarify an issue and reduce ambiguity, then use them. But if they become an end in themselves, do not become a slave to them. That many concepts such as "ecosystem health" are still not readily measurable does not mean they should not be management goals or something we strive to monitor. Other great ideas and aspirations such as peace, equality, liberty, and democracy started out as qualitative concepts. Most remain largely that, but few people say they have no meaning because we cannot measure them precisely.

Some very effective monitoring can be done simply by taking pictures at specified photo-stations. If, for example, an objective is to stabilize stream banks, taking a systematic series of pictures may be far easier than trying to describe bank stability quantitatively. Visual monitoring results can be quite convincing and easy to communicate to the public and others not comfortable with numbers.

Determining the goals and objectives of monitoring is an essential step that is often dealt with superficially. Spellerberg (1991) writes:

> What seems to be an obvious first step, that of defining the objectives of a . . . monitoring programme, needs to be considered very carefully because all other components of the monitoring plan will be dependent on the objectives. After having written these comments, I was pleased to hear that a colleague in industry . . . when discussing monitoring as a management tool in industry, noted the following: "It is the objectives . . . that should discipline the sampling programmes that are undertaken . . . this seems to state the obvious, but I am alarmed at the number of monitoring schemes which I am asked to assess in which the objectives are either not stated at all or are so woolly that they are meaningless."

Without clear objectives, the rest of the adaptive management cycle will be clouded with uncertainty. Throughout this book we have emphasized that all facets of conservation rely on explicit and ambitious goals.

Inventory

The next stage involves gathering existing information, which may involve estimating baseline conditions of vegetation communities or animal numbers, determining hot spots of high species richness, or summarizing historical data from the literature and from museum/herbarium records. Some guidelines for biodiversity inventories were presented in Chapter 4.

Experimental Design and Indicator Selection

Now it is time for the design of the monitoring program to be laid out in detail. General program goals such as "maintain biodiversity" may be refined into lower level objectives such as "maintain viable populations of all native species" or "protect representative natural communities."

Indicators. Since biodiversity cannot be measured in its totality, a certain representative subset of surrogates for biodiversity, termed *indicators,* must be chosen and measured. Indicator selection may be the most critical step in the entire monitoring or adaptive management process. For example, clean water may be an indicator of a properly functioning hydrologic cycle. Similarly, a relatively stable abundance of mountain lions or some other top carnivore may be an indication of ecosystem integrity. The terms *clean water* and *relatively stable abundance* may need to be defined and quantified, but you should not let quantifiability be the initial criterion for reasons already discussed.

Indicators are used in conjunction with biodiversity goals and objectives. In some cases, choice of indicators may be fairly straightforward. For example, if the objective at the species level is "to maintain historic abundance and distribution of vermillion flycatchers," and you know what the historic level was, then you do not need a surrogate. But even abundance can be measured in many ways, such as number of birds, density of birds, number of nests, number of nesting pairs, number of calls/hour, or number of territories. Also, functional indicators such as reproductive success or population growth rate are often better indicators than abundance measures. Landres *et al.* (1988) have described many of the pitfalls of using vertebrate species as indicators and have recommended some remedies.

When we move to the more complex components or correlates of biodiversity such as disturbance regimes, hydrological cycles, connectivity, or ecosystem health, the need for surrogates is imperative. Selection of indicators for such elusive processes and concepts is still in its infancy. A list of indicators for biodiversity at four levels of organization (genetic, population–species, community–ecosystem, regional landscape) including composition, structure, and function is shown in Table 9.3. This list should provide some help in identifying potential indicators, but do not be constrained by what is listed. For each goal or objective, one or more indicators should be selected. If the indicators chosen are not sensitive or representative enough to serve truly as surrogates for the elements of biodiversity at issue, then the process will likely fail.

We suggest several criteria for indicators. First, an indicator must be a good measure of or surrogate for the element of biodiversity that you are concerned with. The size of a salmon run may be a good indicator of the health of the anadromous fishery, but it may be a poor indicator of stream habitat conditions since offshore fishing could depress the runs in the absence of any change in stream habitat. Second, an indicator should detect a problem before it is too late to solve it. The numbers or biomass of a fish population may not be a good indicator if the fish are all 35 to 40 years old and are not reproducing successfully, as in the case of the Colorado River squawfish. A look at the age distribution of the fish would clearly show that there is a problem. Third, whenever possible we should select indicators for which experimental controls are available. However, controls or benchmarks cannot always be found. Regional landscapes and populations of narrow endemic species are inherently unreplicated.

Another consideration is the way in which a change in an indicator beyond a certain threshold will stimulate change in management. Other things being equal, use a flagship species as an indicator. Managers, politicians, and the public are more likely to relate to the decline of a bighorn sheep herd than of a population of kangaroo rats, much less a carabid beetle or lichen. Furthermore, the relationship between the indicator and the element of biodiversity of interest should be well documented and defensible; ideally the relationship can be easily portrayed or explained. If you propose to shut down a multi-million-dollar mine on the basis of macroinvertebrate sampling of downstream pools, then you better be prepared to defend the results. Macroinvertebrates are highly sensitive indicators of pollution (Mangum 1986), but if the relationship is untested or not well established, then another indicator would be better.

Other considerations for indicator selection, as discussed by Noss (1990a) and Spellerberg (1991), include such things as cost-effectiveness to measure, collect, or assay; but the most important determinant in choosing an indicator is the goal or objectives of management and monitoring. In a survey of BLM wildlife and endangered species monitoring, 70 percent of stated objectives could not be monitored because they were too vague, unmeasurable, or had other problems. If goals and objectives are unclear, choosing indicators will be hopelessly muddled, and so will the monitoring program.

Indicators should be selected at all levels of biodiversity from genetic to regional landscape, including compositional, structural, and functional variables. Although finding suitable indicators at the lower levels (gene, species, community) may be fairly straightforward, finding them at higher levels may be difficult for scientists and managers unaccustomed to dealing with broader spatiotemporal scales.

Design. At the experimental design stage, one needs to consult with a statistician. Much time, energy, and money have been wasted on monitoring studies that produced inconclusive results because treated sites were not replicated or controlled and there were many confounding variables. Consultation with a statistician before data collection could have avoided these problems.

Table 9.3 Indicator Variables for Monitoring Biodiversity at Four Levels of Organization[a]

Level	Indicators			
	Composition	Structure	Function	Monitoring tools
Genetic	Allelic diversity; presence of rare alleles, deleterious recessives, or karyotypic variants	Census and effective population size; heterozygosity; chromosomal or phenotypic polymorphism; generation overlap; heritability	Inbreeding depression; outbreeding rate; rate of genetic drift; gene flow; mutation rate; selection intensity	Electrophoresis; karyotypic analysis; DNA sequencing; offspring–parent regression; sib analysis; morphological analysis
Population/species	Absolute or relative abundance; frequency; importance or cover value; biomass; density	Dispersion (microdistribution); range (macrodistribution); population structure (sex ratio, age ratio); habitat variables (see community-ecosystem structure below); within-individual morphological variability	Demographic processes (fertility, recruitment rate, survivorship, mortality); metapopulation dynamics; population genetics (see above); population fluctuations; physiology; life history; phenology; growth rate (of individuals); acclimation; adaptation	Censuses (observations, counts, captures, signs, radiotracking); remote sensing; habitat suitability index (HSI); species–habitat modeling; population viability analysis
Community–ecosystem	Identity, relative abundance, frequency, richness, evenness, and diversity of species and guilds; proportions of endemic, exotic, threatened, and endangered species; dominance–diversity curves; life form proportions; similarity coeffi-	Substrate and soil variables; hydrologic variables; slope and aspect; stream gradient; vegetation biomass and physiognomy; foliage density and layering; horizontal patchiness; canopy openness and gap proportions; pool/riffle/run	Biomass and resource productivity; herbivory, parasitism, and predation rates; colonization and local extinction rates; patch dynamics (fine-scale disturbance processes); nutrient cycling rates; human intrusion rates and intensities	Aerial photographs and other remote sensing data; ground-level photo stations; time series analysis; physical habitat measures and resource inventories; habitat suitability indices (HSI, multispecies); instream flow assessments;

	cients; C_4:C_3 plant species ratios	ratios; abundance, density, and distribution of key physical features (e.g., cliffs, outcrops, sinks) and structural elements (snags, down logs, woody material in water); water and resource (e.g., mast) availability; snow cover; water quality		hydrologic measurements (streamflow, channel stability, sediment transport, etc.); observations, censuses and inventories, captures, and other sampling methodologies; mathematical indices (e.g. of diversity, heterogeneity, layering dispersion, biotic integrity)
Regional landscape	Identity, distribution, richness, and proportions of patch (habitat) types and multipatch landscape types; collective patterns of species distributions (richness, endemism)	Heterogeneity; connectivity; spatial linkage; patchiness; porosity; contrast; grain size; fragmentation; configuration; juxtaposition; patch size frequency distribution; perimeter–area ratio; pattern of habitat layer distribution	Disturbance processes (areal extent, frequency or return interval, rotation period, predictability, intensity, severity, seasonality); nutrient cycling rates; energy flow rates; patch persistence and turnover rates; rates of erosion and geomorphic and hydrologic processes; human land-use trends	Aerial photographs (satellite and conventional aircraft) and other remote sensing data; Geographic Information System (GIS) technology; time series analysis; spatial statistics; mathematical indices (of pattern, heterogeneity, connectivity, layering, diversity, edge, morphology, autocorrelation, fractal dimension)

[a] Modified from Noss 1990a.

As with any experimental design, control (untreated) areas are desirable in a monitoring program. Having true control areas is not always possible, however, particularly at larger geographic scales. If one is interested in maintaining biodiversity in a managed forest landscape, unlogged watersheds might provide some sort of baseline or control, although precious few of them are left in most regions. When no controls are available, such as for regional-scale monitoring, more creative experimental designs must be combined with extrapolation, intuition, and common sense. A P-value is not always needed to demonstrate convincingly that an effect has occurred.

Thresholds. The issue of thresholds also needs thorough consideration at the design stage. The potential for threshold phenomena is especially important to recognize when monitoring human activities suspected to be detrimental to biodiversity. Ideally, monitoring should detect a possible sudden shift of an ecosystem to a new condition well before that shift occurs. Restoration programs also carry risks related to threshold phenomena.

Two types of thresholds should be distinguished: management thresholds and biological thresholds. A biological threshold is a point at which an irreversible change in a population or ecosystem may occur. In earlier chapters, we referred to several biological thresholds such as the point in retrogression on rangelands beyond which a plant community can no longer naturally revert to a previous stable state. This is the "site conservation threshold" (SCT) of the Society for Range Management (1992). Similarly, conservation biologists once tried to determine, at least in theory, a "minimum viable population" level, below which a species was likely to decline to extinction. Agencies typically have minimum standards for wildlife habitat components such as snags or down logs. For the most part biological thresholds are poorly understood.

Managing down to a biological threshold or minimum standard is exceedingly dangerous, especially when that standard is poorly documented. Therefore, management thresholds need to be identified. Management thresholds are points where management must be changed to avoid an unacceptable risk to some element of biodiversity. Said another way, they are points at which the risk of reaching biological thresholds is unacceptably high. For example, consider a coastal salmon stream where summer temperatures below 58°F (14°C) are needed to maintain a viable aquatic system, including suitable conditions for juvenile salmon. Above 75°F the salmon and many other aquatic fauna will die; 75°F is thus a lethal temperature. But above 58°F salmon cannot reproduce well. Thus, the temperature 58°F is a biological threshold beyond which population viability is not possible. To avoid the risk of getting near this temperature, a management threshold of say 55°F might be selected as a reasonable threshold for summer temperature in the stream.

Since most biological thresholds are not well understood, we should ensure that management thresholds provide a sizeable margin for error. Over the years conservation biologists became more cautious about using the term *minimum viable population* because some agency people attached greater reliance on the number than scientists intended and wanted to treat it as a management threshold, assuming that they could safely allow populations to drop to such a level. Given pressures to "get the cut out" or meet other resource production targets, the tendency of many managers to accept minimum standards close to or exceeding biological thresholds must be considered a significant threat to biodiversity.

Threshold values, even if only approximations, need to be taken seriously. If a management threshold of 55°F maximum summer stream temperature has been determined, and if monitoring results show that this threshold is being exceeded, then the management activity (e.g., logging in riparian areas) responsible for this warming should be stopped. In some cases, management thresholds are codified in laws and more commonly regulations such as EPA's water quality criteria. But in most cases, they are not. Agencies should determine and adopt management thresholds as policy or standards at the proper geographic or organizational level before monitoring begins. Decisions about management thresholds that are deferred until after some activity such as logging is causing damage will usually be controversial and highly politicized. By adopting management thresholds beforehand, managers will have a defensible basis for making difficult decisions.

One of the best-known management thresholds is the listing of a species under the Endangered Species Act. At this point federal managers must change their management to ensure that they do not further endanger the species. Furthermore, they must take proactive steps to assist recovery. That they do not always do so has been the basis for many a lawsuit and court injunction.

We emphasize management thresholds because the experimental design of a monitoring program should be focused not only on detecting these points but on determining them before it is too late. If a system has great time lags, then management thresholds need to be set more conservatively. For example, effects of logging on erosion and water quality are often not detected fully for 10 years or more after the activity. Major slumping of soils does not usually occur until roots of cut trees have rotted away—a process that may take 7 or 8 years or even longer. Thus, concluding after 5 years of monitoring that logging is not causing increased erosion and sedimentation would be spurious.

Sampling

Sampling is the process of collecting data according to the experimental design. If the previous steps have been well followed, sampling can be straightforward. Unfortunately, many biologists, in their zeal to get to this stage (and out to the field), tend to take shortcuts in the previous steps. The results may be disastrous. The sampling portion of the adaptive management cycle is often the one labeled "monitoring," but without the preceding stages sampling can be a waste of time. Many agencies have employees who spend a lot of time haphazardly collecting data—shocking a stream here or occasionally measuring stream temperature. Such information is of little value and not defensible scientifically, biologically, or legally.

Validation of Models

Validation—seeing how well our indicators correspond to the real ecological phenomena of interest—should be an ongoing process. Remember that indicators are chosen as surrogates of some element of biodiversity. For example, we have mentioned that cool water (below 58°F) is an indicator of effective coastal stream function (such as shading and adequate flow). The indicator may be based mainly on salmonid fish, and its relevance to other aquatic organisms is unknown. In that case, validation studies may be needed to determine if a similar temperature threshold applies to other elements of aquatic biodiversity. Validation is often conducted independently of monitoring efforts although results of monitoring may provide insight about what models are suspect or need validation.

Data Analysis/Management Adjustment

Periodically, monitoring data need to be analyzed and management changed to reflect knowledge gained. In this way, management can be continually fine tuned by feedback of information on effects of management actions. For an adaptive management program at the watershed level, the human activity causing loss of biodiversity may be pinpointed and controlled. If the indicator is sedimentation and the watershed is being actively logged, then the required management action may be obvious. But at the regional level, causes of decline in biodiversity may be more complex. For example, fire frequency, intensity, and spatial patterns may be functional indicators of biodiversity at the regional landscape level; that is, a natural fire pattern upon which many species and ecosystem functions are dependent has been characterized. If monitoring data show that fire frequency or intensity are increasing or decreasing, or that natural spatial patterns have been disrupted, reasons may not be obvious.

Some effects of human activity at the regional landscape level will be related to larger scale (continental, global) processes, such as global warming or ozone depletion. To determine these effects, larger scale monitoring is needed. The U.S. EPA's Environmental Monitoring and Assessment Program (EMAP)(U.S. Environmental Protection Agency 1990b) is designed to answer questions at this scale, but it is unknown whether this project will ever be fully functional. Many millions of dollars have been spent, but little accomplished.

An Example of Monitoring Program Design: The Blue Mountains of Eastern Oregon

We know of no completely operational monitoring programs for biodiversity at the regional landscape level, although the monitoring program for the Channel Islands (Davis 1990, 1991) may be close. Noss (1992d), however, has outlined an ambitious program for monitoring biodiversity in a regional landscape—the Blue Mountains of eastern Oregon. This is an area whose vertebrate wildlife is relatively well studied and documented (Thomas 1979) and that also has severe forest health problems (Gast *et al.* 1991, Wickman 1991, and see Chapter 6). A key part of the monitoring design process as we have discussed is scoping, during which questions that need to be answered by monitoring are identified. Noss (1992d) provided examples of the types of biodiversity questions to be addressed by research and monitoring at gene, species, community, and ecosystem levels for the Blue Mountains:

Genetic Conservation Questions

1. For species of interest (e.g., commercially important trees, rare species, salmonids, game species, keystone species), is most of the genetic diversity distributed between populations or within populations?
2. For trees, do different branches vary in their genetic composition (i.e., somatic mutations in different meristems can result in genetic variation within individuals of long-lived organisms such as trees; gametes produced in different branches may be distinct genetically)?
3. Has fitness declined in ponderosa pine due to high-grading (removing superior individuals)? To answer this question, a number of high-graded and virgin stands should be assessed genetically and monitored for survivorship, fecundity, and other fitness parameters.

4. Is resistance to insect pests, fungal diseases, drought, or other stresses a heritable genetic trait?

5. What remains of the native genetic diversity of salmonid stocks? Has this diversity been polluted (through introgression) by hatchery stock?

Species Conservation Questions

1. Which species are most sensitive to forest management practices? Conversely, which species tolerate or thrive with human disturbance?

2. Can sensitivity to management practices be predicted from life-history (autecological) characteristics?

3. Are abundance, fertility, survivorship, and other characteristics of sensitive and other special-interest species strongly tied to specific habitat attributes?

4. How should the lists of management indicator species (MIS) for each national forest in the region be revised?

5. What are the population trends of indicator species on each national forest and region wide?

6. What are the trends of threats to sensitive species (road density, rate of timber cutting, livestock grazing, etc.) in each national forest and region wide?

Community and Ecosystem Conservation Questions

1. What is the representation in protected areas, relative to total cover, of each vegetation type in the Blue Mountains region? Which types are most poorly represented?

2. What are the trends in areal cover and condition of each vegetation type?

3. What are the trends in habitat structure (density, size classes, and dispersion of trees; snags and down logs; shrub and grass density; etc.) in each community type?

4. What are the trends in proportions of exotic to native species abundance in various areas?

5. What are the trends in abundance and distribution of animal species assemblages of different "life forms" (feeding and reproductive habitat requirements; Thomas 1979) across the region?

6. What are the trends in natural disturbances and other ecological processes (fire frequency and intensity, insect epidemics, flood, etc.) in each vegetation type and landscape, and across the entire region?
7. What are the trends in soil fertility, tree growth rates, and other measures of ecological health across the region?
8. What are the trends of threats to natural communities (road density, rate of timber cutting, livestock grazing, etc.) in each national forest and region wide?

Landscape Conservation Questions

1. What are the trends in spatial arrangement of vegetation types across various landscapes in the Blue Mountains?
2. Is the landscape becoming more fragmented?
3. What are the trends (frequency and extent, historic and current) in human uses of the regional landscape?
4. How have human land uses affected landscape spatial patterns and disturbance regimes?
5. What are the trends in the movement of materials and processes (propagules, nutrients, fire, etc.) between patches in the landscape? For example, has the rate of sediment movement from slopes to streams been increased?

Monitoring plans for regional landscapes can provide the basis for hierarchical organization of monitoring studies at lower levels. In this way, every study at the ground level can be related to higher level objectives, as exemplified in Table 9.4. Potential indicators, management thresholds, and threshold actions for some monitoring goals for the Blue Mountains are shown in Table 9.5.

Guidelines for Successful Monitoring

A monitoring program involves much more than just data collection. A good monitoring program requires systematic planning, collection and analysis of data, and validation. Furthermore, a good program must be integrated with management so that monitoring results affect management decisions. Yet at the same time, those doing the monitoring must be independent enough from management so that they can "tell it as it is" and not feel compelled to manipulate results to make management look good. Finally, monitoring must be done at proper scales in time and space. Different

TABLE 9.4 Monitoring Goals Tiered from Regional Landscape to Watershed

Monitoring goals and subgoals for regional landscape	Monitoring goals for watershed
Goal: Detect changes in freshwater habitat detrimental to resident and anadromous salmonids	
Subgoal: Detect deleterious increase in siltation in key watersheds	Detect sediment loads in the watershed in excess of 110% of loads in control watersheds
Goal: Detect changes in abundance or viability of old-growth dependent species	
Subgoal: Detect changes in abundance of fishers within existing habitat	Detect decrease in track counts of fishers in watershed in excess of 10% of five-year average
Goal: Detect factors threatening rare and endemic plant species	
Subgoal: Detect excessive disturbance to or loss of natural vegetation on serpentine patches within region	Detect changes in plant species composition (richness and diversity) on serpentine patches within watershed

questions, as indicated above, require monitoring and analysis at different scales. For many elements of biodiversity, the temporal scale may be long (e.g., decades) and the spatial scale may be large (e.g., regional, continental, global). Monitoring such elements will require long-term planning and stable programs as well as integrated efforts between agencies and jurisdictions. Since few efforts are being made along these lines at present, we cannot point to any ideal model to follow at the regional scale or higher, and precious few even at lower levels. From experience in monitoring at lower levels, we suggest the following guidelines for developing a program at the regional landscape level.

1. **Have clear goals and objectives.** Lack of clear goals and objectives is probably the greatest stumbling block for a monitoring program. Many agencies operate with two sets of goals. The set of goals for conserving resources is codified in legislation and agency regulations and manuals. The other set is largely unwritten and relates to satisfying political pressure, commodity users, special interest groups, and others. The newly hired biologist often gets confused in an agency, assuming that the first set of objectives is the only one. For agencies to succeed in biodiversity

conservation, they may need to subordinate their unwritten rules and objectives to the agency mission. This does not mean that agencies will not be influenced by political pressure, or that political pressure is always bad. It simply means that if the agency has a statutory conservation mission, then it needs to keep focused on that mission.

When agencies have unclear or inconsistent goals, then all goals and objectives that they develop for land management are equally muddled. We have no simple solution to this problem. However, developing clear biodiversity goals and objectives and linking them with indicators, management thresholds, and management actions as outlined above may help to clear the air. If managers, staff, and the public can develop a consensus on what these should be, then a major step toward successful monitoring and adaptive management will have been achieved.

2. **Plan and replan.** Agency employees and citizens often tire of seemingly endless planning. Planning is often just "busy-work," and the experience of one of us working for the EPA suggests that planning may be used to avoid taking substantive steps toward solving environmental problems. Yet lack of planning is also one of the most common causes of bureaucratic failures. Typically, agency managers (and often politicians and the public) are eager to get a monitoring program started but will not commit the time and resources to plan that program properly. If you are caught in this situation, resist. An inadequately planned monitoring program can not only cause ecological damage but also waste thousands or even millions of dollars. To plan the program, you may need outside resources or help such as from statisticians or specialists in particular aspects of ecology. Do not try to do the job without the right tools or expertise.

   A well-designed plan should then be followed, but not inflexibly. To ensure comparable data, monitoring protocols, once established, should generally not be changed. But portions of a program may need to be modified as the program proceeds. For example, preliminary analyses may suggest the need to intensify sampling in some watersheds that are at greater risk of degradation. Or new information from the scientific literature may reveal the need to monitor another element of biodiversity, such as soil biota.

3. **Plan for things to go wrong.** A monitoring program should be robust in the sense that it will still get the job done even though many things go wrong. Accidents, mishaps, lost data, malfunctioning equipment, and a plethora of other events are the norm in biological field sampling. Technicians quit during the peak of the field season. Forest fires burn

TABLE 9.5 Goals, Indicators, Thresholds, and Threshold Actions for Blue Mountains at Various Levels of Biodiversity

Level	Monitoring goal	Indicator(s)	Management threshold	Threshold action
Gene	Detect loss of genetic diversity and fitness of ponderosa pine	(1) Heterozygosity of ponderosa pine in cutover stands as opposed to virgin stands; (2) decrease in recruitment of seedlings in managed versus virgin stands with similar structure and soils	Heterozygosity or recruitment significantly less in managed stands	Change selection practices in timber sales; stop cutting ponderosa pine
Species	Maintain historic abundance and distribution of beaver (a keystone species)	Abundance as measured by number of active lodges per mile of stream; distribution as measured by distribution of lodges from aerial flights or photos	A 20% decrease in abundance or areal distribution	Limit trapping; reduce water diversion
	Detect decrease in abundance or viability of bird species threatened by cowbird invasion	Ratio of cowbirds to yellow warblers	Cowbird abundance in excess of 5% of yellow warbler abundance	Eliminate livestock grazing from affected areas

Community/ecosystem	Detect increase in nutrient or sediment loss in streams	Total dissolved solids (TDS), N, and P content in second-order streams sampled at monthly intervals and after peak storm events	Increase of over 5% in TDS, N, or P in logged watersheds compared to control (unlogged/virgin) watersheds	Change in logging practices; elimination of logging on affected watersheds
	Detect loss of vegetation communities or successional stages	Change in areal extent of community or successional stage	Change of over 10% in areal extent of any vegetation community or successional stage	Dependent upon identity of community or successional stage being lost and prior knowledge of that community; may require follow-up study
Landscape	Detect loss of connectivity between critical old-growth areas	Existing corridors are maintained with at least a 4-mile width within which average canopy closure is at least 80% on every section	Corridor width less than 4 miles or with canopy closure below 80%	Eliminate logging in areas where connectivity is inadequate; possible increased fire suppression

control plots. Radiocollars stop working. Film is lost at the lab. A monitoring program must have enough redundancy so that it does not depend on any one set of data or even one study.

4. **Ensure that monitoring is being conducted at the proper spatial and temporal scale.** In assessing long-term phenomena the monitoring must be extended over a long time period. If desert tortoises need 500 years to recover population viability after a collapse, then a 500-year monitoring program should be designed. If tamarisk (an exotic riparian tree species in the Southwest) is invading an entire watershed, then the watershed is the proper scale for monitoring. Agencies often get locked into monitoring on some arbitrary bureaucratic timeframe such as 5 or 10 years. But nature follows its own time scale. Similarly, agencies often do not look beyond their administrative boundaries, even when the problems they are trying to manage and monitor are occurring on a much larger scale. This narrow-mindedness must change if biodiversity protection is to be effective.

5. **When monitoring long-term phenomena, monitor trends or progress in the meantime.** If some phenomenon, such as recovery of desert tortoises or restoration of mature forests, will take hundreds of years, it is prudent to monitor progress toward that end in the meantime. This can be done by measuring expected change of parameters such as size or age class distributions. With forests or tortoises, incremental progress may be fairly easy to recognize and measure. With other phenomena, detecting and measuring progress may be complex and difficult.

6. **Don't confuse implementation monitoring with validation monitoring.** For years natural resource managers have been confusing implementation and validation monitoring. For example, a wildlife management agency would plant bitterbrush to provide increased winter browse to have more mule deer to hunt. The agency would report that 600 acres of bitterbrush had been successfully established but would never census the deer population to see if it had increased as predicted. This is a fairly straightforward and obvious example of confusing the two types of monitoring.

   Sometimes, however, confusion is more subtle, as ends and means get mixed. Consider snags, a common element of the old-growth forest upon which many cavity-nesting birds and other vertebrates and invertebrates depend. Most timber sales on federal lands are designed these days to leave a certain number or density of snags. Furthermore, cutover areas are often monitored to determine if an adequate snag density persists through time. Is this validation or effectiveness monitoring? Is snag density the end or the means? One could argue that snag density is only the means and that we must measure density or, better yet, reproductive success of cavity nesting birds to determine effectiveness.

7. **Choose indicators carefully; don't manage or deliberately manipulate them.** We noted earlier that choice of indicators is critical for many reasons and suggested some criteria for choosing them. For the most part, however, indicators should not be the object of management, or we risk falling back into the halfway technologies discussed in Chapter 8. For example, if an abundant trout population is chosen as an indicator of stream condition, we could simply stock large trout in the stream every April and achieve our annual objectives. Manage to achieve objectives of true conservation value; choose several independent indicators to determine if objectives are being met.

8. **Use all the tools of modern science and statistics as appropriate, including properly trained statisticians and biologists.** People who insist that their bodies be checked by a qualified physician every year and want their bridges and buildings inspected by licensed engineers seem to be content to let their environment be monitored by temporary undergraduate technicians with inadequate direction, supervision, and training. In planning monitoring programs, statisticians and graduate-level conservation biologists should be consulted. No monitoring plan should be approved without review by a statistician. Similarly, trained biologists, hydrologists, and other specialists should be used as appropriate in designing and conducting monitoring.

9. **Separate monitoring from management at the same level of organization.** Monitoring provides a measure of how effective management is at achieving biodiversity goals. Business learned years ago to have certain monitoring functions, particularly the management of money, checked independently by auditors. Asking government managers to devise monitoring schemes to determine how well they are achieving objectives is asking for a biased assessment. Monitoring should be done either by a higher level management unit in the organization or by an independent agency or organization. Hiring an independent contractor or conservation organization may be the best approach, since a higher level unit of an organization is still somewhat responsible and accountable for the lower-level units.

10. **Build an institutional structure for a sustainable monitoring program.** This is one of the most critical needs in most agencies.

Monitoring is usually only one of a large set of responsibilities for most organizational subunits and for most individuals in agencies. Monitoring under these conditions is usually the first task to be dropped with budget or personnel cuts or with new demands on time and money. The best way to have effective long-term monitoring is to have organizational subunits whose sole responsibility is monitoring. Sustained monitoring will also require educating Congress, agency heads, and managers at all levels about the need for predictability and stability of funding for such programs. All budget requests and justification documents should reinforce this idea. A lower, but stable level of funding is likely to produce more useful results than a boom-and-bust funding program.

11. **Use monitoring as a tool to change management or human behavior.** This point has been touched on several times, but deserves one more mention. Monitoring for biodiversity should be issue-driven, and the issue is always to determine if specific human activities are compatible with or damaging biodiversity. Increasingly, we need to monitor our approaches to restoring biodiversity. Monitoring should be designed to detect damage or improvement and the results incorporated into efforts to change human behavior. Unless it leads to positive changes in management and other human behaviors, monitoring is pointless.

CHAPTER TEN

THE TASK AHEAD

> Conservation is a state of harmony between men and land. Despite nearly a century of propaganda, conservation still proceeds at a snail's pace; progress still consists largely of letterhead pieties and convention oratory. On the back forty we still slip two steps backward for each forward stride.
>
> Aldo Leopold (1949), *A Sand County Almanac*

Technocratic optimists may wonder why a book like ours was written. For those indefatigable defenders of the status quo, biodiversity is under little threat today in the United States because we have learned from past mistakes (if mistakes are even admitted) and now know how to manage nature wisely. If what the technocrats tell us is true, this book is unnecessary. The biodiversity crisis is a myth. Or, if there is a problem, it is limited to those underdeveloped countries that lack the human and financial resources and the effective government of the United States.

Unfettered optimism is not limited to politicians and their appointees. Even some highly trained professionals in land-managing agencies are sanguine about the present condition and future of biodiversity in America. For example, three Forest Service scientist/managers recently concluded:

> There is overwhelming evidence that, while some problems remain and others have emerged in the last few years, on balance, multiple use and conservation have worked. The conditions of United States forests, wildlife, rangelands, agricultural lands, and related resources have improved dramatically during the last century. These trends continue. (Salwasser, MacCleery, and Snellgrove 1992)

We are not so convinced that biodiversity is on the upswing. The symptoms of poor land health are everywhere, while signs of recovery are few and far between. Studies on species endangerment, habitat loss, and other forms of biotic impoverishment in the United States show severe and continuing degradation of entire ecosystems and their associated species (Noss *et al.*

1994). Many of the studies documenting biotic impoverishment were conducted by government scientists and are published in government reports. Their conclusions, and those of scientists outside the government, have been echoed by reports issued by the Congressional research arms, the Office of Technology Assessment (OTA) and the General Accounting Office (GAO). Yet, the official position of most agencies (e.g., USDI–USDA 1992) is that environmental problems are minor and biodiversity is well taken care of. Why the discrepancy?

The tendency of agency administrators to hide real problems while glorifying agency programs is widespread. Agencies charged with protecting biodiversity have often developed close ties with industries exploiting biodiversity. Similar ties between agencies and conservationists have been lacking. Not surprisingly, when compromises have been made, those seeking to exploit resources have fared well. The natural environment has been the loser. The pro-commodity bias of agencies reached its peak, we hope, in the 1980s and early 1990s, but it is not limited to Republicans or conservatives. It continues today, for example, in the willingness of both Democrats and Republicans to compromise with the timber industry in the Pacific Northwest. This industry has already cut 90 percent of the old growth in the Northwest—and over 95 percent nationwide—and wants to do as it pleases with what little remains. Underlying a political compromise here is a naive faith that nature can always adjust to our manipulations, and that new technologies will allow us to exploit forests without destroying them.

Past land use on public and private lands nationwide has demonstrably been a disaster. We remain dangerously ignorant of how nature works, even though our knowledge of natural ecosystems and how to manage them has increased over the years. Improvements in knowledge and technology have simply not kept pace with the forces of environmental destruction. These forces—fundamentally human population, resource consumption, anthropocentric arrogance, greed, selfishness, ignorance, and disrespect for other living things—have encouraged higher rates of habitat destruction in spite of better science, stronger laws, and fancier technology. In fact, new technology has often created new problems.

The final chapter of a book like this usually attempts to answer the "Where do we go from here?" question. We do not want to let our readers down. But if you are expecting something like "Ten easy steps to conserving biodiversity," you will be disappointed. This book has presented and reviewed strategies, tactics, actions, and techniques that will probably be useful for protecting and restoring biodiversity, if indeed our people and our government embrace this as a primary policy goal. We have emphasized the fundamental lack of knowledge on many issues, the desirability of an experimental and adaptive approach, and the need to be conservative (to err on the side of protection) in the face of uncertainty. Our caution is undergirded by the belief that wild things and wild places have value for their own sake. Many scientists, managers, and landowners share this belief, and we hope these people will prevail over the exploiters and technocrats in the long run. But it will not be an easy road to travel.

Barriers to Conserving Biodiversity

The strategy we advocate is ambitious. Nothing short of radical changes in the way lands are allocated and managed, and indeed in the fundamental relationship between humans and nature, will get the job done. Many barriers block implementation of this ambitious conservation strategy. By recognizing the nature of these barriers, how and why they were erected, who benefits from them and who does not, we have a better chance of tearing them down.

Impediments to fulfilling the goals and objectives offered in this book are of many kinds: philosophical, institutional, political, educational, and technical, among others. Each of these classes of barriers may warrant a book of its own. Our treatment will be brief, emphasizing barriers we feel are most prominent and problematic.

Philosophical

We were recently sent an editorial from a small city newspaper in Tennessee, written by the Editor Emeritus. It exemplifies an attitude that is all too common in our society—indifference to extinction. The editorial stated:

> We read an article recently that indicated that frogs are becoming an endangered species. The information in the article indicated that scientists (interested in frogs) were unable to explain what has happened. Personally, we have not missed the frogs as we have little contact with them. We often see frogs legs mentioned on menus in fancy restaurants, but we have never ordered them. On one occasion when we took several of our grandsons to a ball game, one of the boys (to the astonishment of his cousins) ordered frog legs, but he couldn't persuade them to join him in eating them. He seemed to enjoy the frogs legs, saying they were better than fried chicken. (Wells 1993)

If an educated individual can ridicule the concern of scientists over the worldwide decline of amphibians, what hope can we have for society in general to take the biodiversity crisis seriously?

Philosophical and ideological factors are primary among the barriers to conservation of biodiversity. Some people are simply not interested in other living things. Saving biodiversity is not likely to be high on the agenda for someone who believes the earth is a smorgasbord set out for people to feast upon. "Who cares if we lose species?" these folks ask. Extinction is natural. The earth is not our real home, anyway. We have a better place to go after we die. The animals and plants have no souls. They were not made in the image of God as we were. They were made for us to use as we please. After Armageddon, they will be left in the dirt while we find glory in heaven.

We have all heard these anti-Earth sentiments; they sprout from the lips of fundamentalists across the country. Yet, not all Christians and other religious people strive for dominion over nature in the sense of selfish exploitation. Some, like Vice-President Al Gore in his book *Earth in the Balance,* emphasize the responsibility of humans for stewardship of the earth. Gore believes that the purpose of life is "to glorify God." "How can one glorify the Creator," he asks, "while heaping contempt on the creation?" (Gore 1992). Similarly, theologian John Cobb believes that "the meaning of the dominion given to us is much better expressed in servanthood and stewardship than in exploitation" (Cobb 1988). One value of religious teaching is that it concerns right and wrong, something often absent from secular thought. If people, through religious insights or teachings, can come to understand that destroying biodiversity is fundamentally wrong, perhaps they will want to save what is left (Ehrenfeld 1988).

We suspect, however, that a more benign interpretation of dominion will not move us far enough toward conservation of all living things and natural processes. Stewardship still involves control of nature. It may be necessary in the short term but fails to satisfy as an ultimate ideal. A long-term relationship with nature depends on a realization that "the greatest human dignity follows from a respectfulness of everything that is as meaningful as ourselves, that is, the entire living world" (Kozlovsky 1974). Biodiversity conservation ultimately requires a rejection of humanism or anthropocentrism, the pernicious tenet that humans are the center of the universe. It requires a biocentric embrace of all life. Protecting only what is easy and convenient to protect is not good enough.

How does one set about rejecting humanism? The answer is straightforward in principle: By affirming our evolutionary kinship and ecological interdependence with other life forms. This is a task for education.

Institutional

The institutional environment for protecting biodiversity in the United States is fragmented. Instead of a coordinated system of databases, regulations, land managers, and citizens all oriented toward a common goal, we have a panoply of agencies and interests, each pursuing its special agenda (Office of Technology Assessment 1987, Blockstein 1989). Even within a government department, agencies often clash in objectives. In the Pacific Northwest, for example, the Fish and Wildlife Service and the BLM—both bureaus of the Department of the Interior—clashed head-on over plans to protect the northern spotted owl. While the Fish and Wildlife Service, with little support from the Secretary of the Interior, attempted to develop a recovery plan that would restore viable populations of owls, the BLM preferred a plan that allowed extinction of owl populations over large portions of the owl's range.

Perhaps the best way to get the agencies working together for biodiversity is to assure that they are guided by an overall national policy. Such a policy should be set by the highest levels of government and include explicit goals and objectives of the type offered in this book. Agencies would then be directed to develop detailed plans and timetables for meeting these goals, and to cooperate with other agencies and landowners in their implementation. Because implementation is often most effective when organized regionally, "bioregional councils" of agencies, landowners, citizens, and scientists (representing biodiversity) should be intimately involved in developing region-specific strategies, goals, plans, and actions. As emphasized throughout this book, neither top-down nor bottom-up planning alone is sufficient. Top-down guidance provides context, while bottom-up involvement provides caring and local knowledge of particular areas.

Political

The political leaders of the United States, from local to national levels, have shown little interest in biodiversity. There are hopeful signs that this situation is changing, but conservationists must be vigilant and persuasive if this interest is to be more than lip service. A persistent problem plaguing implementation of conservation measures is interference from special interests. This interference was rampant during the 1980s and early 1990s. However, it continues today, particularly in the western states, where many elected officials (Republican and Democrat alike) have close ties to the timber, mining, oil and gas, and livestock industries. They don't want to let their friends and donors down.

Political maneuvering extends down through all levels of public agencies. It takes the form of manipulation and suppression of scientific findings and punishment and censure of whistle blowers, biologists, and other professionals who put the resource or the land above agency glory. Repression of conscientious professionals must not only stop, it must be turned around.

Employees should be rewarded, rather than punished, for defending biodiversity. They should be given certificates of merit. Acting on behalf of natural ecosystems should be seen as a professional obligation. To this end, a group of 22 prominent scientists wrote a letter to Vice-President Gore in February 1993, charging that the Forest Service routinely suppresses scientific information and intimidates professionals who criticize logging practices. The scientists urged that the burden of proof be shifted to those who would extract resources, to demonstrate that projects will not cause unacceptable damage. If such a shift occurs at all levels of management decision making, political manipulation will not be so easy to achieve.

Educational

Another barrier to conservation is simple ignorance. We are all ignorant about nature and its myriad processes. As ecologist Frank Egler (1977) once remarked, "Ecosystems are not only more complex than we think, but more complex than we *can* think." (Egler probably based his comment on a 1927 quip from evolutionary biologist J.B.S. Haldane, who said "My suspicion is that the universe is not only queerer than we suppose, but queerer than we *can* suppose"; see Gould 1993.) Nevertheless, the tremendous growth of knowledge in natural science cannot be denied. Ecology and evolutionary biology have blossomed over the last century, leading to new insights into the generation and maintenance of biodiversity at all levels of organization. Every year we know more than we did the year before about the relationships of organisms to their environments, the interactions between organisms, the evolution of new life forms, and the factors that lead to extinction or persistence of life forms. Some people have made a serious attempt to familiarize themselves with this body of knowledge, while others have not. Those who have not still dominate the higher echelons of natural resources agencies and government.

To their credit, land-managing agencies have sincerely tried in recent years to upgrade their staffs with knowledgeable scientists and to educate existing staff about new ideas, knowledge, and techniques relevant to biodiversity and its conservation. The BLM, through its Training Center in Phoenix, has produced several high-quality interagency courses on biodiversity for managers and staff. These courses have brought in top scientists from universities and other institutions to train agency employees about biodiversity and the effects of various management practices. We have helped design, organize, and teach interagency shortcourses to administrators and staffs of federal agencies over the last several years. Ethics, politics, and other value considerations are discussed openly. Opinions have ranged widely among participants in these courses, making for lively discussions.

But it is fair to say that everyone, including instructors, has come away with a better understanding of biodiversity, how to conserve it, and what kinds of barriers stand in the way.

But more can be done. Salaries and opportunities can be upgraded to attract more qualified professionals to the agencies. For many years, Ph.D.-level scientists have ended up working for agencies only when they could not find jobs in academia. If the intellectual environment within agencies were improved and competent biologists and other scientists had more influence on management decisions, agency jobs might be the first choice for many graduates. If these professionals were placed in policy-making positions, we might begin to see a real translation of knowledge to management. The biodiversity training courses now being conducted reach only a small fraction of agency people; they should be expanded to reach more.

Placing highly trained scientists in agencies will not automatically lead to more enlightened land management. A major problem is that most university training is irrelevant to real-life conservation work. Many or most graduates with advanced degrees have had little experience outside the university, where they were taught by professors who also have known nothing but academia. They may have had many courses in ecology, botany, or zoology, but all were heavy in theory and facts but lacking in application. Alternately, they may have taken courses in wildlife, fisheries, forestry, or range management, where emphasis was on management all right, but management toward consumptive goals: more fish to hook, deer to shoot, trees to cut, or grass for cows. With either applied or basic science degrees alone, graduates are unprepared for modern conservation work.

Fortunately, the academic situation is improving. We credit the new interdisciplinary field of conservation biology for stimulating new curricula in many universities that foster the merging of pure and applied science, theory and practice, as well as better cooperation among departments. Foresters and wildlifers are actually talking to each other, and sometimes even to botanists and zoologists.

Perhaps most distressing of the educational problems today is the increasing separation of people from exposure to nature. Ecological and environmental education has largely become an indoor activity—using books, magazines, videos, and computers. We note with trepidation that the glut of articles, books, television shows, and videos on falcons, bears, and other charismatic wildlife seems to be inversely related to the numbers of these creatures in the wild. Is this merely an expression of society's increasing interest in wildlife and biodiversity—a complementary activity for those who cannot always be out in the woods? Or is it society's way of substituting an artificial experience for a real but vanishing one? Is it another "technological fix," another "halfway technology"? This raises the threshold question: Is viewing the most polished video or movie on grizzly bears or a desert pupfish or a glow worm (or any other aspect of nature) an adequate substitute for spending time in a wilderness where there are real grizzlies, seeing a pupfish in its hot desert habitat, or spotting a glow worm on a rainy night?

The separation of humans from nature cannot be remedied by technology or reliance on indoor learning. Exposure to nature will be necessary to foster an appreciation and desire to conserve it. An educational activity that emphasizes first-hand experience with nature, and that has a relatively long history in this country, is the art of nature interpretation. Interpretation is "an educational activity which aims to reveal meanings and relationships through the use of original objects, by firsthand experience, and by illustrative media, rather than simply to communicate factual information" (Tilden 1967). Thus interpretation combines direct experience with education about ecological relationships. This is probably the best way to learn about biodiversity.

Noss (1992e) describes guidelines for interpreting biodiversity, emphasizing the "big picture" of how local natural history and environmental problems can be related to patterns and problems on a global scale. With this interpretive philosophy, local observations provide a window of direct experience through which a bigger picture can be seen. The bigger picture will not always be a pleasant one, but we hope it will stimulate action.

Technical

Even with the best educated staffs in the world, and the most knowledgeable and supportive public, agencies will find themselves facing technical barriers to conserving biodiversity. Throughout this book we have discussed ecological problems that are insoluble given the current understanding of ecology and existing methods of managing landscapes, communities, and species. Examples include restoring aquatic and terrestrial ecosystems and controlling introduced organisms. No one knows how to restore the Intermountain West's rangelands that have been taken over by cheatgrass and other exotics. No one knows what to do about the rapidly spreading infestation of zebra mussels throughout the Great Lakes and river systems in the East and Midwest. No one has any idea how many green trees, logs, and snags must be retained on sites with New Forestry harvests in order to replicate natural disturbances and ensure site fertility over hundreds of years.

The most sensible response to our lack of knowledge and appropriate technology in these cases is adaptive management, where we test management treatments experimentally and monitor ecosystem responses over long periods. Through these experiments, we may eventually learn how to control cheatgrass and restore healthy stands of bunchgrass, how to rid lakes and streams of exotic pests, how to manage forests sustainably, and how to retain many other elements of native biodiversity. But if the past is any guide to the future, we suspect that for every technical problem solved in conservation, several more will arise, many of them direct consequences of failed management experiments. Global stressors such as climate change and stratospheric ozone depletion may make our very best management techniques futile. Ultimately, we may have to face the fact that only a spectacular reduction in the scale of human activities on earth will allow biodiversity to recover, and even then it may take centuries or millennia.

Priorities

We feel that the best hope of overcoming the barriers reviewed above and providing for recovery of biodiversity is an aggressive and ambitious strategy that outlines what needs to be done and how to do it, to the best of current knowledge and expert opinion. Scientists, managers, and citizens must join together to advocate these goals vigorously. If they do, goals that seem impossibly utopian today will be attainable tomorrow. Although no one knows exactly what needs to be done to accomplish the goals espoused in this book, a reasonable consensus exists among conservation biologists on many issues. For instance, we know that populations of a few dozen individuals rarely persist for very long, so it is a good idea to provide for populations of hundreds to thousands of individuals. Every consensus will shift with new knowledge, but the collective opinion of those who study biodiversity problems is a reasonable foundation for a conservation program.

In this book, we have drawn liberally from the work of many colleagues. We have sought to recognize areas of broad agreement, although our interpretations are colored by our experiences, values, and preferences. Without venturing to speak for all our colleagues, we offer the following as priority actions for conserving biodiversity at several spatial scales.

Global

- Encourage rapid stabilization, followed by decline, of the human population. This can only be accomplished equitably by promoting sustainable economies, reducing poverty, and offering strong incentives for small families. If we fail to reduce our population, nature will find a way to do it for us.

- Develop international strategies for controlling substances that threaten the environment at a global scale, including greenhouse gases (carbon dioxide, methane, etc.), chlorofluorocarbons (responsible for ozone thinning), and other stressors.
- Identify and protect biodiversity hot spots in all countries worldwide. Although this book has focused on conservation in the United States, global conservation is a national issue here because most of the funds for land protection in less developed nations must come from the United States and other rich countries. So far, funds that have financed dams and other destructive developments in the Third World have come from here. Our wealth and expertise should be used to help poorer countries establish economic development programs that do not degrade biodiversity.
- Set a good example. For better or worse, people in Third World nations often aspire to emulate the life styles and consumption patterns of the United States, Europe, and Japan. They want what we have. If we reduce our rate of resource consumption, positive effects may be felt worldwide. Similarly, Americans are hypocritical to demand that tropical countries save their forests when we are cutting down our own, a fact that has been pointed out by the president of Brazil, among others.

Continental

- Look beyond national borders. Nationalism is as much a threat to biodiversity conservation as it is to world peace. The continents are connected ecological systems and must be viewed as such. Reserve networks and land management plans should cross international boundaries and follow natural boundaries such as mountain ranges, basins, and watersheds. As a positive example, grassroots activists in the state of Washington and province of British Columbia are developing conservation strategies for the bioregions that cross their international border.

National

- Provide incentives (tax credits, deductions) and other bonuses for land donations to conservation agencies, energy and resource conservation practices, sound stewardship of private lands, and other actions that benefit biodiversity. Provide tax burdens, penalties, and other disincentives for profligate consumption of energy and resources, poor land use, and other destructive behaviors.

- Establish a national biodiversity center and implement a national biological survey to inventory nationwide the distribution and status of species, ecosystems, and genetic material.
- Complete the U.S. Fish and Wildlife Service Gap Analysis, which will provide a nationwide overview of the extent to which ecosystem types and species are represented in protected areas. Gap Analysis must be funded at an order of magnitude higher than it is today.
- Upgrade the Natural Heritage programs (Conservation Data Centers) in each state. These databases are often the best single source of information on rare species and natural communities in each state, but many programs are not as effective as they might be due to insufficient funding and staffing. The national biological survey and Gap Analysis programs will depend on vigorous Heritage programs for information.
- Determine which ecosystem types are most endangered (i.e., those that have declined or been degraded most since European settlement and are at greatest risk of further loss). This process can begin with Gap Analysis data but must involve higher resolution mapping of specific habitats, seral stages, and plant associations; documentation of presettlement vegetation and historical trends; and identification and mapping of threats to ecosystems nationwide.
- Pass legislation that will set recovery goals for endangered and threatened ecosystems and standards for representing all ecosystem types in protected areas, regardless of current endangerment. We specifically propose a Native Ecosystems Act (Noss 1991a, 1991h).
- Strengthen the Endangered Species Act and its implementation by providing equal protection for plants and invertebrates and greater and more explicit protection of habitats, by quickly completing the listing process for all candidate species that qualify, and by establishing aggressive recovery plans that will restore viable populations of threatened and endangered species throughout their historical ranges. The use of ecosystem classifications and maps to group candidates for listing, and listed species for recovery planning, will streamline the process and tie the Endangered Species Act to a Native Ecosystems Act that will help prevent the need for future listings.
- Require assessment of direct, indirect, and cumulative effects of federal actions on biodiversity in all NEPA documents (environmental impact statements and environmental assessments). Require rigorous monitoring to assure that biodiversity protection measures are implemented effectively.

- Interconnect regional reserve systems to form a continuous U. S. network tied to similar networks in Canada and Mexico.
- Pass legislation that places biodiversity conservation as the highest management objective for all federal lands and that prohibits degradation of natural or near-natural habitats.
- Phase out intensive commodity production, off-road vehicle use, and other harmful activities on federal lands. A good place to start would be removing subsidies that encourage below-cost timber sales, roadbuilding, grazing, mining, and subsidized irrigation. Use the money saved for inventory, protection, and restoration of biodiversity.
- Establish a new Civilian Conservation Corps, this time oriented toward scientifically defensible restoration projects on federal lands (removal of roads and fences, prescribed burning, etc.). This program could employ workers who lose their jobs as resource extraction activities are phased out.
- Establish incentive programs for conservation leadership in public service.
- Establish a national program for environmental education, emphasizing biodiversity interpretation in the classroom and outdoors. Provide guidelines, curricula, training, major grants, and other assistance to state and local school districts, park agencies, nature centers, outdoor schools, museums, and camps for biodiversity education.

Regional

- Develop a biogeographically sensible but flexible classification of bioregions across the country, for purposes of conservation planning. The classification should be hierarchical, recognizing regions within regions (for example, Intermountain West—Great Basin—Northern Basin and Range—Steens Mountain/Alvord Desert/Malheur Basin).
- Identify hot spots of biodiversity (e.g., centers of endemism and high species richness), remaining wild areas, unrepresented or underrepresented ecosystem types, and other areas requiring protection in each region.
- Develop a map-based plan for establishing core reserves, buffer zones, and connectivity adequate to meet the goals of representing all ecosystem types across their natural range of variation, maintaining viable populations of all native species, sustaining ecological and evolutionary processes, and being adaptable to environmental change.

- Develop land management and restoration plans for the reserve system and surrounding matrix of public and private lands that will allow these goals to be met. Proceed from general (entire region) to specific (small watershed) with increasingly detailed plans that can be interpreted and integrated within a regional context.
- Determine specific measures needed to implement the plan, including federal, state, and local legislation for land use and zoning; land acquisition by public or private conservation agencies (e.g., The Nature Conservancy); designations of protected areas on public lands; conservation easements and landowner agreements on private lands; and public education programs.

Local

- Put the regional conservation plan on the ground. This will require implementing measures such as land acquisitions, land use and zoning regulations, conservation easements, and public education programs in each local area.
- Provide incentives (for example, reduced property taxes, public recognition, and awards) to landowners who are managing their properties effectively for biodiversity, and to teachers, environmental activists, scientists, and other citizens who advance the cause of conservation. Create bad publicity for those engaged in destructive activities.
- Remember to consider the regional, national, and global context in each local decision so that local actions can be coordinated and will contribute to higher conservation goals.
- Implement each action in an adaptive management framework, establishing a scientifically defensible design, experimenting cautiously, monitoring, and adjusting management to new information.

Personal

- Limit your reproduction to at or below replacement level.
- Limit your use of resources, both renewable and nonrenewable, and especially those that require much energy to produce. An occasional sirloin steak or case of cold beer (if in returnable bottles) does little harm, but squandering paper, gasoline, plastic, and other materials only contributes to demands for increased logging, mining, and other resource extraction. Recycle or, better yet, reuse containers and other materials.

- Enjoy this marvelous planet. Get to know your local natural areas. Explore the untamed wilds beyond. Learn the names, habits, and needs of other creatures. Include them among your closest friends.
- Defend the wild. If a development or agency action threatens a natural area, do all you can to thwart destruction. But because habitat destruction is so prevasive today, establish clear priorities. Channel your energies toward the most critical sites. For example, it is more important to save the sole habitat of an endemic isopod than to rescue a few old trees in a parking lot or a rabbit in a laboratory.
- Tell others about all this. Share your affection for nature, your knowledge about it, and what must be done to protect it. Learning from a friend or neighbor may be more effective in changing behavior than reading a book.

One Final Word

> One final paragraph of advice. Do not burn yourselves out. Be as I am—a reluctant enthusiast—a part-time crusader, a half-hearted fanatic. Save the other half of yourselves and your lives for pleasure and adventure. It is not enough to fight for the land; it is even more important to enjoy it. While you can. While it's still here. So get out there and hunt and fish and mess around with your friends, ramble out yonder and explore the forests, encounter the grizzly, climb the mountains, bag the peaks, run the rivers, breathe deep of that yet sweet and lucid air, sit quietly for awhile and contemplate the precious stillness, that lovely, mysterious and awesome space. Enjoy yourselves, keep your brain in your head and your head firmly attached to the body, the body active and alive, and I promise you this one sweet victory over our enemies, over those desk-bound people with their hearts in a safe deposit box and their eyes hypnotized by desk calculators. I promise you this: you will outlive the bastards.
>
> Edward Abbey

EPILOGUE

> Will the things that are being lost—the wilderness, the plants and animals, the skills, and all the others—leave too vast a gap in the continuity of life to be bridged even by the human spirit? This is the unanswerable question. In the meantime, we must live in our century and wait, enduring somehow the unavoidable sadness.
>
> David Ehrenfeld (1978), *The Arrogance of Humanism*

Here we sit in yet another meeting to discuss the fate of biodiversity. Another conference room without windows, speakers devoid of zest, listeners wishing they were somewhere else. Questions are raised but never really answered. Solutions are proposed to insurmountable problems, but no one pays attention. Ah, the life of the conference biologist.

Loss of biodiversity is a serious matter, to be sure—a worthy topic for a thousand conferences. But do any of our efforts make a difference? By all signs we are powerless. The people who care profoundly about toads and liverworts may always be a minority in our society, a fact Aldo Leopold reminded us of throughout his writings. We are a minority even in this room and more surely out there in the street. To the rest of humanity, concerned with the "real" world of business, television, and mortgage payments, we are truly a bunch of weirdos and zealots.

The mind wanders in these meetings. If a mind really stayed in conference rooms and lecture halls like it is supposed to, it would soon be lost. The mind is more comfortable in the context in which it evolved millions of years ago. Out in the hot savannas, by the forest edge, by the riverside with the screeching birds and hungry crocodiles. That is where the mind naturally returns when the lectures become unbearable. It returns to the trees—real trees, not Latin names, not genotypes, transpiration machines, or even components of ecosystems. Nature satisfies best when raw, unclassified, unlabeled; when experienced directly through the senses rather than distorted by the intellect. But we are so busy with our professional responsibilities, with writing and reviewing countless manuscripts, that we hardly ever experience the *real* world. Who has time?

We were trained as biologists to classify, pigeonhole, hypothesize, experiment, and analyze. It is our job to go to conferences, give lectures, and write obscure papers. Now we have written a book. Can such actions do any good? Maybe. Science is a kind of understanding, albeit incomplete, and a kind of problem solving. The number of problems that need solving in the field of conservation are not becoming any fewer. Science in the form of conservation biology is a set of principles, empirical generalizations, and techniques for protecting nature—a toolbox for saving the earth. It is not a perfect toolbox (some tools are rusty, others lost, and still others yet to be invented) but it is the best we have. We feel confident enough about the science of conservation biology that we want to share it with others. And we want to see it applied in the real world, not just pontificated about in lecture halls, journals, or books.

Science is only the toolbox, of course; ultimately it provides no reason at all for conservation. Something deeper puts us on this quest. Maybe it is just the simple caring for our fellow creatures and our one common home. Caring, it seems, leads to ethical commitments. The only science worth doing is one firmly grounded in an ethic and an emotional commitment. Without values and commitments, science is perilous or at least irrelevant. But together, feeling and thinking might accomplish great things. As Aldo Leopold [1990, (1936)] said about his field of wildlife management: "There seem to be few fields of research where the means are so largely of the brain, but the ends so largely of the heart."

Strangely enough, we remain hopeful about the long-term future of biodiversity. This book is full, perhaps excessively so, of proposals, recommendations, and suggestions for how biodiversity might be conserved—conserved in its entirety, not just in small pieces where we find it convenient. We are by no means certain our proposals will work, but we think they have half a chance if implemented quickly and energetically. And we are tired of equivocation and excuses of not having enough data to say what needs to be done. Some people will label our proposals premature. Others, jaded by long years in government service or environmental politics, will call them utterly impractical and utopian. Still others may find what we have to offer misanthropic. But despite many years of government service between us and countless rejections of our earlier proposals, we remain guardedly optimistic. We believe that the trends of destruction can be turned around. The emergence of bioregional wilderness groups, increased activism within the scientific community, strong coalitions of biologists and other citizens who care about their environments give us hope. And why not? The alternative is despair.

Back to the conference. Someone must have said something mildly humorous, for some chortles have snapped the mind out of its daydream. Our fellow scientists and bureaucrats still sit uncomfortably around us (our bodies were not designed for these chairs) as another speaker drones on. The slide projector is still out of focus. The moderator is still holding up his one minute sign (for the last 15 minutes?). Everyone is looking forward to cocktail hour. And somewhere outside this windowless room—somewhere in the real world of forests, prairies, rivers, and seashores—the birds still sing, mate, and feed their young.

LITERATURE CITED

Abbey, E. 1982. "Down the River." E.P. Dutton, New York.

Abele, L.G., and E.F. Connor. 1979. Application of island biogeographic theory to refuge design: Making the right decision for the wrong reasons. Pages 89–94 in R.M. Linn, ed. "Proceedings of the First Conference on Scientific Research in the National Parks." Vol. I. USDI National Park Service, Washington, DC.

Abrahamson, W.G., and D.C. Hartnett. 1990. Pine flatwoods and dry prairies. Pages 103–149 in R.L. Myers and J.J. Ewel, eds. "Ecosystems of Florida." University of Central Florida Press, Orlando, FL.

Adams, A.W. 1982. Migration. Pages 301–321 in J.W. Thomas and D.E. Toweill, eds. "Elk of North America: Ecology and Management." Stackpole Books, Harrisburg, PA.

Adams, L.W., and A.D. Geis. 1983. Effects of roads on small mammals. *Journal of Applied Ecology* 20: 403–415.

Allan, J.D., and A.S. Flecker. 1993. Biodiversity conservation in running waters. *BioScience* 43: 32–43.

Allen, E.B. (Ed.). 1988. "The Reconstruction of Disturbed Arid Ecosystems." Westview Press, Boulder, CO.

Allen, E.B. 1989. The restoration of disturbed arid landscapes with special reference to mycorrhizal fungi. *Journal of Arid Environments* 17:279–286.

Allen, E.B. and M.F. Allen. 1986. Water relations of xeric grasses in the field: interactions of Mycorrhizas and competition. *New Phytologist* 104:559–571.

Allen, E.B. and L.L. Jackson. 1992. The arid West. *Restoration and Management Notes* 10(1):56–59.

Allendorf, F.W., R.B. Harris, and L.H. Metzgar. 1991. Estimation of effective population size of grizzly bears by computer simulation. Pages 650–654 in E.C. Dudley, ed. "The Unity of Evolutionary Biology: Proceedings of the Fourth International Congress of Systematic and Evolutionary Biology." Dioscorides Press, Portland, OR.

Alverson, W.S., D.M. Waller, and S.L. Solheim. 1988. Forests too deer: Edge effects in northern Wisconsin. *Conservation Biology* 2: 348–358.

Anderson, A.E. 1983. "A Critical Review of Literature on Puma (*Felis concolor*)." Special Report No. 54. Colorado Division of Wildlife, Denver, CO.

Anderson, J.E. 1991. A conceptual framework for evaluating and quantifying naturalness. *Conservation Biology* 5: 347–352.

Anderson, L., C.E. Carlson, and R.H. Wakimoto. 1987. Forest fire frequency and western spruce budworm outbreaks in western Montana. *Forest Ecology and Management* 22: 251–260.

Andren, H., and P. Angelstam. 1988. Elevated predation rates as an edge effect in habitat islands: Experimental evidence. *Ecology* 69: 544–547.
Archer, S. 1989. Have southern Texas savannas been converted to woodlands in recent history? *American Naturalist* 134: 545–561.
Archer, S. and F.E. Smeins. 1991. Ecosystem-level processes. Pages 109–139 in R.K. Heitschmidt and J.W. Stuth, eds. "Grazing Management: An Ecological Perspective." Timber Press, Portland, OR.
Archer, S. and F.E. Smeins. 1992. Non-linear dynamics in grazed ecosystems: Thresholds, multiple steady states and positive feedbacks. In "Is the Range Condition Concept Compatible with Ecosystem Dynamics?" Society for Range Management, Spokane, WA.
Askins, R.A., M.J. Philbrick, and D.S. Sugeno. 1987. Relationship between the regional abundance of forest and the composition of forest bird communities. *Biological Conservation* 39: 129–152.
Asquith, A., J.D. Lattin, and A.R. Moldenke. 1990. Arthropods: The invisible diversity. *Northwest Environmental Journal* 6: 404–405.
Bailey, J.A. 1982. Implications of "muddling through" for wildlife management. *Wildlife Society Bulletin* 10:363–369.
Baker, W.L. 1989. Landscape ecology and nature reserve design in the Boundary Waters Canoe Area, Minnesota. *Ecology* 70: 23–35.
Baker, W.L. 1992a. The landscape ecology of large disturbances in the design and management of nature reserves. *Landscape Ecology* 7: 181–194.
Baker, W.L. 1992b. Effects of settlement and fire suppression on landscape structure. *Ecology* 73: 1879–1887.
Ballou, J.D., T.J. Foose, R.C. Lacy, and U.S. Seal. 1989. "Florida Panther (*Felis concolor coryi*) Population Viability Analysis and Recommendations." Captive Breeding Specialist Group, Species Survival Commission, IUCN, Apple Valley, MN.
Barrett, T.S. and P. Livermore. 1983. "The Conservation Easement in California." Island Press, Covelo, CA.
Bartolome, J.W. 1993. Application of herbivore optimization theory to rangelands of the western United States. *Ecological Applications* 3: 27–29.
Batisse, M. 1990. Development and implementation of the biosphere reserve concept and its applicability to coastal regions. *Environmental Conservation* 17: 111–116.
Bean, M.J. 1988. The 1973 Endangered Species Act: Looking back over the first 15 years. *Endangered Species Update* 5(10):4–6.
Bean, M. 1991. Issues and controversies in the forthcoming reauthorization battle. *Endangered Species Update* 9(1 and 2): 1–4.
Bedward, M., R.L. Pressey, and D.A. Keith. 1992. A new approach for selecting fully representative reserve networks: Addressing efficiency, reserve design and land suitability with an iterative analysis. *Biological Conservation* 62: 115–125.
Behan, R.W. 1990. Multiresource forest management: a paradigmatic challenge to professional forestry. *Journal of Forestry* 88(4):12–18.

Beier, P. 1993. Determining minimum habitat areas and habitat corridors for cougars. *Conservation Biology* 7: 94–108.

Beier, P., and R.H. Barrett. 1991. "Orange County Cooperative Mountain Lion Study." Department of Forestry and Resource Management, University of California, Berkeley, CA.

Beier, P., and S. Loe. 1992. A checklist for evaluating impacts to wildlife movement corridors. *Wildlife Society Bulletin* 20: 434–440.

Belsky, A.J. 1986. Does herbivory benefit plants? A review of the evidence. *American Naturalist* 127: 870–892.

Benke, A.C. 1990. A perspective on America's vanishing streams. *Journal of the North American Benthological Society* 91: 77–88.

Bennett, A.F. 1990. "Habitat Corridors: Their Role in Wildlife Management and Conservation." Arthur Rylah Institute for Environmental Research, Department of Conservation and Environment, Victoria, Australia.

Bennett, A.F. 1991. Roads, roadsides, and wildlife conservation: A review. Pages 99–118 in D.A. Saunders and R.J. Hobbs, eds. "Nature Conservation 2: The Role of Corridors." Surrey Beatty and Sons, Chipping Norton, NSW , Australia.

Berger, J.J. (ed.). 1990. "Environmental Restoration—Science and Strategies for Restoring the Earth." Island Press, Covelo, CA.

Berry, K.H. 1978. Livestock grazing and the desert tortoise. *Transactions of the North American Wildlife and Natural Resources Conference* 43: 505–519.

Betz, R.F. 1992. The tallgrass prairie. *Restoration and Management Notes* 10(1): 33–35.

Blackburn, W.H. 1984. Impacts of grazing intensity and specialized grazing systems on watershed characteristics and responses. Pages 927–983 in "Developing Strategies for Rangeland Management." National Research Council/National Academy of Sciences. Westview Press, Boulder, CO.

Blake, J.G., and W.G. Hoppes. 1986. Influence of resource abundance on use of treefall gaps by birds. *Auk* 103: 328–430.

Bleich, V.C., J.D. Wehausen, and S.A. Holl. 1990. Desert-dwelling mountain sheep: Conservation implications of a naturally fragmented distribution. *Conservation Biology* 4: 383–390.

Blockstein, D.E. 1989. Toward a federal plan for biological diversity. *Issues in Science and Technology* 4(4): 63–67.

Bock, C.E., J.H. Bock, W.R. Kenney, and V.M. Hawthorne. 1984. Responses of birds, rodents, and vegetation in livestock enclosure in a semidesert grassland site. *Journal of Range Management* 37: 239–242.

Boecklen, W.J. 1986. Effects of habitat heterogeneity on the species–area relationships of forest birds. *Journal of Biogeography* 13: 59–68.

Bohning-Gaese, K., M.L. Taper, and J.H. Brown. 1993. Are declines in North American insectivorous songbirds due to causes on the breeding range? *Conservation Biology* 7: 76–86.

Bormann, F.H., and G.E. Likens. 1979. "Pattern and Process in a Forested Ecosystem." Springer-Verlag, New York.

Botkin, D.B. 1990. "Discordant Harmonies: A New Ecology for the Twenty-first Century." Oxford University Press, New York.

Bourgeron, P.S., and L. Engelking (eds.). 1992. Preliminary compilation of a series level classification of the vegetation of the western United States using a physiognomic framework. "Report to the Idaho Cooperative Fish and Wildlife Research Unit." Western Heritage Task Force, The Nature Conservancy, Boulder, CO.

Bowers, C.A. 1993. "Education, Cultural Myths, and the Ecological Crisis—Toward Deep Changes." State University of New York Press, Albany, NY.

Boyce, M.S. 1992. Population viability analysis. *Annual Review of Ecology and Systematics* 23: 481–506.

Brady, W.W., M.R. Stromberg, E.F. Aldon, C.D. Bonham, and S.H. Henry. 1989. Response of a semidesert grassland to 16 years of rest from grazing. *Journal of Range Management* 42: 284–288.

Branson, F.A. 1985. "Vegetation Changes on Western Rangelands." Society for Range Management, Range Monograph No. 2, Denver, CO.

Branson, F.A., G.F. Gifford, K.G. Renard, and R.F. Hadley. 1981. "Rangeland Hydrology." 2nd Ed. Society for Range Management, Range Science Series No. 1, Kendall/Hunt, Dubuque, IA.

Bratton, S.P., and P.S. White. 1980. Rare plant management—after preservation what? *Rhodora* 82: 49–75.

Braun, E.L. 1950. "Deciduous Forests of Eastern North America." Macmillan, New York.

Bray, J.R. 1956. Gap phase replacement in a maple-basswood forest. *Ecology* 37: 598–600.

Brittingham, M.C., and S.A. Temple. 1983. Have cowbirds caused forest songbirds to decline? *BioScience* 33: 31–35.

Brody, A.J. 1984. Habitat use by black bears in relation to forest management in Pisgah National Forest, North Carolina. M.S. Thesis. University of Tennessee, Knoxville.

Brody, A.J., and M.P. Pelton. 1989. Effects of roads on black bear movements in western North Carolina. *Wildlife Society Bulletin* 17: 5–10.

Brokaw, N.V.L. 1985. Treefalls, regrowth, and community structure in tropical forests. Pages 53–69 in S.T.A. Pickett and P.S. White, eds. "The Ecology of Natural Disturbance and Patch Dynamics." Academic Press, Orlando, FL.

Brown, J.H. 1988. Alternative conservation priorities and practices. Paper presented at 73rd Annual Meeting, Ecological Society of America. Davis, CA.

Brown, J.H. and A.C. Gibson. 1983. "Biogeography." C.V. Mosby, St. Louis, MO.

Brown, J.H., and A. Kodric-Brown. 1977. Turnover rates in insular biogeography: Effect of immigration on extinction. *Ecology* 58: 445–449.

Brown, M.T., J.M. Schaefer, K.H. Brandt, S.J. Doherty, C.D. Dove, J.P. Dudley, D.A. Eifler, L.D. Harris, R.F. Noss, and R.W. Wolfe. 1987. "An Evaluation of the Applicability of Upland Buffers for the Wetlands of the Wekiva Basin." Center for Wetlands, University of Florida, Gainesville, FL.

Browne, J. 1983. "The Secular Ark: Studies in the History of Biogeography." Yale University Press, New Haven, CT.

Brubaker, L.B. 1991. Climate change and the origin of old-growth Douglas-fir forests in the Puget Sound Lowland. Pages 17–24 in L.F. Ruggiero, K.B. Aubry, A.B. Carey, and M.H. Huff, tech. coordinators. "Wildlife and Vegetation of Unmanaged Douglas-fir Forests." USDA Forest Service, Pacific Northwest Research Station, Portland, OR.

Brussard, P.F. 1991. The role of ecology in biological conservation. *Ecological Applications* 1: 6–12.

Buechner, M. 1987. Conservation in insular parks: Simulation models of factors affecting the movement of animals across park boundaries. *Biological Conservation* 41: 57–76.

Burgess, R.L., and D.M. Sharpe, eds. 1981. "Forest Island Dynamics in Man-Dominated Landscapes." Springer-Verlag, New York.

Callicott, J.B. 1982. Traditional American Indian and western European attitudes toward nature: An overview. *Environmental Ethics* 4: 293–318.

Callicott, J.B. 1991. The wilderness idea revisited: The sustainable development alternative. *Environmental Professional* 13: 235–247.

Campbell, F.T. 1988. The desert tortoise. Pages 567–581 in W.J. Chandler, ed. "Audubon Wildlife Report 1988/1989." Academic Press, New York.

Canham, C.D. 1989. Different responses to gaps among shade-tolerant tree species. *Ecology* 70: 548–550.

Carey, A.B. 1989. Wildlife associated with old-growth forests in the Pacific Northwest. *Natural Areas Journal* 9: 151–162.

Carey, H.H. 1989. Forest trust: two-year report, 1987–1988. The Forest Trust, Santa Fe, NM.

Chadwick, D. 1990. The biodiversity challenge. *Defenders* 65(3): 19–30.

Charley, J.L. 1977. Mineral cycling in rangeland ecosystems. Pages 215–256 in R.E. Sosebee, ed. "Rangeland Plant Physiology." Society for Range Management, Range Science Series, Number 4, Denver, CO.

Chen, J., and J.F. Franklin. 1990. Microclimatic pattern and basic biological responses at the clearcut edges of old-growth Douglas-fir stands. *Northwest Environmental Journal* 6: 424–425.

Chen, J., J.F. Franklin, and T.A. Spies. 1992. Vegetation responses to edge environments in old-growth Douglas-fir forests. *Ecological Applications* 2: 387–396.

Christensen, N.L., J.K. Agee, P.F. Brussard, J. Hughes, D.H. Knight, G.W. Minshall, J.M. Peek, S.J. Pyne, F.J. Swanson, J.W. Thomas, S. Wells, S.E. Williams, and H.A. Wright. 1989. Interpreting the Yellowstone fires of 1988. *BioScience* 39: 678–685.

Clark, T.W., and D. Zaunbrecher. 1987. The Greater Yellowstone Ecosystem: The ecosystem concept in natural resource policy and management. *Renewable Resources Journal* 5(3):8–16.

Clawson, M. 1979. Forests in the long sweep of American history. *Science* 204: 1168–1174.

Clebsch, E.E.C., and R.T. Busing. 1989. Secondary succession, gap dynamics, and community structure in a southern Appalachian cove forest. *Ecology* 70: 728–735.

Clewell, A.F. 1986. "Natural Setting and Vegetation of the Florida Panhandle." U.S. Army Corps of Engineers. Mobile, AL.

Clewell, A.F. 1989. Natural history of wiregrass (*Aristida stricta* Michx., Gramineae). *Natural Areas Journal* 9: 223–233.

Clout, M.N., and P.D. Gaze. 1984. Effects of plantation forestry on birds in New Zealand. *Journal of Applied Ecology* 21: 795–815.

Clutton-Brock, J. 1981. "Domesticated Animals from Early Times." University of Texas Press, Austin, TX.

Cobb, J.B. 1988. A Christian view of biodiversity. Pages 481–485 in E.O. Wilson, ed. "Biodiversity." National Academy Press, Washington, DC.

Coblentz, B. 1990. Exotic organisms: A dilemma for conservation biology. *Conservation Biology* 4: 261–265.

Commoner, B. 1971. "The Closing Circle." Alfred A. Knopf, New York.

Connell, J.H. 1978. Diversity in tropical rain forests and coral reefs. *Science* 199:1302–1310.

Connor, E.F., and E.D. McCoy. 1979. The statistics and biology of the species-area relationship. *American Naturalist* 113: 791–833.

Cook, R.E. 1969. Variation in species density of North American birds. *Systematic Zoology* 18: 63–84.

Cooper, W.S. 1913. The climax forest of Isle Royale, Lake Superior, and its development. *Botanical Gazette* 55: 1–44, 115–140, 189–235.

Cooperrider, A.Y. 1985. The desert bighorn. Pages 472–485 in R.L. Di Silvestro, ed. "Audubon Wildlife Report 1985." National Audubon Society, New York.

Cooperrider, A.Y. 1986. Food habits. Pages 699–710 in A.Y. Cooperrider, R.J. Boyd, and H.R. Stuart, eds. "Inventory and Monitoring of Wildlife Habitat." USDI Bureau of Land Management, Washington, DC.

Cooperrider, A. 1993. Ecological quality in the National Parks. Paper presented at "Diamonds or Dust?—A forum on the quality of our national parks," San Francisco, CA, May, 1991.

Cooperrider, A.Y. and J.A. Bailey. 1982. Analysis of competition between domestic livestock and wild ungulates with computer simulation. Pages 421–440 in J.M. Peek and P.D. Dalke, eds. "Wildlife–Livestock Relationships Symposium: Proceedings 10." University of Idaho, Forest, Wildlife and Range Experiment Station, Moscow, ID.

Cooperrider, A.Y. and J.A. Bailey. 1984. A simulation approach to forage allocation. Pages 525–559 in "Developing Strategies for Rangeland Management." National Research Council/National Academy of Sciences. Westview Press, Boulder, CO.

Courtenay, W.R. Jr. 1990. Fish conservation and the enigma of introduced species. Pages 11–20 in D.A. Pollard, ed. "Introduced and Translocated Fishes and Their Ecological Effects." *Proceedings of the Bureau of Rural Resources* 8, Canberra, Australia.

Courtenay, W.R. Jr., and P.B. Moyle. 1992. Crimes against biodiversity: the lasting legacy of fish introductions. *Transactions of the North American Wildlife and Natural Resources Conference* 56: 365–372.

Courtenay, W.R. Jr., and P.B. Moyle. 1994. Biodiversity, fishes, and the introduction paradigm. *In* R. Szaro ed. "Biodiversity in Managed Landscapes," Oxford University Press, New York.

Courtenay, W.R. Jr., and C.R. Robins. 1989. Fish introductions: good management, mismanagement, or no management? *Reviews in Aquatic Science* 1: 159–172.

Cowell, E.B. 1978. Ecological monitoring as a management tool in industry. *Ocean Management* 4: 273–285.

Craig, G. 1986. Peregrine falcon. Pages 806–824 in R.L. DiSilvestro, ed., "Audubon Wildlife Report 1986." The National Audubon Society, New York.

Crumpacker, D.W., S.W. Hodge, D.F. Friedley, and W.P. Gregg, Jr. 1988. A preliminary assessment of the status of major terrestrial and wetland ecosystems on federal and Indian lands in the United States. *Conservation Biology* 2: 103–115.

Currie, D.J. 1991. Energy and large-scale patterns of animal- and plant-species richness. *American Naturalist* 137: 27–49.

Currie, D.J., and V. Paquin. 1987. Large-scale biogeographical patterns of species richness in trees. *Nature* 329: 326–327.

Curtin, C.G. 1993. The evolution of the U.S. National Wildlife Refuge system and the doctrine of compatibility. *Conservation Biology* 7: 29–38.

Curtis, J.T. 1956. The modification of mid-latitude grasslands and forests by man. Pages 721–736 in W.L. Thomas, ed. "Man's Role in Changing the Face of the Earth." University of Chicago Press, Chicago, IL.

Dahl, T.E. 1990. "Wetland Losses in the United States 1780s to 1980s." USDI Fish and Wildlife Service, Washington, DC.

Daniel, J. 1990. Old growth on the dry side. *High Country News* 22(22): 27–28.

Dasmann, R.F. 1988. Biosphere reserves, buffers, and boundaries. *BioScience* 38: 487–489.

Davis, G.D. 1988. Preservation of natural diversity: The role of ecosystem representation within wilderness. Paper presented at National Wilderness Colloquium, Tampa, Florida, January 1988.

Davis, G.E. 1989. Design of a long-term ecological monitoring program for Channel Islands National Park, California. *Natural Areas Journal* 9: 80–89.

Davis, M.B. 1981. Quaternary history and the stability of forest communities. Pages 132–153 in D.C. West, H.H. Shugart, and D.B. Botkin, eds. "Forest Succession." Springer-Verlag, New York.

Day, G.M. 1953. The Indian as an ecological factor in the northeastern forest. *Ecology* 34: 329–346.

Deacon, J.E. 1988. The endangered woundfin and water management in the Virgin river, Utah, Arizona, Nevada. *Fisheries* 13(1): 18–24.

Deacon, J.E., and J.E. Williams. 1984. Annotated list of the fishes of Nevada. *Proceedings of the Biological Society of Washington* 97: 103–118.

Defenders of Wildlife. 1992. "Putting Wildlife First: Recommendations for Reforming Our Troubled Refuge System." Defenders of Wildlife, Washington, DC.

Deloria, V. Jr. 1992. Prospect for restoration on tribal lands. *Restoration and Management Notes* 10(1):48–50.

Den Boer, P.J. 1981. On the survival of populations in a heterogeneous and variable environment. *Oecologia* 50: 39–53.

Den Boer, P.J. 1990. The survival value of dispersal in terrestrial arthropods. *Biological Conservation* 54: 175–192.

Devall, B., and G. Sessions. 1985. "Deep Ecology: Living as if Nature Mattered." Peregrine Smith, Salt Lake City, UT.

Dewberry, T.C. 1992. Protecting the biodiversity of riverine and riparian ecosystems: the National River Public Land Policy Development Project. *Transactions of the North American Wildlife and Natural Resources Conference* 57: 424–432.

Diamond, J.M. 1972. Biogeographic kinetics: Estimation of relaxation times for avifaunas of Southwest Pacific Islands. *Proceedings of the National Academy of Science* 69: 3199–3203.

Diamond, J.M. 1975. The island dilemma: Lessons of modern biogeographic studies for the design of natural preserves. *Biological Conservation* 7: 129–146.

Diamond, J.M. 1976. Island biogeography and conservation: Strategy and limitations. *Science* 193: 1027–1029.

Diamond, J.M. 1982. Man the exterminator. *Nature* 298: 787–789.

Diamond, J.M. 1984. Historic extinctions: A Rosetta stone for understanding prehistoric extinctions. Pages 824–862 in P.S. Martin and R.G. Klein, eds. "Quaternary Extinctions: A Prehistoric Revolution." University of Arizona Press, Tucson, AZ.

Diamond, J.M. 1990. Playing dice with megadeath. *Discover* (April): 55–59.

Diamond, J. 1992. Must we shoot deer to save nature? *Natural History* 8/92: 2–8.

Diamond, J.M., and R.M. May. 1976. Island biogeography and the design of natural reserves. Pages 163–186 in R.M. May, ed. "Theoretical Ecology: Principles and Applications." W.B. Saunders, Philadelphia, PA.

Diamondback. 1990. Ecological effects of roads (or, the road to destruction). Pages 1–5 in J. Davis, ed. "Killing Roads: A Citizens' Primer on the Effects and Removal of Roads." Earth First! Biodiversity Project Special Publication. Tucson, AZ.

Dortignac, E.J. 1956. Watershed resources and problems of the Upper Rio Grande Basin. U.S. Forest Service, Rocky Mountain Forest and Range Experiment Station. Unpublished Paper.

Dregne, H.E. 1977. Desertification of arid lands. *Economic Geography* 3: 322–331.

Driscoll, R.S., D.L. Merkel, D.L. Radloff, D.E. Snyder, and J.S. Hagihara. 1984. "An Ecological Land Classification Framework for the United States." USDA Forest Service, Misc. Pub. 1439, Washington, DC.

Duever, L.C., and R.F. Noss. 1990. A computerized method of priority ranking for natural areas. Pages 22–33 in R.S. Mitchell, C.J. Sheviak, and D.J. Leopold, eds. "Ecosystem Management: Rare Species and Significant Habitats." Proceedings of the 15th Annual Natural Areas Conference. Bulletin No. 471. New York State Museum, Albany, NY.

Duffy, D.C., and A.J. Meier. 1992. Do Appalachian herbaceous understories ever recover from clearcutting? *Conservation Biology* 6: 196–201.

Dunlap, T.R. 1988. "Saving America's Wildlife." Princeton University Press, Princeton, NJ.

Durning, A. 1992. "How Much is Enough?" W.W. Norton and Company, New York.

Dyer, M.I., and M.M. Holland. 1991. The biosphere-reserve concept: Needs for a network design. *BioScience* 41: 319–325.

Egan, T. Satellite view shows NW forest damage. *New York Times* News Service, June 12, 1992.

Egler, F. 1977. "The Nature of Vegetation: Its Management and Mismanagement." Aton Forest, Norfolk, CT.

Ehrenfeld, D. 1978. "The Arrogance of Humanism." Oxford University Press, New York.

Ehrenfeld, D. 1988. Why put a value on biodiversity. Pages 212–216 in E.O. Wilson, ed. "Biodiversity." National Academy Press, Washington, DC.

Ehrlich, A.H., and P.R. Ehrlich. 1986. Needed: An endangered humanity act? *Amicus Journal.* Reprinted on pages 298–302 of K.A. Kohm, ed. 1991. "Balancing on the Brink of Extinction: The Endangered Species Act and Lessons for the Future." Island Press, Washington, DC.

Ehrlich, P.R. 1986. "The Machinery of Nature." Simon and Schuster, New York.

Ehrlich, P.R., and A.H. Ehrlich. 1981. "Extinction: The Causes and Consequences of the Disappearance of Species." Random House, New York.

Ehrlich, P.R., and E.O. Wilson. 1991. Biodiversity studies: Science and policy. *Science* 253: 758–762.

Elton, C.S. 1958. "The Ecology of Invasions by Animals and Plants." Methuen, London.

Enge, K.M., and W.R. Marion. 1986. Effects of clearcutting and site preparation on herpetofauna of a north Florida flatwoods. *Forest Ecology and Management* 14: 177–192.

Errington, P.L., and F.N. Hamerstrom. 1937. The evaluation of nesting losses and juvenile mortality of the ring-necked pheasant. *Journal of Wildlife Management* 1: 3–20.

Erwin, T.L. 1982. Tropical forests: Their richness in Coleoptera and other arthropod species. *Coleopterists' Bulletin* 36: 74–75.

Fahey, T.J., and D.H. Knight. 1986. Lodgepole pine ecosystems. *BioScience* 36: 610–617.

Fahrig, L., and G. Merriam. 1985. Habitat patch connectivity and population survival. *Ecology* 66: 1762–1768.

Fahrig, L., and J. Paloheimo. 1988. Effect of spatial arrangement of habitat patches on local population size. *Ecology* 69: 468–475.

FAO (Food and Agricultural Organization of the United Nations). 1988. Current fisheries statistics. FAO, Rome.

Ferguson, D. and N. Ferguson. 1983. "Sacred Cows at the Public Trough." Maverick Publications, Bend, OR.

Ferreras, P., J.J. Aldama, J.F. Beltran, and M. Delibes. 1992. Rates and causes of mortality in a fragmented population of Iberian lynx *Felis pardina* Temminck, 1824. *Biological Conservation* 61: 197–202.

Findley, R. 1990. Will we save our own? *National Geographic* 178(3): 106–136.

Findley, R.W. and D.A. Farber. 1992. "Environmental Law in a Nutshell," Third Edition. West Publishing Co., St. Paul, MN.

Fleischner, T.L. 1992. Preservation is not enough—the need for courage in wilderness management. Pages 236–253 in S.I. Zeveloff, L.M. Vause, and W.H. McVaugh, eds., "Wilderness Tapestry: An Eclectic Approach to Preservation." University of Nevada Press, Reno, NV.

Foreman, D. 1991. The new conservation movement. *Wild Earth* 1(2): 6–12.

Foreman, D., and H. Wolke. 1989. "The Big Outside." Ned Ludd Books, Tucson, AZ.

Foreman, D., J. Davis, D. Johns, R. Noss, and M. Soulé. 1992. The Wildlands Project mission statement. *Wild Earth* (Special Issue): 3–4.

Forman, R.T.T. 1981. Interactions among landscape elements: A core of landscape ecology. Pages 35–48 in S.P. Tjallingii and A.A. de Veer, eds. "Perspectives in Landscape Ecology." Center for Agricultural Publication and Documentation, Wageningen, The Netherlands.

Forman, R.T.T., and M. Godron. 1981. Patches and structural components for a landscape ecology. *BioScience* 31: 733–740.

Forman, R.T.T., and M. Godron. 1986. "Landscape Ecology." John Wiley and Sons, New York.

Forman, R.T.T., A.E. Galli, and C.F. Leck. 1976. Forest size and avian diversity in New Jersey woodlots with some land use implications. *Oecologia* 26: 1–8.

Foster, M.L., and S.R. Humphrey. 1991. Effectiveness of wildlife crossing structures on Alligator Alley (I–75) for reducing animal/auto collisions. Report to Florida Game and Fresh Water Fish Commission and Florida Department of Transportation. Tallahassee, FL.

Fox, S.R. 1981. "John Muir and His Legacy: The American Conservation Movement." Little, Brown and Co., Boston, MA.

Frankel, O.H., and M.E. Soulé. 1981. "Conservation and Evolution." Cambridge University Press. Cambridge, UK.

Franklin, J.F. 1992. Scientific basis for New Perspectives in forests and streams. Pages 25–72 in R.J. Naiman, ed. "Watershed Management: Balancing Sustainability and Environmental Change." Springer-Verlag, New York.

Franklin, J.F., and R.T.T. Forman. 1987. Creating landscape patterns by forest cutting: Ecological consequences and principles. *Landscape Ecology* 1: 5–18.

Franklin, J.F., and T.A. Spies. 1991. Composition, function, and structure of old-growth Douglas-fir forests. Pages 71–80 in L.F. Ruggiero, K.B. Aubry, A.B. Carey, and M.H. Huff, tech. coordinators. "Wildlife and Vegetation of Unmanaged Douglas-fir Forests." USDA Forest Service, Pacific Northwest Research Station, Portland, OR.

Franklin, J.F., K. Cromack, W. Denison, A. McKee, C. Maser, J. Sedell, F. Swanson, and G. Juday. 1981. "Ecological Characteristics of Old-growth Douglas-Fir Forests." Gen. Tech. Rep. PNW-118. USDA Forest Service, Pacific Northwest Forest and Range Experiment Station, Portland, OR.

Franklin, J.F., D.A. Perry, T.D. Schowalter, M.E. Harmon, A. McKee, and T.A. Spies. 1989. Importance of ecological diversity in maintaining long-term site productivity. Pages 82–97 in D.A. Perry, R. Meurisse, B. Thomas, R. Miller, J. Boyle, J. Means, C.R. Perry, and R.F. Powers, eds. "Maintaining Long-term Productivity of Pacific Northwest Forest Ecosystems." Timber Press, Portland, OR.

Frazer, N. 1992. Sea turtle conservation and halfway technology. *Conservation Biology* 6: 179–184.

Freemark, K.E., and H.G. Merriam. 1986. Importance of area and habitat heterogeneity to bird assemblages in temperate forest fragments. *Biological Conservation* 36: 115–141.

Frelich, L.E., and C.G. Lorimer. 1985. Current and predicted long-term effects of deer browsing in hemlock forests in Michigan, USA. *Biological Conservation* 34: 99–120.

Friedel, M.H. 1991. Range condition assessment and the concept of thresholds. A Viewpoint. *Journal of Range Management* 44:422–426.

Frissell, C.A. 1992. New strategies for watershed restoration and recovery of salmon in the Pacific Northwest. Unpublished report. Pacific Rivers Council, Eugene, OR.

Frissell, C.A., R.K. Nawa, and R. Noss. 1992. Is there any conservation biology in "New Perspectives?": A response to Salwasser. *Conservation Biology* 6: 461–464.

Futuyma, D.J. 1979. "Evolutionary Biology." Sinauer Associates, Sunderland, MA.

Garland, T., and W.G. Bradley. 1984. Effects of a highway on Mojave Desert rodent populations. *American Midland Naturalist* 111: 47–56.

Gast, W.R., D.W. Scott, C. Schmitt, D. Clemens, S. Howes, C.G. Johnson, R. Mason, F. Mohr, and R.A. Clapp. 1991. "Blue Mountains Forest Health Report: New Perspectives in Forest Health." USDA Forest Service, Portland, OR.

Gates, J.E., and L.W. Gysel. 1978. Avian nest dispersion and fledgling success in field-forest ecotones. *Ecology* 59: 871–883.

Geist, V. 1971. "Mountain Sheep: A Study in Behavior and Evolution." University of Chicago Press, Chicago, IL.

Gentry, A.W. 1986. Endemism in tropical versus temperate plant communities. Pages 153–181 in M.E. Soulé, ed. "Conservation Biology: The Science of Scarcity and Diversity." Sinauer, Sunderland, MA.

Gentry, A.H. 1992. Tropical forest biodiversity: Distributional patterns and their conservational significance. *Oikos* 63: 19–28.

Gifford, G.F. 1984. Vegetation allocation for meeting site requirements. Pages 35–116 in "Developing Strategies for Rangeland Management." National Research Council/National Academy of Sciences. Westview Press, Boulder, CO.

Gilbert, L.E. 1980. Food web organization and the conservation of neotropical diversity. Pages 11–33 in M.E. Soulé and B.A. Wilcox, eds. "Conservation Biology: An Evolutionary–Ecological Perspective." Sinauer, Sunderland, MA.

Gilpin, M.E., and I. Hanski (eds.). 1991. "Metapopulaton Dynamics: Empirical and Theoretical Investigations." Linnaean Society of London and Academic Press, London.

Gilpin, M.E., and M.E. Soulé. 1986. Minimum viable populations: Processes of species extinction. Pages 19–34 in M.E. Soulé ed. "Conservation Biology: The Science of Scarcity and Diversity." Sinauer, Sunderland, MA.

Gleason, H.A. 1926. The individualistic concept of the plant association. *Bulletin of the Torrey Botanical Club* 43: 463–481.

Glick, D., M. Carr, and B. Harting. 1991. "An Environmental Profile of the Greater Yellowstone Ecosystem." Greater Yellowstone Coalition, Bozeman, MT.

Godfrey, D. 1993. Opinion. *The Monitor* 12(4): 2.

Godron, M., and R.T.T. Forman. 1983. Landscape modification and changing ecological characteristics. Pages 12–28 in H.A. Mooney and M. Godron, eds. "Disturbance and Ecosystems." Springer-Verlag, Berlin.

Goldstein, B.E. 1992. Can ecosystem management turn an administrative patchwork into a Greater Yellowstone Ecosystem? *Northwest Environmental Journal* 8: 285–324.

Goodland, R., H.E. Daly, and S.E. Serafy (eds.). 1992. "Population, Technology, and Lifestyle." Island Press, Covelo, CA.

Goodson, N.J. 1982. Effects of domestic sheep grazing on bighorn sheep populations: a review. *Proceedings Biennial Symposium of Northern Wild Sheep and Goat Council* 3: 287–313.

Gore, A. 1992. "Earth in the Balance: Ecology and the Human Spirit." Houghton Mifflin, Boston, MA.

Gould, S.J. 1993. A special fondness for beetles. *Natural History* 102(1): 4–12.

Graecen, S. (ed.). 1991. A citizen's guide to the timber industry and a profile of U.S. forests. *Forest Watch* 12(1): 1–38.

Graham, R.W. 1986. Response of mammalian communities to environmental changes during the Late Quaternary. Pages 300–313 in J. Diamond and T.J. Case, eds. "Community Ecology." Harper and Row, New York.

Gray, J.G. 1984. "Re-thinking American Education." Wesleyan University Press, Middletown, CT.

Gregory, S.V., F.J. Swanson, W.A. McKee, and K.W. Cummins. 1991. An ecosystem perspective on riparian zones. *BioScience* 41: 540–551.

Grimm, E.C. 1984. Fire and other factors controlling the Big Woods vegetation of Minnesota in the mid-nineteenth century. *Ecological Monographs* 54: 291–311.

Grove, R.H. 1992. Origins of western environmentalism. *Scientific American*, July 1992: 42–47.

Grumbine, R.E. 1990a. Viable populations, reserve size, and federal lands management: A critique. *Conservation Biology* 4: 127–134.

Grumbine, R.E. 1990b. Protecting biological diversity through the greater ecosystem concept. *Natural Areas Journal* 10: 114–120.

Grumbine, R.E. 1991. Cooperation or conflict? Interagency relationships and the future of biodiversity for U.S. parks and forests. *Environmental Management* 15: 27–37.

Grumbine, R.E. 1992. "Ghost Bears: Exploring the Biodiversity Crisis." Island Press, Washington, DC.

Grumbine, R.E. 1994. What is ecosystem management? *Conservation Biology* 8.

Habeck, J.R. 1990. Old-growth ponderosa pine-western larch forests in western Montana: Ecology and management. *Northwest Environmental Journal* 6: 271–292.

Hall, E. R. 1981. "The Mammals of North America." Wiley and Sons, New York.

Hammer, K.J. 1986. An on-site study of the effectiveness of the U.S. Forest Service road closure program in Management Situation One grizzly bear habitat, Swan Lake Ranger District, Flathead National Forest, Montana. Unpublished Report.

Hammer, K.J. 1988. Roads revisited: A travelway inventory of the Upper Swan and Lower Swan Geographic Units, Swan Lake Ranger District, Flathead National Forest. Stage 2, Report No. 4.

Hammer, K.J. 1990. A road ripper's guide to the national forests. Pages 6–8 in J. Davis, ed. "Killing Roads: A Citizens' Primer on the Effects and Removal of Roads." Earth First! Biodiversity Project Special Publication. Tucson, AZ.

Hansen, A.J., T.A. Spies, F.J. Swanson, and J.L. Ohmann. 1991. Conserving biodiversity in managed forests. *BioScience* 41: 382–392.

Hansen, A.J., and D.L. Urban. 1992. Avian response to landscape pattern: The role of species' life histories. *Landscape Ecology* 7: 163–180.

Hardin, E.D., and D.L. White. 1989. Rare vascular plant taxa associated with wiregrass (*Aristida stricta*) in the Southeastern United States. *Natural Areas Journal* 9: 234–245.

Hardin, Garrett. 1986. "Filters Against Folly." Penguin Books, New York.

Harris, L.D. 1984. "The Fragmented Forest: Island Biogeography Theory and the Preservation of Biotic Diversity." University of Chicago Press, Chicago, IL.

Harris, L.D. 1989. The faunal significance of fragmentation of southeastern bottomland forests. Pages 126–134 in D.D. Hook and R. Lea, eds. Proceedings of the Symposium: "The Forested Wetlands of the Southern United States." General Technical Report SE-50. USDA Forest Service, Southeastern Forest Experiment Station, Asheville, NC.

Harris, L.D. 1990. An Everglades regional biosphere reserve. *Florida Fish and Wildlife News* 4(2): 12–15.

Harris, L.D., and P.B. Gallagher. 1989. New initiatives for wildlife conservation: The need for movement corridors. Pages 11–34 in G. MacKintosh, ed. "Preserving Communities and Corridors." Defenders of Wildlife, Washington, DC.

Harris, T. 1991. "Death in the Marsh." Island Press, Covelo, CA.

Harrison, R.L. 1992. Toward a theory of inter-refuge corridor design. *Conservation Biology* 6: 293–295.

Harrison, S. 1987. Treefall gaps versus forest understory as environments for a defoliating moth on a tropical forest shrub. *Oecologia* 72: 65–68.

Hartshorn, G.S. 1978. Treefalls and tropical forest dynamics. Pages 617–628 in P.B. Tomlinson and M.H. Zimmerman, eds. "Tropical Trees as Living Systems." Cambridge University Press, New York.

Harvey, H. T., H.S. Shellhammer, and R.E. Stecker. 1980. "Giant Sequoia Ecology—Fire and Reproduction." Scientific Monograph Series No. 12. U.S. Department of the Interior, National Park Service, Washington, DC.

Haskell, B.D., B.G. Norton, and R. Constanza. 1992. Introduction: What is ecosystem health and why should we worry about it? Pages 3–20 in R. Costanza, B.G. Norton, and B.D. Haskell, eds. "Ecosystem Health—New Goals for Environmental Management." Island Press, Covelo, CA.

Hastings, J.R. and R.M. Turner. 1965. "The Changing Mile: An Ecological Study of Vegetation Changes with Time in the Lower Mile of an Arid and Semi-arid Region." University of Arizona Press, Tucson, AZ.

Haynes, R.W. 1990. An Analysis of the Timber Situation in the United States: 1989–2040. Gen. Tech. Rep. RM-199. USDA Forest Service, Rocky Mountain Forest and Range Experiment Station, Ft. Collins, CO.

Heady, H.F. (ed.) 1988. The Vale Rangeland Rehabilitation Program: An Evaluation. USDA Forest Service, Resource Bulletin PNW-RB-157, 151 pp.

Heady, H.F., and J. Bartolome. 1977. The Vale Rangeland Rehabilitation Program: The Desert Repaired in Southeastern Oregon. USDA Forest Service, Resource Bulletin PHW–70, 139 pp.

Heinselman, M.L. 1973. Fire in the virgin forests of the Boundary Waters Canoe Area, Minnesota. *Quaternary Research* 3: 329–382.

Hellgren, E.C., M.R. Vaughan, and D.F. Stauffer. 1991. Microhabitat use by black bears in a southeastern wetland. *Journal of Wildlife Management* 55: 442–448.

Henderson, M.T., G. Merriam, and J. Wegner. 1985. Patchy environments and species survival: Chipmunks in an agricultural mosaic. *Biological Conservation* 31: 95–105.

Henein, K., and G. Merriam. 1990. The elements of connectivity where corridor quality is variable. *Landscape Ecology* 4: 157–170.

High Country News. 1987. "Western Water Made Simple." Island Press, Covelo, CA.

Hobbs, R.J., and L.F. Huenneke. 1992. Disturbance, diversity, and invasion: implications for conservation. *Conservation Biology* 6(3):324–337.

Hoffecker, J.F., W.R. Powers, and T. Goebel. 1993. The colonization of Beringia and the peopling of the New World. *Science* 259:46–53.

Holechek, J.L., R.D. Pieper, and C.H. Herbel. 1989. "Range Management—Principles and Practices." Prentice Hall, Englewood Cliffs, NJ.

Holland, R.F. 1987. Is *Quercus lobata* a rare plant? Approaches to conservation of rare plant communities that lack rare plant species. Pages 129–132 in T.S. Elias, ed. "Conservation and Management of Rare and Endangered Plants." California Native Plant Society, Sacramento, CA.

Holling, C.S. (ed.). 1978. "Adaptive Environmental Assessment and Management." John Wiley and Sons, New York.

Holsinger, K.E., and L.D. Gottlieb. 1991. Conservation of rare and endangered plants: Principles and prospects. Pages 195–208 in D.A. Falk and K.E. Holsinger, eds. "Genetics and Conservation of Rare Plants." Oxford University Press, New York.

Hopwood, D. 1991. "Principles and Practices of New Forestry." Land Management Report 71. Ministry of Forests, Victoria, British Columbia.

Hough, A.F. 1965. A twenty-year record of understory vegetation change in a virgin Pennsylvania forest. *Ecology* 46: 370–373.

Hough, J. 1988. Biosphere reserves: myth and reality. *Endangered Species Update* 6(1 and 2): 1–4.

Howe, R.W., G.J. Davis, and V. Mosca. 1991. The demographic significance of "sink" populations. *Biological Conservation* 57: 239–255.

Hudson, W.E. (ed.). 1991. "Landscape Linkages and Biodiversity." Defenders of Wildlife and Island Press, Washington, DC.

Hummel, M. 1990. "A Conservation Strategy for Large Carnivores in Canada." World Wildlife Fund Canada, Toronto.

Humphreys, W.F., and D.J. Kitchener. 1982. The effect of habitat utilization on species-area curves: Implications for optimal reserve area. *Journal of Biogeography* 9: 391–396.

Hunt, C.E. 1989. Creating an Endangered Ecosystems Act. *Endangered Species Update* 6(3–4): 1–5.

Hunter, M.L. 1990. "Wildlife, Forests, and Forestry." Prentice Hall, Englewood Cliffs, NJ.

Hunter, M.L. 1991. Coping with ignorance: The coarse-filter strategy for maintaining biodiversity. Pages 266–281 in K.A. Kohm, ed. "Balancing on the Brink of Extinction: The Endangered Species Act and Lessons for the Future." Island Press, Washington, DC.

Hunter, M.L., and A. Calhoun. 1994. A triad approach to land use allocation. In R. Szaro, ed. "Biodiversity in Managed Landscapes." Oxford University Press, New York.

Hunter, M.L., G.L. Jacobson, and T. Webb. 1988. Paleoecology and the coarse-filter approach to maintaining biological diversity. *Conservation Biology* 2: 375–385.

Hurst, J. 1990. Education for social change. Pages 198–201 in B. Erickson, ed. "Call to Action—Handbook for Ecology, Peace and Justice." Sierra Club Books, San Francisco, CA.

Huston, M. 1979. A general hypothesis of species diversity. *American Naturalist* 113: 81–101.

Hutchinson, G.E. 1957. "A Treatise on Limnology," Wiley, New York.

Hutto, R.L. S. Reel, and P.B. Landres. 1987. A critical evaluation of the species approach to biological conservation. *Endangered Species Update* 4(12): 1–4.

Ingram, P. 1991. "Monitoring Summaries, Siuslaw National Forest, 1987–1989." Oregon Department of Fish and Wildlife, Corvallis, OR.

International Union for the Conservation of Nature and Natural Resources (IUCN). 1980. "World Conservation Strategy." Gland, Switzerland.

Jackson, D.R., and E.G. Milstrey. 1989. The fauna of gopher tortoise burrows. Pages 86–98 in J.E. Diemer, D.R. Jackson, J.L. Landers, J.N. Layne, and D.A. Wood, eds. "Gopher Tortoise Relocation Symposium Proceedings." Nongame Wildlife Program Technical Report No. 5. Florida Game and Fresh Water Fish Commission, Tallahassee, FL.

Jackson, J.A. 1986. Biopolitics, management of federal lands, and the conservation of the red-cockaded woodpecker. *American Birds* 40: 1162–1168.

Jacobs, L. 1991. "Waste of the West: Public Lands Ranching." Lynn Jacobs, PO Box 5784, Tucson, AZ 85703.

Jacobson, S.K. 1990. Graduate education in conservation biology. *Conservation Biology* 4: 431–440.

Janzen, D.H. 1983. No park is an island: Increase in interference from outside as park size decreases. *Oikos* 41: 402–410.

Janzen, D.H. 1986. The eternal external threat. Pages 286–303 in M.E. Soulé, ed. "Conservation Biology: The Science of Scarcity and Diversity." Sinauer, Sunderland, MA.

Jenkins, R.E. 1985. Information methods: Why the heritage programs work. *Nature Conservancy News* 35(6): 21–23.

Jenkins, R.E. 1988. Information management for the conservation of biodiversity. Pages 231–239 in E.O. Wilson, ed. "Biodiversity." National Academy Press, Washington, DC.

Jensen, D.B., M. Torn, and J. Harte. 1990. In our own hands: a strategy for conserving biological diversity in California. California Policy Seminar, Research Report.

Jensen, M.N., and P.R. Krausman. 1993. *Conservation Biology's* literature: New wine or just a new bottle? *Wildlife Society Bulletin* 21: 199–203.

Johnson, A.S. 1989. The thin green line—riparian corridors and endangered species in Arizona and New Mexico. Pages 35–46 in G. Mackintosh, ed. "Preserving Communities and Corridors." Defenders of Wildlife, Washington, DC.

Johnson, H.B., and H.S. Mayeux. 1992. Viewpoint: a view on species additions and deletions and the balance of nature. *Journal of Range Management* 45(4): 322–333.

Johnson, J.E., and B.L. Jensen. 1991. Hatcheries for endangered freshwater fishes. Pages 199–217 in W.L. Minckley and J.E. Deacon, eds. "Battle Against Extinction: Native Fish Management in the American West." University of Arizona Press, Tucson, AZ.

Johnson, K.N., J.F. Franklin, J.W. Thomas, and J. Gordon. 1991. Alternatives for management of late-successional forests of the Pacific Northwest: A report to the U.S. House of Representatives. Oregon State University, Corvallis, OR.

Johnson, M.P., and P.H. Raven. 1973. Species number and endemism: The Galapagos Archipelago revisited. *Science* 179: 893–895.

Johnson, M.P., and D.S. Simberloff. 1974. Environmental determinants of island species numbers in the British Isles. *Journal of Biogeography* 1: 149–154.

Johnson, N.K. 1975. Controls on number of bird species on montane islands in the Great Basin. *Evolution* 29: 545–567.

Johnson, R.G., and S.A. Temple. 1990. Nest predation and brood parasitism of tallgrass prairie birds. *Journal of Wildlife Management* 54: 106–111.

Johnson, W.C., and C.S. Adkisson. 1985. Dispersal of beech nuts by blue jays in fragmented landscapes. *American Midland Naturalist* 113: 319–324.

Johnston, D.W., and E.P. Odum. 1956. Breeding bird populations in relation to plant succession on the piedmont of Georgia. *Ecology* 37: 50–62.

Johnston, V.R. 1947. Breeding birds of the forest edge in east-central Illinois. *Condor* 49: 45–53.

Jones, B. 1986. Inventory and monitoring process. Pages 1–10 in A.Y. Cooperrider, R.J. Boyd, and H.R. Stuart, eds. "Inventory and Monitoring of Wildlife Habitat." USDI Bureau of Land Management, Washington, DC.

Jontz, J. 1993. The Sustainable Ecosystems Act. Draft report. Silver Lake, IN.

Jordan, D.B. 1991. A Proposal to Establish a Captive Breeding Population of Florida Panthers. Draft Supplemental Environmental Assessment. U.S. Fish and Wildlife Service, Gainesville, FL.

Jordan, W.R., III, M.E. Gilpin, and J.D. Aber. 1987. "Restoration Ecology—A Synthetic Approach to Ecological Research." Cambridge University Press, New York.

Judy, R.D., Jr. , T.M. Murray, S.C. Svirsky, M.R. Whitworth, and L. S. Ischinger. 1984. 1982 national fisheries survey. Volume 1. FWS/OBS-84/06. U.S. Fish and Wildlife Service, Washington, DC.

Karr, J.R. 1976. Within- and between-habitat avian diversity in African and neotropical lowland habitats. *Ecological Monographs* 46: 457–481.

Karr, J.R. 1982. Population variability and extinction in the avifauna of a tropical land bridge island. *Ecology* 63: 1975–1978.

Karr, J.R., and R.R. Roth. 1971. Vegetation structure and avian diversity in several New World areas. *American Naturalist* 105: 423–435.

Kartesz, J.T. 1992. Preliminary counts for native vascular plant species of U.S. states and Canadian provinces. *Biodiversity Network News* 5(3): 6.

Keiter, R.B. 1989. Taking account of the ecosystem on the public domain: Law and ecology in the Greater Yellowstone Ecosystem. *University of Colorado Law Review* 60: 923–1007.

Kellogg, E. (ed.). 1992. "Coastal Temperate Rain Forests: Ecological Characteristics, Status, and Distribution Worldwide." Ecotrust and Conservation International, Portland, OR, and Washington, DC.

Kendeigh, S.C., H.I. Baldwin, V.H. Cahalane, C.H.D. Clarke, C. Cottam, I.M. Cowan, P. Dansereau, J.H. Davis, F.W. Emerson, I.T. Haig, A. Hayden, C.L. Hayward, J.M. Linsdale, J.A. MacNab, and J.E. Potzger. 1950–51. Nature sanctuaries in the United States and Canada: A preliminary inventory. *The Living Wilderness* 15(35): 1–45.

Kepler, C.B., and J.M. Scott. 1985. Conservation of island ecosystems. Pages 255–271 in P.O. Moors, ed. "Conservation of Island Birds." International Council for Bird Preservation Technical Publication 3.

Kessler, W.B., H. Salwasser, C.W. Cartwright, and J.A. Caplan. 1992. New perspectives for sustainable natural resources management. *Ecological Applications* 2: 221–225.

Keystone Center, The. 1991. "Final Consensus Report of the Keystone Policy Dialogue on Biological Diversity on Federal Lands." The Keystone Center, Keystone, CO.

Kiester, A.R. 1971. Species density of North American amphibians and reptiles. *Systematic Zoology* 20: 127–157.

Kirkpatrick, J.B. 1983. An iterative method for establishing priorities for the selection of nature reserves: An example for Tasmania. *Biological Conservation* 25: 127–134.

Koch, P. 1991. Wood vs. non-wood materials in U.S. residential construction: Some energy-related international implications. Working Paper 36. Center for International Trade in Forest Products. University of Washington, Seattle, WA.

Kohm, K.A. (ed.). 1991. "Balancing on the Brink of Extinction: The Endangered Species Act and Lessons for the Future." Island Press, Washington, DC.

Kosztarab, M. 1986. Prefatory comments: Some of the activities leading to this symposium. Pages 23–27 in K.C. Kim and L. Knutson, eds. "Foundations for a National Biological Survey." Association of Systematic Collections, Lawrence, KS.

Kovalchik, B.L., and W. Elmore. 1992. Effects of cattle grazing systems on willow-dominated plant associations in central Oregon. Pages 111–119 in "Proceedings—Symposium on Ecology and Management of Riparian Shrub

Communities." U.S. Department of Agriculture, Forest Service, Intermountain Research Station, General Technical Report INT-289.
Kozlovsky, D.G. 1974. "An Ecological and Evolutionary Ethic." Prentice-Hall, Englewood Cliffs, NJ.
Kufeld, R.C., O.C. Wallmo, and C. Feddema. 1973. Foods of the Rocky Mountain mule deer. USDA, Forest Service, Research Paper RM-111.
Kushlan, J.A. 1979. Design and management of continental wildlife reserves: Lessons from the Everglades. *Biological Conservation* 15: 281–290.
Kushlan, J.A. 1983. Special species and ecosystem preserves: Colonial water birds in US national parks. *Environmental Management* 7: 201–207.
LaBounty, J. 1986. Lakes. Pages 237–253 in A.Y. Cooperrider, R.J. Boyd, and H.R. Stuart, eds. "Inventory and Monitoring of Wildlife Habitat." USDI Bureau of Land Management, Washington, DC.
Lack, D. 1976. "Island Biology: Illustrated by the Land Birds of Jamaica." Blackwell, Oxford, England.
Lalo, J. 1987. The problem of road kill. *American Forests* Sept./Oct.: 50–53,72.
Lande, R. 1988. Genetics and demography of biological conservation. *Science* 241: 1455–1460.
Landres, P.B., J. Verner, and J.W. Thomas. 1988. Ecological uses of vertebrate indicator species: a critique. *Conservation Biology* 2:316–328.
Langton, T.E.S. (ed.). 1989. "Amphibians and Roads." ACO Polymer Products, Ltd., Shefford, UK.
Lattin, J.D. 1990. Arthropod diversity in Northwest old-growth forests. *Wings* 15(2): 7–10.
Lauenroth, W.K. and D.P. Coffin. 1992. Belowground processes and the recovery of semiarid grassland from disturbance. Pages 131–150 in M.K. Wali, ed. "Ecosystem Rehabilitation: Preamble to Sustainable Development." Volume 2: "Ecosystem Analysis and Synthesis." SPB Academic Publishing, The Hague, The Netherlands.
Lawrence, N., and D. Murphy. 1992. New perspectives or old priorities? *Conservation Biology* 6: 465–468.
Lay, D. 1938. How valuable are woodland clearings to birdlife? *Wilson Bulletin* 50: 254–256.
Laycock, W.A. 1991. Stable states and thresholds of range condition on North American rangelands: a viewpoint. *Journal of Range Management* 44(5):427–433.
Ledig, F.T. 1986. Heterozygosity, heterosis, and fitness in outbreeding plants. Pages 77–104 in M.E. Soulé, ed. "Conservation Biology: The Science of Scarcity and Diversity." Sinauer, Sunderland, MA.
Leopold, A. 1933. "Game Management." Charles Scribners Sons, New York.
Leopold, A. 1941a. Wilderness as a land laboratory. *Living Wilderness* 6(July): 3.
Leopold, A. 1941b. Lakes in relation to terrestrial life patterns. Pages 17–22 in J.G. Needham and others, A Symposium on Hydrobiology.
Leopold, A. 1949. "A Sand County Almanac." Oxford University Press, New York.

Leopold, A. 1953. "Round River." Oxford University Press, New York.

Leopold, A. 1990 (1936). Means and ends in wildlife management. *Environmental Ethics* 12: 329–332.

Leopold, L. 1990. Ethos, equity, and the water resource. *Environment* 32(2): 16–19, 37–42.

Levin, S.A. 1993. Forum: Grazing theory and rangeland management. *Ecological Applications* 3: 1.

Levins, R. 1970. Extinction. Pages 77–107 in M. Gerstenhaber, ed. "Some Mathematical Questions in Biology. Lectures on Mathematics in the Life Sciences," Vol. 2. American Mathematical Society, Providence, RI.

Linhart, Y.B., P. Feinsinger, J.H. Beach, W.H. Busby, K.G. Murray, W.Z. Pounds, S. Kinsman, C.A. Guindon, and M. Kooiman. 1987. Disturbance and predictability of flowering patterns in bird-pollinated cloud forest plants. *Ecology* 68: 1696–1710.

Lopez, B. 1992. The rediscovery of North America. *Orion* 11(3): 10–16.

Lorimer, C.G. 1989. Relative effects of small and large disturbances on temperate forest structure. *Ecology* 70: 565–567.

Lowe, D.W., J.R. Matthews, and C.J. Moseley. 1990. "The Official World Wildlife Fund Guide to Endangered Species." Beacham Publishing, Inc., Washington, DC.

Lynch, J.F. 1987. Responses of breeding bird communities to forest fragmentation. Pages 123–140 in D.A. Saunders, G.W. Arnold, A.A. Burbidge, and A.J.M. Hopkins, eds. "Nature Conservation: The Role of Remnants of Native Vegetation." Surrey Beatty and Sons, Chipping Norton, NSW , Australia.

Lyon, L.J. 1983. Road density models describing habitat effectiveness for elk. *Journal of Forestry* 81: 592–595.

MacArthur, R.H. 1972. "Geographical Ecology: Patterns in the Distribution of Species." Princeton University Press, Princeton, NJ.

MacArthur, R.H., and J.W. MacArthur. 1961. On bird species diversity. *Ecology* 42: 594–598.

MacArthur, R.H., J.W. MacArthur, and J. Preer. 1962. On bird species diversity—Prediction of bird censuses from habitat measurements. *American Naturalist* 96: 167–174.

MacArthur, R.H., and E.O. Wilson. 1963. An equilibrium theory of insular zoogeography. *Evolution* 17: 373–387.

MacArthur, R.H., and E.O. Wilson. 1967. "The Theory of Island Biogeography." Princeton University Press, Princeton, NJ.

Machlis, G.E., and D.L. Tichnell. 1985. "The State of the World's Parks: An International Assessment for Resource Management, Policy, and Research." Westview Press, Boulder, CO.

Mack, R.N. 1986. Alien plant invasions into the Intermountain West: A case history. Pages 191–213 in H.A. Mooney and J.A. Drake, eds. "Ecology of

Biological Invasions of North America and Hawaii." Springer-Verlag, New York.

Mack, R.N. and J.N. Thompson. 1982. Evolution in steppe with few large, hooved mammals. *American Naturalist* 119: 757–773.

MacMahon, J.A. 1986. Disturbed lands and ecological theory: An essay about a mutualistic association. Pages 221–237 in W.R. Jordan III, M.E. Gilpin, and J.D. Aber, eds. "Restoration Ecology: A Synthetic Approach to Ecological Research." Cambridge University Press, New York.

Mader, H.J. 1984. Animal habitat isolation by roads and agricultural fields. *Biological Conservation* 29: 81–96.

Mader, H.J., C. Schell, and P. Kornacker. 1990. Linear barriers to movements in the landscape. *Biological Conservation* 54: 209–222.

Madson, C. 1987. Down from the mountains. *Nature Conservancy Magazine* 37(4): 5–11.

Maehr, D.S. 1990. The Florida panther and private lands. *Conservation Biology* 4: 167–170.

Mangum, F. 1986. Macroinvertebrates. Pages 661–675 in A.Y. Cooperrider, R.J. Boyd, and H.R. Stuart, eds. "Inventory and Monitoring of Wildlife Habitat." USDI Bureau of Land Management, Washington, DC.

Mann, C.C., and M.L. Plummer. 1993. The high cost of biodiversity. *Science* 260: 1868–1871.

Margules, C.R. 1989. Introduction to some Australian developments in conservation evaluation. *Biological Conservation* 50: 1–11.

Margules, C.R., A.J. Higgs, and R.W. Rafe. 1982. Modern biogeographic theory: Are there any lessons for nature reserve design? *Biological Conservation* 24: 115–128.

Margules, C.R., A.O. Nicholls, and R.L. Pressey. 1988. Selecting networks of reserves to maximize biological diversity. *Biological Conservation* 43: 63–76.

Margules, C., and M.B. Usher. 1981. Criteria used in assessing wildlife conservation potential: A review. *Biological Conservation* 24: 115–128.

Marsh, G.P. 1864. "Man and Nature." (reprint 1965). Harvard University Press, Cambridge, MA.

Marston, E. 1987. Reworking the Colorado River Basin. Pages 199–210 in E. Marston, ed. "Western Water Made Simple." Island Press, Covelo, CA.

Martin, P.S. 1967. Prehistoric overkill. Pages 75–120 in P.S. Martin and H.E. Wright, eds. "Pleistocene Extinctions: The Search for a Cause." Yale University Press, New Haven, CT.

Martin, P.S. 1970. Pleistocene niches for alien animals. *BioScience* 20: 218–221.

Martin, P.S. 1973. The discovery of America. *Science* 180: 969–974.

Martin, P.S., and R.G. Klein, eds. 1984. "Quaternary Extinctions: A Prehistoric Revolution." University of Arizona Press, Tucson, AZ.

Martin, T.E., and J.R. Karr. 1986. Patch utilization by migrating birds: Resource oriented? *Ornis Scandinavica* 17: 165–174.

Martin, W.H. 1992. Characteristics of old-growth mixed mesophytic forests. *Natural Areas Journal* 12: 127–135.

Maser, C. 1986. "The Redesigned Forest." R. & E. Miles, San Pedro, CA.

Maser, C. 1989. "Forest Primeval: The Natural History of an Ancient Forest." Sierra Club Books, San Francisco, CA.

Maser, C., and J.M. Trappe. 1984. The fallen tree: A source of diversity. Pages 335–339 in "New Forests for a Changing World." Proceedings of the 1983 SAF Convention, Portland, OR, October 16–20. Society of American Foresters, Washington, DC.

Maser, C., R.F. Tarrant, J.M. Trappe, and J.F. Franklin, eds. 1988. "From the Forest to the Sea: A Story of Fallen Trees." General Technical Report PNW-GTR-229. USDA Forest Service. Pacific Northwest Research Station, Portland, OR.

Master, L. 1990. The imperiled status of North American aquatic animals. *Biodiversity Network News* 3(3):1–2,7–8.

Master, L.L. 1991. Assessing threats and setting priorities for conservation. *Conservation Biology* 5: 559–563.

Matthiessen, P. 1959. "Wildlife in America." Viking Press, New York.

Mattson, D.J. 1990. Human impacts on bear habitat use. *International Conference on Bear Research and Management* 8: 33–56.

Mattson, D.J., and R.R. Knight. 1991. Effects of access on human-caused mortality of Yellowstone grizzly bears. USDI National Park Service Interagency Grizzly Bear Study Team Report 1991B.

Mattson, D.J., and M.M. Reid. 1991. Conservation of the Yellowstone grizzly bear. *Conservation Biology* 5: 364–372.

Mattson, D.J., R.R. Knight, and B.M. Blanchard. 1987. The effects of developments and primary roads on grizzly bear habitat use in Yellowstone National Park, Wyoming. *International Conference on Bear Research and Management* 7: 259–273.

Mayer, K.E., and W.F. Laudenslayer. 1988. "A Guide to Wildlife Habitats of California." California Department of Forestry and Fire Protection, Sacramento, CA.

Mayr, E. 1970. "Populations, Species, and Evolution." Belknap Press of Harvard University Press, Cambridge, MA.

McAllister, D.E., S.P. Platania, F.W. Schueler, M.E. Baldwin, and D.S. Lee. 1986. Ichthyofaunal patterns on a geographic grid. Pages 17–51 in C.H. Hocutt and E.O. Wiley, eds. "The Zoogeography of North American Freshwater Fishes." John Wiley and Sons, New York.

McDonald, J.N. 1981. "North American Bison: Their Classification and Evolution." University of California Press, Berkeley, CA.

McIver, J.D., A.R. Moldenke, and G.L. Parsons. 1990. Litter spiders as bioindicators of recovery after clearcutting in a western coniferous forest. *Northwest Environmental Journal* 6: 410–412.

McLellan, B.N., and R.D. Mace. 1985. "Behavior of Grizzly Bears in Response to Roads, Seismic Activity, and People." British Columbia Ministry of Environment, Cranbrook, B.C.

McLellan, B.N., and D.M. Shackleton. 1988. Grizzly bears and resource extraction industries: Effects of roads on behavior, habitat use, and demography. *Journal of Applied Ecology* 25: 451–460.

McNaughton, S.J. 1993. Grasses and grazers, science and management. *Ecological Applications* 3: 17–20.

McNeely, J.A., K.R. Miller, W.V. Reid, R.A. Mittermeier, T.B. Werner. 1990. "Conserving the World's Biological Diversity." IUCN, WRI, CI, WWF-US, World Bank. Gland, Switzerland and Washington, DC.

Mech, L.D., S.H. Fritts, G.L. Radde, and W.J. Paul. 1988. Wolf distribution and road density in Minnesota. *Wildlife Society Bulletin* 16: 85–87.

Meffe, G.K. 1987. Conserving fish genomes: philosophies and practices. *Environmental Biology of Fishes* 18(1):3–9.

Meffe, G.K. 1990. Genetic approaches to conservation of rare fishes: examples from North American desert species. *Journal of Fish Biology* 37 (Supplement A): 105–112.

Meffe, G.K. 1992. Techno-arrogance and halfway technologies: salmon hatcheries on the Pacific Coast of North America. *Conservation Biology* 6: 350–354.

Meine, C. 1988. "Aldo Leopold: His Life and Work." University of Wisconsin Press, Madison, WI.

Meine, C. 1992. Keeper of the cogs. *Defenders* 67(6): 9–17.

Merriam, G. 1988. Landscape dynamics in farmland. *Trends in Ecology and Evolution* 3: 16–20.

Merriam, G. 1991. Corridors and connectivity: Animal populations in heterogeneous environments. Pages 133–142 in D.A. Saunders and R.J. Hobbs, eds. "Nature Conservation 2: The Role of Corridors." Surrey Beatty and Sons, Chipping Norton, NSW , Australia.

Meslow, E.C., C. Maser, and J. Verner. 1981. Old-growth forests as wildlife habitat. *Transactions of the North American Wildlife and Natural Resources Conference* 46: 329–335.

Metzgar, L.H., and M. Bader. 1992. Large mammal predators in the northern Rockies: Grizzly bears and their habitat. *Northwest Environmental Journal* 8: 231–233.

Milchunas, D.G., O.E. Sala, and W.K. Lauenroth. 1988. A generalized model of the effects of grazing by large herbivores on grassland community structure. *American Naturalist* 132(1):87–106.

Miller, G.T., Jr. 1982. "Living in the Environment." Wadsworth Publishing Company, Belmont, CA.

Miller, R.M. 1987. Mycorrhizae and succession. Pages 205–219 in W.R. Jordan III, M.E. Gilpin, and J.D. Aber, eds. "Restoration Ecology: A Synthetic Approach to Ecological Research." Cambridge University Press, New York.

Miller, R.R., J.D. Williams, and J.E. Williams. 1989. Extinctions of North American fishes during the past century. *Fisheries* 14(6):22–38.
Miller, S.G., S.P. Bratton, and J. Hadidian. 1992. Impacts of white-tailed deer on endangered and threatened vascular plants. *Natural Areas Journal* 12: 67–74.
Mills, L.S., M.E. Soulé, and D.F. Doak. 1993. The keystone species concept in ecology and conservation. *BioScience* 43: 219–224.
Milstein, M. 1991. A fading Yellowstone "vision." *High Country News* 23(10): 1, 10–11.
Minckley, W.L., 1991. Native fishes of the Grand Canyon region: an obituary? Pages 124–177 in "Colorado River Ecology and Dam Management." National Academy Press, Washington, DC.
Minckley, W.L., and J.E. Deacon (eds.). 1991. "Battle Against Extinction: Native Fish Management in the American West." University of Arizona Press, Tucson, AZ.
Mittermeier, R.A. 1988. Primate diversity and the tropical forest: Case studies from Brazil and Madagascar and the importance of the megadiversity countries. Pages 145–154 in E.O. Wilson, ed. "Biodiversity." National Academy Press, Washington, DC.
Moldenke, A.R., and J.D. Lattin. 1990. Dispersal characteristics of old-growth soil arthropods: The potential for loss of diversity and biological function. *Northwest Environmental Journal* 6: 408–409.
Mooney, H.A. 1988. Lessons from Mediterranean-climate regions. Pages 157–165 in E.O. Wilson, ed. "Biodiversity." National Academy Press, Washington, DC.
Mooney, H.A., and J. Drake. (eds.). 1986. "The Ecology of Biological Invasions of North America and Hawaii." Springer-Verlag, New York.
Mooney, H.A., S.P. Hamburg, and J.A. Drake. 1986. Pages 250–272 in H.A. Mooney and J.A. Drake, eds. "Ecology of Biological Invasions of North America and Hawaii." Springer-Verlag, New York.
Moore, W.H., B.F. Swindel, and W.S. Terry. 1982. Vegetative response to clearcutting and chopping in a north Florida flatwoods forest. *Journal of Range Management* 35: 214–218.
Morris, L.A., W.L. Pritchett, and B.F. Swindel. 1983. Displacement of nutrients into windrows during site preparation of a flatwoods forest. *Journal of the Soil Science Society of America* 47: 591–594.
Morrison, P.H. 1990. "Ancient Forests on the Olympic National Forest: Analysis from a Historical and Landscape Perspective." The Wilderness Society, Washington, DC.
Morrison, P.H., and F.J. Swanson. 1990. "Fire History and Pattern in a Cascade Range Landscape." PNW-GTR-254. USDA Forest Service, Portland, OR.
Moyle, P.B. 1976. Fish introductions in California: history and impact on native fishes. *Biological Conservation* 9:101–118.
Moyle, P.B., and G.M. Sato. 1991. On the design of preserves to protect native fishes. Pages 155–169 in W.L. Minckley and J.E. Deacon, eds. "Battle Against

Extinction: Native Fish Management in the American West." University of Arizona Press, Tucson, AZ.

Moyle, P.B., and J.E. Williams. 1990. Biodiversity loss in the temperate zone: Decline of the native fish fauna of California. *Conservation Biology* 4: 475–484.

Mueller-Dombois, D. 1987. Natural dieback in forest. *BioScience* 37: 575–583.

Murray, D.F. 1992. Vascular plant diversity in Alaskan arctic tundra. *Northwest Environmental Journal* 8: 29–52.

Mutch, R.W. 1970. Wildland fires and ecosystems: A hypothesis. *Ecology* 51: 1046–1051.

Myers, R.L. 1985. Fire and the dynamic relationship between Florida sandhill and sand pine scrub vegetation. *Bulletin of the Torrey Botanical Club* 112: 241–252.

Myers, R.L. 1990. Scrub and high pine. Pages 150–193 in R.L. Myers and J.J. Ewel, eds. "Ecosystems of Florida." University of Central Florida Press, Orlando, FL.

Nabhan, G.P. 1982. "The Desert Smells Like Rain: A Naturalist in Papago Country." North Point Press, San Francisco, CA.

Naess, A. 1989. "Ecology, Community and Lifestyle: Outline of an Ecosophy." Translated and revised by David Rothenberg. Cambridge University Press, Cambridge, UK.

Naiman, R.J., C.A. Johnston, and J.C. Kelley. 1988. Alteration of North American streams by beaver. *BioScience* 38: 753–762.

Nash, R.F. 1982 (1967). "Wilderness and the American Mind." Yale University Press, New Haven, CT.

Nash, R.F. 1989. "The Rights of Nature: A History of Environmental Ethics." University of Wisconsin Press, Madison, WI.

National Research Council. 1992. "Restoration of Aquatic Ecosystems: Science, Technology, and Public Policy." Committee on Restoration of Aquatic Ecosystems. National Academy Press, Washington, DC.

Natural Areas Association. 1993. Compendium on Exotic Species (43 articles from the *Natural Areas Journal*). Natural Areas Association, Mukwonago, WI.

Nature Conservancy, The. 1992. Extinct vertebrate species in North America. Unpublished draft list, March 4, 1992. The Nature Conservancy, Arlington, VA.

Nehlsen, W., J.E. Williams, and J.A. Lichatowich. 1991. Pacific salmon at the crossroads: Stocks at risk from California, Oregon, Idaho, and Washington. *Fisheries* 16(2):4–21.

Nelson, J. 1989. Agriculture, wetlands, and endangered species. The Food Security Act of 1985. *Endangered Species Technical Bulletin* 14(5): 1, 6–8.

Neves, R.J. and P.L. Angermeier. 1990. Habitat alteration and its effects on native fishes in the upper Tennessee River system, east-central U.S.A. *Journal of Fish Biology* 37 (Suppl. A):45–52.

Newman, B., H. Irwin, K. Lowe, A. Mostwill, S. Smith, and J. Jones. 1992. Southern Appalachian wildlands proposal. *Wild Earth* (Special Issue): 46–60.

Newmark, W.D. 1985. Legal and biotic boundaries of western North American national parks: A problem of congruence. *Biological Conservation* 33: 197–208.

Newmark, W.D. 1987. A land-bridge island perspective on mammalian extinctions in western North American parks. *Nature* 325: 430–432.

Nichols, G.E. 1913. The vegetation of Connecticut. II. Virgin forests. *Torreya* 13: 199–215.

Niering, W.A. 1987. Vegetation dynamics (succession and climax) in relation to plant community management. *Conservation Biology* 1:287–295.

Nilsson, S.G. 1979. Effect of forest management on the breeding bird community in southern Sweden. *Biological Conservation* 15: 135–143.

Noon, B.R., V.P. Bingman, and J.P. Noon. 1979. The effects of changes in habitat on northern hardwood forest bird communities. Pages 33–48 in R.M. DeGraaf and K.E. Evans, eds. "Management of Northcentral and Northeastern Forests for Nongame Birds." Gen. Tech. Rep. NC-51. USDA Forest Service, St. Paul, MN.

Norris, R. 1992. A prescription for refuges. *Defenders* 67(3): 15–19.

Norse, E.A. 1990. "Ancient Forests of the Pacific Northwest." The Wilderness Society and Island Press, Washington, DC.

Norse, E.A., K.L. Rosenbaum, D.S. Wilcove, B.A. Wilcox, W.H. Romme, D.W. Johnston, and M.L. Stout. 1986. "Conserving Biological Diversity in Our National Forests." The Wilderness Society, Washington, D.C.

Norton, B.G., ed. 1986. "The Preservation of Species: The Value of Biological Diversity." Princeton University Press, Princeton, NJ.

Noss, R.F. 1981. The birds of Sugarcreek, an Ohio nature reserve. *Ohio Journal of Science* 81: 29–40.

Noss, R.F. 1983. A regional landscape approach to maintain diversity. *BioScience* 33: 700–706.

Noss, R.F. 1985. Wilderness recovery and ecological restoration: An example for Florida. *Earth First!* 5(8): 18–19.

Noss, R.F. 1987a. Protecting natural areas in fragmented landscapes. *Natural Areas Journal* 7: 2–13.

Noss, R.F. 1987b. From plant communities to landscapes in conservation inventories: A look at The Nature Conservancy (USA). *Biological Conservation* 41: 11–37.

Noss, R.F. 1987c. Corridors in real landscapes: A reply to Simberloff and Cox. *Conservation Biology* 1: 159–164.

Noss, R.F. 1988. The longleaf pine landscape of the Southeast: Almost gone and almost forgotten. *Endangered Species Update* 5(5): 1–8.

Noss, R.F. 1989. Longleaf pine and wiregrass: Keystone components of an endangered ecosystem. *Natural Areas Journal* 9: 211–213.

Noss, R.F. 1990a. Indicators for monitoring biodiversity: A hierarchical approach. *Conservation Biology* 4: 355–364.

Noss, R.F. 1990b. What can wilderness do for biodiversity? Pages 49–61 in P. Reed, ed. "Preparing to Manage Wilderness in the 21st Century." General

Technical Report SE-66. USDA Forest Service, Southeastern Forest Experiment Station, Asheville, NC.

Noss, R.F. 1991a. From endangered species to biodiversity. Pages 227–246 in K.A. Kohm, ed. "Balancing on the Brink of Extinction: The Endangered Species Act and Lessons for the Future." Island Press, Washington, DC.

Noss, R.F. 1991b. Sustainability and wilderness. *Conservation Biology* 5: 120–121.

Noss, R.F. 1991c. Landscape connectivity: Different functions at different scales. Pages 27–39 in W.E. Hudson, ed. "Landscape Linkages and Biodiversity." Defenders of Wildlife and Island Press, Washington, DC.

Noss, R.F. 1991d. Wilderness recovery: Thinking big in restoration ecology. *Environmental Professional* 13: 225–234.

Noss, R.F. 1991e. Landscape conservation priorities in the Greater Yellowstone Ecosystem. Report to The Nature Conservancy, Arlington, VA and Boulder, CO.

Noss, R.F. 1991f. A critical review of the U.S. Fish and Wildlife Service's proposal to establish a captive breeding population of Florida panthers, with emphasis on the population reestablishment issue. Report to the Fund for Animals, Washington, DC.

Noss, R.F. 1991g. Effects of edge and internal patchiness on avian habitat use in an old-growth Florida hammock. *Natural Areas Journal* 11: 34–47.

Noss, R.F. 1991h. A Native Ecosystems Act. *Wild Earth* 1(1): 24.

Noss, R.F. 1991/92. Biologists, biophiles, and warriors. *Wild Earth* 1(4): 56–60.

Noss, R.F. 1992a. Issues of scale in conservation biology, Pages 239–250 in P.L. Fiedler and S.K. Jain, eds. "Conservation Biology: The Theory and Practice of Nature Conservation, Preservation, and Management." Chapman and Hall, New York.

Noss, R.F. 1992b. The Wildlands Project: Land conservation strategy. *Wild Earth* (Special Issue): 10–25.

Noss, R.F. 1992c. A preliminary conservation plan for the Oregon Coast Range. Coast Range Association, Newport, OR.

Noss, R.F. 1992d. Biodiversity in the Blue Mountains: A framework for monitoring and assessment. Blue Mountains Biodiversity Conference, May 26–29, 1992, Whitman College, Walla Walla, WA.

Noss, R.F. 1992e. Interpreting biodiversity. Pages 11–37 in W.E. Hudson, ed. "Nature Watch: A Resource for Enhancing Wildlife Viewing Areas." Defenders of Wildlife and Falcon Press, Helena, MT.

Noss, R.F. 1993a. A bioregional conservation plan for the Oregon Coast Range. *Natural Areas Journal* 13: 276–290.

Noss, R.F. 1993b. Wildlife corridors. Pages 43–68 in D. Smith and P. Hellmund, eds. "Ecology of Greenways." University of Minnesota Press, Minneapolis, MN.

Noss, R.F. 1993c. Sustainable forestry or sustainable forests? In J.T. Olson, ed. "Defining Sustainable Forestry." Island Press, Washington, DC.

Noss, R.F., and B. Csuti. 1994. Habitat fragmentation. In G.K. Meffe and C.R. Carroll, eds. "An Introduction to Conservation Biology." Sinauer, Sunderland, MA.

Noss, R.F., and L.D. Harris. 1986. Nodes, networks, and MUMs: Preserving diversity at all scales. *Environmental Management* 10: 299–309.

Noss, R.F., and L.D. Harris. 1990. Habitat connectivity and the conservation of biological diversity: Florida as a case history. Pages 131–135 in Proceedings of the 1989 Society of American Foresters National Convention, Spokane, WA, Sept. 24–27. Society of American Foresters, Bethesda, MD.

Noss, R.F., and R.F. Labisky. 1988. Sensitivity of Vertebrates to Development in Four Upland Community Types in Northern Peninsular Florida. Report to Florida Game and Fresh Water Fish Commission, Tallahassee, FL.

Noss, R.F., S.P. Cline, B. Csuti, and J.M. Scott. 1992. Monitoring and assessing biodiversity. Pages 67–85 in E. Lykke, ed. "Achieving Environmental Goals: The Concept and Practice of Environmental Performance Review." Belhaven Press, London, UK.

Noss, R.F., E.T. LaRoe, and J.M. Scott. 1994. Endangered Ecosystems of the United States: A Preliminary Assessment of Loss and Degradation. U.S. Fish and Wildlife Service Report, Washington, DC.

Noy-Meir, I. 1993. Compensating growth of grazed plants and its relevance to the use of rangelands. *Ecological Applications* 3: 32–34.

O'Callaghan, K. 1992. Whose agenda for America? *Audubon* 94(5): 80–91.

Odum, E.P. 1970. Optimum population and environment: A Georgia microcosm. *Current History* 58: 355–359.

Odum, E.P. 1989. Input management of production systems. *Science* 243: 177–182.

Odum, E.P., and H.T. Odum. 1972. Natural areas as necessary components of Man's total environment. *Proceedings North American Wildlife and Natural Resources Conference* 37: 178–189.

Office of Technology Assessment. 1987. "Technologies to Maintain Biological Diversity." U.S. Government Printing Office, Washington, DC.

Ogden (Environmental and Energy Services Company, Inc.) 1992. Task 3.5a-1. Accounts of MSCP Target Species. Unpublished draft report to Clean Water Program, City of San Diego, CA.

Ogden, J.C., D.A. McCrimmon, G.T. Bancroft, and B.W. Patty. 1987. Breeding populations of Wood Stork *Mycteria americana* in southeastern United States. *Condor* 89: 752–759.

Opdam, P., D. van Dorp, and C.J.F. ter Braak. 1984. The effect of isolation on the number of woodland birds in small woods in the Netherlands. *Journal of Biogeography* 11: 473–478.

Oregon Natural Heritage Data Base. 1989. "Rare, Threatened, and Endangered Plants and Animals of Oregon." Oregon Natural Heritage Data Base, Portland, OR.

Orians, G.H. 1986. Site characteristics favoring invasions. Pages 133–148 in H.A. Mooney and J.A. Drake, eds. "Ecology of Biological Invasions of North America and Hawaii." Springer-Verlag, New York.

Oxley, D.J., M.B. Fenton, and G.R. Carmody. 1974. The effects of roads on populations of small mammals. *Journal of Applied Ecology* 11: 51–59.

Padley, W.D. 1991. Mountain lion ecology in the southern Santa Ana Mountains, California. California State Polytechnic University, Pomona, CA.

Painter, E.L., and A.J. Belsky. 1993. Application of herbivore optimization theory to rangelands of the western United States. *Ecological Applications* 3: 2–9.

Palmer, S. 1985. Some extinct mollusks of the U.S.A. *Atala* 13: 1–7.

Parker, G.R. 1989. Old-growth forests of the central hardwood region. *Natural Areas Journal* 9: 5–11.

Patten, D.T. 1993. Herbivore optimization and overcompensation: does native herbivory on western ranagelands support these theories? *Ecological Applications* 3: 35–36.

Payne, N.F., and F. Copes. 1986. "Wildlife and Fisheries Habitat Improvement Handbook." U.S. Department of Agriculture, Forest Service, Wildlife and Fisheries Administrative Report. (Unnumbered).

Pearson, D.L., and F. Cassola. 1992. World-wide species richness patterns of tiger beetles (Coleoptera: Cicindelidae): Indicator taxon for biodiversity and conservation studies. *Conservation Biology* 6: 376–391.

Peek, J.M., and P.D. Dalke (eds.). 1982. "Wildlife–Livestock Relationships," Symposium: Proceedings 10. University of Idaho, Forest, Wildlife and Range Experiment Station, Moscow, ID.

Peet, R.K. 1988. Forests of the Rocky Mountains. Pages 63–101 in M.G. Barbour and W.D. Billings, eds. "North American Terrestrial Vegetation." Cambridge University Press, New York.

Perry, D.A., R. Meurisse, B. Thomas, R. Miller, J. Boyle, J. Means, C.R. Perry, and R.F. Powers (eds.). 1989a. "Maintaining Long-term Productivity of Pacific Northwest Forest Ecosystems." Timber Press, Portland, OR.

Perry, D.A., M.P. Amaranthus, J.G. Borchers, S.L. Borchers, and R.E. Brainerd. 1989b. Bootstrapping in ecosystems. *BioScience* 39: 230–237.

Perry, D.A., J.G. Borchers, S.L. Borchers, and M.P. Amaranthus. 1990. Species migrations and ecosystem stability during climate change: The belowground connection. *Conservation Biology* 4: 266–274.

Peterken, G. 1981. "Woodland Conservation and Management." Chapman and Hall, New York.

Peters, R.L. 1988. Effects of global warming on species and habitats: An overview. *Endangered Species Update* 5(7):1–8.

Peters, R.L., and J.D.S. Darling. 1985. The greenhouse effect and nature reserves. *BioScience* 35: 707–717.

Peters, R.L., and T.E. Lovejoy (eds.). 1992. "Global Warming and Biological Diversity." Yale University Pres, New Haven, CT.

Petraitis, P.S., R.E. Latham, and R.A. Niesenbaum. 1989. The maintenance of species diversity by disturbance. *Quarterly Review of Biology* 64: 393–418.

Pickett, S.T.A., and J.N. Thompson. 1978. Patch dynamics and the design of nature reserves. *Biological Conservation* 13: 27–37.

Pickett, S.T.A., and P.S. White. 1985. "The Ecology of Natural Disturbance and Patch Dynamics." Acadmic Press, Orlando, FL.

Pickett, S.T.A., V.T. Parker, and P.L. Fiedler. 1992. The new paradigm in ecology: Implications for conservation biology above the species level. Pages 65–88 in P.L. Fiedler and S.K. Jain, eds. "Conservation Biology: The Theory and Practice of Nature Conservation, Preservation, and Management." Chapman and Hall, New York.

Pielou, E.C. 1991. "After the Ice Age." University of Chicago Press, Chicago, IL.

Pinchot, G. 1910. "The Fight for Conservation." University of Washington Press, Seattle, WA.

Platt, W.J., and M. Schwartz. 1990. Temperate hardwood forests. Pages 194–229 in R.L. Myers and J.J. Ewel, eds. "Ecosystems of Florida." University of Central Florida Press, Orlando, FL.

Platt, W.J., G.W. Evans, and S.L. Rathbun. 1988. The population dynamics of a long-lived conifer (*Pinus palustris*). *American Naturalist* 131: 491–525.

Postel, S., and J.C. Ryan. 1991. Reforming forestry. Pages 74–92 in L. Starke, ed. "State of the World 1991: A Worldwatch Institute Report on Progess Toward a Sustainable Society." W.W. Norton, New York.

Povilitis, T. 1993. Applying the biosphere reserve concept to a greater ecosystem: The San Juan Mountain area of Colorado and New Mexico. *Natural Areas Journal* 13: 18–28.

Power, T.M. 1991. Ecosystem preservation and the economy in the Greater Yellowstone area. *Conservation Biology* 5: 395–404.

Prescott-Allen, C., and R. Prescott-Allen. 1986. "The First Resource." Yale University Press, New Haven, CT.

Prescott-Allen, R., and C. Prescott-Allen. 1983. "Genes from the Wild." Earthscan, London, UK.

Pressey, R.L., and A.O. Nicholls. 1989. Application of a numerical algorithm to the selection of reserves in semi-arid New South Wales. *Biological Conservation* 50: 263–278.

Pulliam, H.R. 1988. Sources, sinks, and population regulation. *American Naturalist* 132: 652–661.

Pyle, R.M. 1980. Management of nature reserves. Pages 319–327 in M.E. Soulé and B.A. Wilcox, eds. "Conservation Biology: An Evolutionary-Ecological Perspective." Sinauer Associates, Sunderland, MA.

Pyne, S.J. 1982. "Fire in America: A Cultural History of Wildland and Rural Fire." Princeton University Press, Princeton, NJ.

Ranney, J.W., M.C. Bruner, and J.B. Levenson. 1981. The importance of edge in the structure and dynamics of forest islands. Pages 67–95 in R.L. Burgess and D.M. Sharpe, eds. "Forest Island Dynamics in Man-Dominated Landscapes." Springer-Verlag, New York.

Ratti, J.T., and K.P. Reese. 1988. Preliminary test of the ecological trap hypothesis. *Journal of Wildlife Management* 52: 484–491.

Recher, H. 1969. Bird species diversity and habitat diversity in Australia and North America. *American Naturalist* 103: 75–80.

Reed, D.F. 1981. Mule deer behavior at a highway underpass exit. *Journal of Wildlife Management* 45: 542–543.

Reed, D.F., T.N. Woodard, and T.M. Pojar. 1975. Behavioral response of mule deer to a highway underpass. *Journal of Wildlife Management* 39: 361–367.

Reed, P., G. Haas, F. Beum, and L. Sherrick. 1989. Non-recreational uses of the National Wilderness Preservation System: A 1988 telephone survey. Pages 220–228 in Wilderness Benchmark 1988: Proceedings of the National Wilderness Colloquium. Gen. Tech. Rep. SE-51, Asheville, NC.

Reeves, F.B., and E.F. Redente. 1992. The importance of mutualism in succession. Pages 423–442 in J. Skujins, ed. "Semiarid Lands and Deserts—Soil Resource and Reclamation." Marcel Dekker, New York.

Reeves, G.H., and J.R. Sedell. 1992. An ecosystem approach to the conservation and management of freshwater habitat for anadromous salmonids in the Pacific Northwest. *Transactions of the North American Wildlife and Natural Resources Conference* 57: 408–415.

Reid, W.V., and K.R. Miller. 1989. "Keeping Options Alive—The Scientific Basis for Conserving Biodiversity." World Resources Institute, Washington, DC.

Repenning, R.W., and R.F. Labisky. 1985. Effects of even-age timber management on bird communities of the longleaf pine forest in northern Florida. *Journal of Wildlife Management* 49: 1088–1098.

Reschke, C. 1990. "Ecological Communities of New York State." New York Natural Heritage Program, Latham, NY.

Reynolds, R.V., and A.H. Pierson. 1923. Lumber cut of the United States, 1870–1920. USDA Bulletin No. 1119, Washington, DC.

Rieger, J. 1992. Western riparian and wetland ecosystems. *Restoration and Management Notes* 10(1): 52–55.

Rinne, J.N. 1990. Biodiversity and native fish management: Arizona, Southwestern United States. Pages 152–165 in Proceedings of 70th Annual Conference of the Western Association of Fish and Wildlife Agencies and Western Division, American Fisheries Society, Sun Valley, ID.

Rinne, J.N. 1992. An approach to management and conservation of a declining regional native fish fauna: southwestern United States. Pages 56–60 in Proceedings International Symposium Wildlife, 5th International Congress of Zoology, Yokohama, Japan.

Ripple, W.J., G.A. Bradshaw, and T.A. Spies. 1991. Measuring forest landscape patterns in the Cascade Range of Oregon, USA. *Biological Conservation* 57: 73–88.

Risser, P.G. 1992. Landscape ecology approach to ecosystem rehabilitation. Pages 37–46 in M.K. Wali, ed. "Ecosystem Rehabilitation—Preamble to Sustainable Development." Volume 1: "Policy Issues." SPB Academic Publishing, The Hague, The Netherlands.

Robbins, C.S. 1979. Effect of forest fragmentation on bird populations. Pages 198–212 in R.M. DeGraaf and K.E. Evans, eds. "Management of North Central and Northeastern Forests for Nongame Birds." USDA Forest Service General Technical Report NC-51. Washington, DC.

Robbins, C.S., D.K. Dawson, and B.A. Dowell. 1989. Habitat area requirements of breeding forest birds of the Middle Atlantic states. *Wildlife Monographs* 103: 1–34.

Robertson, W.B. 1965. Inside the Everglades. *Audubon* 67: 274–279.

Robinson, J.G. 1993. The limits to caring: Sustainable living and the loss of biodiversity. *Conservation Biology* 7: 20–28.

Robinson, S.K. 1992. Effects of Forest Fragmentation on Migrant Songbirds in the Shawnee National Forest. Report to Illinois Department of Energy and Natural Resources. Illinois Natural History Survey, Champaign, IL.

Rolston, H. 1988. "Environmental Ethics: Duties to and Values in the Natural World." Temple University Press, Philadelphia, PA.

Romme, W.H., and D.H. Knight. 1982. Landscape diversity: The concept applied to Yellowstone Park. *BioScience* 32: 664–670.

Roth, R.R. 1976. Spatial heterogeneity and bird species diversity. *Ecology* 57: 773–782.

Rudnicky, T.C., and M.L. Hunter. 1993. Avian nest predation in clearcuts, forests, and edges in a forest-dominated landscape. *Journal of Wildlife Management* 57: 358–364.

Ruggiero, L.F., K.B. Aubry, A.B. Carey, and M.H. Huff, tech. coordinators. 1991. "Wildlife and Vegetation of Unmanaged Douglas-fir Forests." USDA Forest Service, Pacific Northwest Research Station, Portland, OR.

Runkle, J.R. 1981. Gap regeneration in some old-growth forests of the eastern United States. *Ecology* 62: 1041–1051.

Runkle, J.R. 1985. Disturbance regimes in temperate forests. Pages 17–33 in S.T.A. Pickett and P.S. White, eds. "The Ecology of Natural Disturbance and Patch Dynamics." Academic Press, Orlando, FL.

Runte, A. 1987. "National Parks: The American Experience." Second Edition. University of Nebraska Press, Lincoln, NE.

Russell, C., and L. Morse. 1992. Extinct and possibly extinct plant species of the United States and Canada. Unpublished report. Review draft, 13 March 1992. The Nature Conservancy, Arlington, VA.

Ryti, R.T. 1992. Effect of the focal taxon on the selection of nature reserves. *Ecological Applications* 2: 404–410.

Sale, K. 1985. "Dwellers in the Land—the Bioregional Vision." Sierra Club Books, San Francisco, CA.

Salley, A.S. (ed.). 1911. "Narrative of Early Carolina 1650–1708." Barnes and Noble, New York.

Salwasser, H. 1990. Sustainability as a conservation paradigm. *Conservation Biology* 4: 213–216.

Salwasser, H. 1991. New perspectives for sustaining diversity in U.S. national forest ecosystems. *Conservation Biology* 5: 567–569.

Salwasser, H. 1992. From New Perspectives to Ecosystem Management: Response to Frissell et al. and Lawrence and Murphy. *Conservation Biology* 6: 469–472.

Salwasser, H., D.W. MacCleery, and T.A. Snellgrove. 1992. New perspectives for managing the U.S. national forest system. Report to the North American Forestry Commission, 16th Session, Cancun, Mexico, February, 1992.

Sampson, A.H., and L.H. Wehl. 1918. Range preservation and its relation to erosion control on western grazing lands. U.S. Department of Agriculture, Bulletin No. 675.

Samson, F.B., and F.L. Knopf. 1982. In search of a diversity ethic for wildlife management. *Transactions of the North American Wildlife and Natural Resources Conference* 47: 421–431.

Saunders, D.A., and R.J. Hobbs (eds.). 1991. "Nature Conservation 2: The Role of Corridors." Surrey Beatty & Sons, Chipping Norton, Australia.

Saunders, D.A., R.J. Hobbs, and C.R. Margules. 1991. Biological consequences of ecosystem fragmentation: A review. *Conservation Biology* 5: 18–32.

Savory, A. 1988. "Holistic Resource Management." Island Press, Covelo, CA.

Schamberger, M.A., and A.H. Farmer. 1978. The habitat evaluation procedures: their application in project planning and impact evaluation. *Transactions of the North American Wildlife and Natural Resources Conference* 43: 274–283.

Schlesinger, W.H., J.F. Reynolds, G.L. Cunningham, L.F. Huenneke, W.M. Jarrell, R.A. Virginia, and W.G. Whitford. 1990. Biological feedbacks in global desertification. *Science* 247:1043–1048.

Schlosser, I.J. 1991. Stream fish ecology: a landscape perspective. *BioScience* 41: 704–712.

Schoen, J.W. 1990. Bear habitat management: A review and future perspective. *International Conference on Bear Research and Management* 8: 143–154.

Schonewald-Cox, C.M. 1983. Conclusions. Guidelines to management: A beginning attempt. Pages 141–145 in C.M. Schonewald-Cox, S.M. Chambers, B. MacBryde, and W.L. Thomas, eds. "Genetics and Conservation: A Reference for Managing Wild Animal and Plant Populations." Benjamin/Cummings, Menlo Park, CA.

Schonewald-Cox, C.M. 1988. Boundaries in the protection of nature reserves. *BioScience* 38: 480–486.

Schonewald-Cox, C.M., and J.W. Bayless. 1986. The boundary model: A geographical analysis of design and conservation of nature reserves. *Biological Conservation* 38: 305–322.

Schonewald-Cox, C.M., S.M. Chambers, B. MacBryde, and W.L. Thomas (eds.). 1983. "Genetics and Conservation: A Reference for Managing Wild Animal and Plant Populations." Benjamin/Cummings, Menlo Park, CA.

Schoonmaker, P., and A. McKee. 1988. Species composition and diversity during secondary succession of coniferous forests in the western Cascade Mountains of Oregon. *Forest Science* 34: 960–979.

Schowalter, T.D. 1988. Forest pest management: A synopsis. *Northwest Environmental Journal* 4: 313–318.

Schowalter, T.D. 1990. Differences and consequences for insects. Pages 91–106 in A.F. Pearson and B.A. Challenger, eds. "Forests, Wild and Managed: Differences and Consequences." Students for Forestry Awareness, University of British Columbia, Vancouver, BC.

Schroeder, M. 1987. The black-footed ferret. Pages 446–455 in R.L. DiSilvestro, ed. "Audubon Wildlife Report 1987," Adademic Press, San Diego, CA.

Schultz, R.P., and L.P. Wilhite. 1975. Changes in a flatwoods site following intensive site preparation. *Forest Science* 10: 230–237.

Scott, B. 1992. The big open. *Restoration and Management Notes* 10(1):51–52.

Scott, J.M., S. Mountainspring, F.L. Ramsey, and C.B. Kepler. 1986. Forest bird communities of the Hawaiian Islands: Their dynamics, ecology, and conservation. *Studies in Avian Biology* No. 9.

Scott, J.M., B. Csuti, J.D. Jacobi, and J.E. Estes. 1987. Species richness: A geographic approach to protecting future biological diversity. *BioScience* 37: 782–788.

Scott, J.M., B. Csuti, K. Smith, J.E. Estes, and S. Caicco. 1991a. Gap analysis of species richness and vegetation cover: An integrated biodiversity conservation strategy. Pages 282–297 in K.A. Kohm, ed. "Balancing on the Brink of Extinction: The Endangered Species Act and Lessons for the Future." Island Press, Washington, DC.

Scott, J.M., B. Csuti, and S. Caicco. 1991b. Gap analysis: Assessing protection needs. Pages 15–26 in W.E. Hudson, ed. "Landscape Linkages and Biodiversity." Defenders of Wildlife and Island Press, Washington, DC.

Scott, J.M., F. Davis, B. Csuti, R. Noss, B. Butterfield, C. Groves, J. Anderson, S. Caicco, F. D'Erchia, T.C. Edwards, J. Ulliman, and R.G. Wright. 1993. Gap analysis: A geographical approach to protection of biological diversity. *Wildlife Monographs* 123: 1–41.

Scott, J.M., A.R. Kiester, B. Csuti, R.F. Noss, and B. Butterfield. (In press). Gap analysis: A spatial approach to identifying representative areas for maintaining biodiversity. *Ecology.*

Sedell, J.R., P.A. Bisson, F.J. Swanson, and S.V. Gregory. 1988. What we know about large trees that fall into streams and rivers. Pages 47–81 in C. Maser, R.F. Tarrant, J.M. Trappe, and J.F. Franklin, eds. "From the Forest to the Sea: A Story of Fallen Trees." U.S. Department of Agriculture, Forest Service, Pacific Northwest Research Station, General Technical Report PNW-GTR-229.

Seegmiller, R.F., and R.D. Ohmart. 1981. Ecological Relationships of feral burros and desert bighorn sheep. *Wildlife Monographs* No. 78.

Sessions, G. 1992. Radical environmentalism in the 90s. *Wild Earth* 2(3): 64–67.

Shafer, C.L. 1990. "Nature Reserves: Island Theory and Conservation Practice." Smithsonian Institution Press, Washington, DC.

Shaffer, M.L. 1981. Minimum population sizes for species conservation. *BioScience* 31: 131–134.

Shaffer, M.L. 1992. "Keeping the Grizzly Bear in the American West: A Strategy for Real Recovery." The Wilderness Society, Washington, DC.

Shannon, C.E., and W. Weaver. 1949. "The Mathematical Theory of Communication." University of Illinois Press, Urbana, IL.

Sheldon, A.L. 1988. Conservation of stream fishes: patterns of diversity, rarity, and risk. *Conservation Biology* 2: 220–245.

Shelford, V.E. (ed.). 1926. "Naturalist's Guide to the Americas." Williams and Wilkins, Baltimore, MD.

Shelford, V.E. 1933. Ecological Society of America: A nature sanctuary plan unanimously adopted by the Society, December 28, 1932. *Ecology* 14: 240–245.

Shen, S. 1987. Biological diversity and public policy. *BioScience* 37: 709–712.

Sheridan, D. 1979. Off-road vehicles on public land. Council on Environmental Quality, U.S. Government Printing Office, Washington, DC.

Sheridan, D. 1981. Desertification of the United States. U.S. Council on Environmental Quality, U.S. Government Printing Office, Washington, DC.

Shindler, B., P. List, and B.S. Steel. 1993. Managing federal forests: Public attitudes in Oregon and nationwide. *Journal of Forestry* 91(7): 36–42.

Shugart, H.H., and S.W. Seagle. 1985. Modeling forest landscapes and the role of disturbance in ecosystems and communities. Pages 353–368 in S.T.A. Pickett and P.S. White, eds. "The Ecology of Natural Disturbance and Patch Dynamics." Academic Press, Orlando, FL.

Shugart, H.H., and D.C. West. 1981. Long-term dynamics of forest ecosystems. *American Scientist* 69: 647–652.

Simberloff, D. 1988. The contribution of population and community biology to conservation science. *Annual Review of Ecology and Systematics* 19: 473–511.

Simberloff, D. 1991. Review of theory relevant to acquiring land. Report to Florida Department of Natural Resources. Florida State University, Tallahassee, FL.

Simberloff, D., and L.G. Abele. 1976. Island biogeography theory and conservation practice. *Science* 191: 285–286.

Simberloff, D., and L.G. Abele. 1982. Refuge design and island biogeographic theory: Effects of fragmentation. *American Naturalist* 120: 41–50.

Simberloff, D., and J. Cox. 1987. Consequences and costs of conservation corridors. *Conservation Biology* 1: 63–71.

Simberloff, D., J.A. Farr, J. Cox, and D.W. Mehlman. 1992. Movement corridors: Conservation bargains or poor investments? *Conservation Biology* 6: 493–504.

Simpson, E.H. 1949. Measurement of diversity. *Nature* 163: 688.

Simpson, G.G. 1964. Species densities of North American mammals. *Systematic Zoology* 13: 361–389.

Small, M.F., and M.L. Hunter. Forest fragmentation and avian nest predation in forested landscapes. *Oecologia* 76: 62–64.

Smeins, F.E. 1983. Origin of the brush problem—a geological and ecological perspective of contemporary distributions. Pages 5–16 in Proceedings Brush Management Symposium, Society for Range Management, Texas Tech University Press, Lubbock, TX.

Smeins, F.E. 1992. Concepts for solving biodiversity problems—rangeland ecosystems. Unpublished handout for BLM Shortcourse.

Snyder, G. 1980. "The Real Work—Interviews and Talks 1964–1979." New Directions Publishing Corporation, New York.

Snyder, G. 1990. "The Practice of the Wild." North Point Press, San Francisco, CA.

Snyder, G. 1992. Coming in to the watershed. *Wild Earth* (Special Issue): 65–70.

Society for Range Management. 1992. Society for Range Management Guidelines for Rangeland Assessment. Denver, CO.

Soulé, M.E. 1983. What do we really know about extinction? Pages 111–124 in C.M. Schonewald-Cox, S.M. Chambers, B. MacBryde, and W.L. Thomas, eds. "Genetics and Conservation: A Reference for Managing Wild Animal and Plant Populations." Benjamin/Cummings, Menlo Park, CA.

Soulé, M.E. 1985. What is conservation biology? *BioScience* 35:727–734.

Soulé, M.E., ed. 1987. "Viable Populations for Conservation." Cambridge University Press, Cambridge, UK.

Soulé, M.E. 1991a. Conservation: Tactics for a constant crisis. *Science* 253: 744–750.

Soulé, M.E. 1991b. Theory and strategy. Pages 91–104 in W.E. Hudson, ed. "Landscape Linkages and Biodiversity." Defenders of Wildlife and Island Press, Washington, DC.

Soulé, M.E. 1992. A vision for the meantime. *Wild Earth* (Special Issue): 7–8.

Soulé, M.E., and M.E. Gilpin. 1991. The theory of wildlife corridor capability. Pages 3–8 in D.A. Saunders and R.J. Hobbs, eds. "Nature Conservation 2: The Role of Corridors." Surrey Beatty and Sons, Chipping Norton, NSW, Australia.

Soulé, M.E., and D. Simberloff. 1986. What do genetics and ecology tell us about the design of nature reserves? *Biological Conservation* 35: 19–40.

Soulé, M.E., and B.A. Wilcox. 1980. Conservation biology: Its scope and challenge. Pages 1–8 in M.E. Soulé and B.A. Wilcox, eds. "Conservation Biology: An Evolutionary-Ecological Perspective." Sinauer, Sunderland, MA.

Soulé, M.E., D.T. Bolger, A.C. Alberts, J. Wright, M. Sorice, and S. Hill. 1988. Reconstructed dynamics of rapid extinctions of chaparral-requiring birds in urban habitat islands. *Conservation Biology* 2: 75–92.

Spellerberg, I.F. 1991. "Monitoring Ecological Change." Cambridge University Press, Cambridge, UK.

Spencer, C.N., B.R. McClelland, and J.A. Stanford. 1991. Shrimp stocking, salmon collapse, and eagle displacement. *BioScience* 41:14–21.

Spies, T.A. 1991. Plant species diversity and occurrence in young, mature, and old-growth Douglas-fir stands in western Oregon and Washington. Pages 111–121 in L.F. Ruggiero, K.B. Aubry, A.B. Carey, and M.H. Huff, tech. coordinators. "Wildlife and Vegetation of Unmanaged Douglas-fir Forests." USDA Forest Service, Pacific Northwest Research Station, Portland, OR.

Spies, T.A., and S.P. Cline. 1988. Coarse woody debris in forests and plantations of coastal Oregon. Pages 5–24 in C. Maser, R.F. Tarrant, J.M. Trappe, and J.F. Franklin, eds. "From the Forest to the Sea: A Story of Fallen Trees." General Technical Report PNW-GTR-229. USDA Forest Service. Pacific Northwest Research Station, Portland, OR.

Stanford, J.A., and J.V. Ward. 1979. Stream regulation in North America. Pages 215–236 in J.V. Ward and J.A. Stanford, eds. "The Ecology of Regulated Streams." Plenum Press, New York.

Stansbery, D.H. 1970. 2. Eastern freshwater mollusks (I) The Mississippi and St. Lawrence River systems. *Malacologia* 10: 9–22.

Stebbins, G.L. 1980. Rarity of plant species: A synthetic viewpoint. *Rhodora* 82: 77–86.

Steuter, A.A. 1992. A practitioner's view. *Restoration and Management Notes* 10(1):45–47.

Stoddard, H.L. 1931. "The Bobwhite Quail: Its Habits, Preservation and Increase." Scribners, New York.

Stoddart, L.A., A.D. Smith, and T.W. Box. 1975. "Range Management," Third Edition. McGraw-Hill Book Co., New York.

Stolzenburg, W. 1992. Silent sirens. *Nature Conservancy* May/June 1992: 8–13.

Stork, N.E. 1992. Insect diversity: Facts, fiction, and speculation. *Biological Journal of the Linnaean Society* 35: 321–337.

Strole, T.A., and R.C. Anderson. 1992. White-tailed deer browsing: Species preferences and implications for central Illinois forests. *Natural Areas Journal* 12: 139–144.

Swanson, F.J., and J.F. Franklin. 1992. New forestry principles from ecosystem analysis of Pacific Northwest forests. *Ecological Applications* 2: 262–274.

Swanson, F.J., J.F. Franklin, and J.R. Sedell. 1990. Landscape patterns, disturbance, and management in the Pacific Northwest, USA. Pages 191–213 in I.S. Zonneveld and R.T.T. Forman, eds. "Changing Landscapes: An Ecological Perspective." Springer-Verlag, New York.

Swihart, R.K., and N.A. Slade. 1984. Road crossing in *Sigmodon hispidus* and *Microtus ochrogaster*. *Journal of Mammalogy* 65: 357–360.

Swindel, B.F., L.F. Conde, and J.E. Smith. 1983. Plant cover and biomass response to clear-cutting, site preparation, and planting in *Pinus elliottii* flatwoods. *Science* 219:1421–1422.

Tangley, L. 1988. A new era for biosphere reserves. *BioScience* 38: 148–155.

Tear, T.H., J.M. Scott, P.H. Hayward, and B. Griffith. 1993. Status and prospects for success of the Endangered Species Act: A look at recovery plans. *Science* 262: 976.

Temple, S.A. 1986. Predicting impacts of habitat fragmentation on forest birds: A comparison of two models. Pages 301–304 in J. Verner, M.L. Morrison, and C.J. Ralph, eds. "Wildlife 2000: Modeling Habitat Relationships of Terrestrial Vertebrates." University of Wisconsin Press, Madison, WI.

Temple, S.A., and J.R. Cary. 1988. Modeling dynamics of habitat-interior bird populations in fragmented landscapes. *Conservation Biology* 2: 340–347.

Temple, S.A., P.F. Brussard, E.G. Bolen, H. Salwasser, M.E. Soulé, and J.G. Teer. 1988. What's so new about conservation biology? *Transactions of the North American Wildlife and Natural Resources Conference* 53:609–612.

Terborgh, J. 1974. Preservation of natural diversity: The problem of extinction prone species. *BioScience* 24: 715–722.

Terborgh, J. 1976. Island biogeography and conservation: Strategy and limitations. *Science* 193: 1029–1030.

Terborgh, J. 1988. The big things that run the world—A sequel to E.O. Wilson. *Conservation Biology* 2: 402–403.

Terborgh, J., and B. Winter. 1983. A method for siting parks and reserves with special reference to Columbia and Ecuador. *Biological Conservation* 27: 45–58.

Thiel, R.P. 1985. Relationship between road densities and wolf habitat suitability in Wisconsin. *American Midland Naturalist* 113: 404–407.

Thomas, C.D. 1990. What do real population dynamics tell us about minimum viable population sizes? *Conservation Biology* 4: 324–327.

Thomas, J.W. (ed.). 1979. "Wildlife Habitats in Managed Forests: The Blue Mountains of Oregon and Washington." U.S. Department of Agriculture, Forest Service, Agriculture Handbook 553, Washington, DC.

Thomas, J.W., E.D. Forsman, J.B. Lint, E.C. Meslow, B.R. Noon, and J. Verner. 1990. "A Conservation Strategy for the Northern Spotted Owl." USDA Forest Service, USDI Bureau of Land Management, USDI Fish and Wildlife Service, and USDI National Park Service, Portland, OR.

Thomas, J.W., M.G. Raphael, R.G. Anthony, E.D. Forsman, A.G. Gunderson, R.S. Holthausen, B.G. Marcot, G.H. Reeves, J.R. Sedell, and D.M. Solis. 1993. Viability Assessments and Management Considerations for Species Associated with Late-Successional and Old-Growth Forests of the Pacific Northwest. USDA Forest Service, Washington, DC.

Thomas, T.R., and L.R. Irby. 1990. Habitat use and movement patterns by migrating mule deer in southeastern Idaho. *Northwest Science* 64: 19–27.

Thompson, F.R., W.D. Dijak, T.G. Kulowiec, and D.A. Hamilton. 1992. Breeding bird populations in Missouri Ozark forests with and without clearcutting. *Journal of Wildlife Management* 56: 23–30.

Thoreau, H.D. 1960 (1854). "Walden or, Life in the Woods," and "On the Duty of Civil Disobediance." New American Library, New York.

Tilden, F. 1967. "Interpreting Our Heritage." University of North Carolina Press, Chapel Hill, NC.

Tilghman, N.G. 1989. Impacts of white-tailed deer on forest regeneration in northwestern Pennsylvania. *Journal of Wildlife Management* 53: 524–532.

Tober, J.A. 1981. "Who Owns the Wildlife? The Political Economy of Conservation in 19th-Century America." Greenwood Press, Westport, CT.

Tscharntke, T. 1992. Fragmentation of *Phragmites* habitats, minimum viable population size, habitat suitability, and local extinction of moths, midges, flies, aphids, and birds. *Conservation Biology* 6: 530–536.

Turner, M.G. 1989. Landscape ecology: The effect of pattern on process. *Annual Review of Ecology and Systematics* 20: 171–197.

Turner, M.G., and C.L. Ruscher. 1988. Changes in landscape patterns in Georgia, USA. *Landscape Ecology* 1: 241–251.

Turner, M.G., W.H. Romme, R.H. Gardner, R.V. O'Neill, and T.K. Kratz. 1993. A revised concept of landscape equilibrium: Disturbance and stability on scaled landscapes. *Landscape Ecology* 8: 213–227.

Turner, R.M. 1990. Long term vegetation change at a fully protected Sonoran Desert site. *Ecology* 71(2):464–477.

Udall, S.L. 1988. "The Quiet Crisis and the Next Generation." Peregrine Smith Books, Salt Lake City, UT.

Udvardy, M.D.F. 1975. "World Biogeographical Provinces." Occasional Paper No. 18. International Union for Conservation of Nature and Natural Resources, Gland, Switzerland.

UNESCO. 1974. Task Force on Criteria and Guidelines for the Choice and Establishment of Biosphere Reserves. Man and the Biosphere Report No. 22. Paris, France.

Urban, D.L., R.V. O'Neill, and H.H. Shugart. 1987. Landscape ecology. *BioScience* 37: 119–127.

U.S. Comptroller General of the United States. 1977. Public rangelands continue to deteriorate. Report to Congress by the Comptroller General of the United States, Report CED-77-88, 28 pp.

U.S. Department of Agriculture. 1936. The Western Range—A great but neglected natural resource. Report from Secretary of Agriculture to U.S. Senate. Senate Document 199, U.S. Government Printing Office, Washington, DC.

U.S. Department of the Interior, Bureau of Land Management. 1962. Adjustments in grazing use—An evaluation of adjustments in grazing use as they occur in the management of the federal range by the bureau of land management. USDI Bureau of Land Management, Washington, DC.

U.S. Department of the Interior, Bureau of Land Management. 1975. Effects of livestock grazing on wildlife, watershed, recreation and other resource values in Nevada. USDI Bureau of Land Management, Washington, DC.

U.S. Department of the Interior, Bureau of Land Management. 1975. Range condition report. Prepared for The Senate Committee on Appropriations.

U.S. Department of Interior, Bureau of Land Management. 1984. Range conditions—then and now. Unauthored article in Your Public Lands, Summer 1984, pages 10–11.

U.S. Department of Interior, Bureau of Land Management. 1990. State of the public rangelands 1990—the range of our vision.

U.S. Department of the Interior, Fish and Wildlife Service. 1992. *Endangered Species Technical Bulletin* 17(9–11):16.

U.S. Department of Interior and U.S. Department of Agriculture (USDI–USDA). 1992. America's Biodiversity Strategy: Actions to Conserve Species and Habitats. Washington, DC.

U.S. Environmental Protection Agency. 1990a. National water quality inventory: 1988 report to Congress. EPA 440-4-90-003, Office of Water. Washington, DC.

U.S. Environmental Protection Agency. 1990b. Environmental monitoring and assessment program—overview. U.S. Environmental Protection Agency, EPA/600/9-90/001.

U.S. Fish and Wildlife Service. 1993. Draft Recovery Plan for the Desert Tortoise (Mojave Population). Portland, OR.

U.S. General Accounting Office. 1988a. Rangeland management—more emphasis needed on declining and overstocked grazing allotments. U.S. General Accounting Office, Report to Congressional Requesters. GAO/RCED-88-80.

U.S. General Accounting Office. 1988b. Public Rangelands—some riparian areas restored but widespread improvement will be slow. U.S. General Accounting Office, Report to Congressional Requesters, GAO/RCED-88-105.

U.S. General Accounting Office. 1991. Rangeland Management—BLM's hot desert grazing program merits reconsideration. U.S. General Accounting Office, Report to the Chairman, Subcommittee on National Parks and Public Lands, Committee on Interior and Insular Affairs, House of Representatives, GAO/RCED-92-12.

Usher, M.B. 1986. "Wildlife Conservation Evaluation." Chapman and Hall, London, UK.

Vale, T.R., K.C. Parker, and A.J. Parker. 1989. Terrestrial vertebrates and vegetation structure in the western United States. *Professional Geographer* 41: 450–464.

Vallentine, J.F. 1971. "Range Development and Improvements." Brigham Young University Press, Provo, UT.

Van Dyke, F.G., R.H. Brocke, H.G. Shaw, B.A. Ackerman, T.H. Hemker, and F.G. Lindzey. 1986. Reactions of mountain lions to logging and human activity. *Journal of Wildlife Management* 50: 95–102.

Van Dyne, G.M., W. Burch, S.K. Fairfax, and W. Huey. 1984. Forage allocation on arid and semiarid public grazing lands: Summary and recommendations. Pages 1–25 in "Developing Strategies for Rangeland Management." National Research Council/National Academy of Sciences. Westview Press, Boulder, CO.

Vane-Wright, R.I., C.J. Humphries, and P.H. Williams. 1991. What to protect?—Systematics and the agony of choice. *Biological Conservation* 55: 235–253.

VanHaveren, B.P., and W.L. Jackson. 1986. Concepts in stream riparian rehabilitation. *Transactions of the North American Wildlife and Natural Resources Conference* 51:280–289.

Vannote, R.L., G. W. Minshall, K.W. Cummins, J.R. Sedell, and C.E. Cushing. 1980. The river continuum concept. *Canadian Journal of Fisheries and Aquatic Sciences* 37:130–137.

Van Vuren, D. 1982. Comparative ecology of bison and cattle in the Henry Mountains, Utah. Pages 449–457 in J.M. Peek and P.D. Dalke, eds. "Wildlife–Livestock Relationships Symposium: Proceedings 10." University of Idaho, Forest, Wildlife and Range Experiment Station, Moscow, ID.

Vickerman, S. 1992. Developing a national network of conservation lands. Pages 1–10 in W. Hudson, ed. "Nature Watch—A Resource for Enhancing Wildlife Viewing Areas." Falcon Press, Helena, MT.

Vitousek, P.M. 1986. Biological invasions and ecosystem properties: can species make a difference? Pages 163–176 in H.A. Mooney and J.A. Drake, eds. "Ecology of Biological Invasions of North America and Hawaii." Springer-Verlag, New York.

Vitousek, P.M. 1988. Diversity and biological invasions of oceanic islands. Pages 81–89 in E.O. Wilson, ed. "Biodiversity." National Academy Press, Washington, DC.

Vitousek, P.M., and P.A. Matson. 1984. Mechanisms of nitrogen retention in forest ecosystems: A field experiment. *Science* 225: 51–52.

Vitousek, P.M., P.R. Ehrlich, A.H. Ehrlich, and P.M. Matson. 1986. Human appropriation of the products of photosynthesis. *BioScience* 36: 368–373.

Voigt, W., Jr. 1976. "Public Grazing Lands—Use and Misuse by Industry and Government." Rutgers University Press, New Brunswick, NJ.

Wagner, F.H. 1978. Livestock grazing and the livestock industry. Pages 121–145 in H.P. Brokaw, ed. "Wildlife and America." Council on Environmental Quality, Washington, DC.

Wald, J., and D. Alberswerth. 1985. "Our Ailing Public Rangelands." National Wildlife Federation and Natural Resources Defense Council, Washington, DC.

Wald, J., and D. Alberswerth. 1989. "Our Ailing Public Rangelands: Still Ailing." National Wildlife Federation and Natural Resources Defense Council, Washington, DC.

Walker, J., and R.K. Peet. 1983. Composition and species diversity of pine-wiregrass savannas of the Green Swamp, North Carolina. *Vegetatio* 55: 163–179.

Wallace, D.R. 1983. "The Klamath Knot." Sierra Club Books, San Francisco, CA.

Walter, H.S. 1990. Small viable population: The red-tailed hawk of Socorro Island. *Conservation Biology* 4: 441–443.

Walters, C.J. 1986. "Adaptive Management of Renewable Resources." McGraw-Hill, New York.

Walters, C.J., and C.S. Holling. 1990. Large-scale management experiments and learning by doing. *Ecology* 71: 2060–2068.

Watkins, T.H., ed. 1989. A special report — Wilderness America: A vision for the future of the Nation's wildlands. *Wilderness* 52(184): 3–64.

Watt, A.S. 1947. Pattern and process in the plant community. *Journal of Ecology* 35: 12–22.

Wayne, P.M., and F.A. Bazzaz. 1991. Assessing diversity in plant communities: The importance of within-species variation. *Trends in Ecology and Evolution* 6: 400–404.

Webb, S. David. 1990. Historical biogeography. Pages 70–100 in R. Myers, and J.J. Ewel, eds. "Ecosystems of Florida," University of Central Florida Press, Orlando, FL.

Wegner, J.F., and G. Merriam. 1979. Movements of birds and small mammals between a wood and adjoining farmland habitat. *Journal of Applied Ecology* 16: 349–357.

Wells, H.V. 1993. Frogs missing? *The Courier-News* (Clinton, TN) April 7, 1993.

West, N.E. (ed.). 1983. "Temperate Deserts and Semi-deserts. Ecosystems of the World 5." Elsevier Scientific Publishing Co., New York.

West, N.E. 1991. Nutrient cycling in soils of semiarid and arid regions. Pages 295–332 in J. Skujins, ed. "Semiarid Lands and Deserts—Soil Resource and Reclamation." Marcel Dekker, Inc., New York.

West, N.E. 1993. Biodiversity of rangelands. *Journal of Range Management* 46: 2–13.

West, N.E. 1994. Strategies for maintenance and repair of biotic community diversity on rangelands. In R. Szaro, ed. "Biodiversity in Managed Landscapes." Oxford University Press, New York.

Westman, W.E. 1990. Managing for biodiversity. *BioScience* 40: 26–33.

Westman, W.E. 1990. Park management of exotic plant species: problems and issues. *Conservation Biology* 4: 251–260.

Westoby, M., B.Walker, and I. Noy-Meir. 1989. Opportunistic management for rangelands not at equilibrium. *Journal of Range Management* 42: 266–274.

Wetzel, R.G. 1975. "Limnology." Saunders College Publishing, Philadelphia, PA.

Whelan, C.J., and M.L. Dilger. 1992. Invasive, exotic shrubs: A paradox for natural area managers? *Natural Areas Journal* 12: 109–110.

Whitcomb, R.F., J.F. Lynch, P.A. Opler, and C.S. Robbins. 1976. Island biogeography and conservation: Strategy and limitations. *Science* 193: 1030–1032.

Whitcomb, R.F., C.S. Robbins, J.F. Lynch, B.L. Whitcomb, K. Klimkiewicz, and D. Bystrak. 1981. Effects of forest fragmentation on avifauna of the eastern deciduous forest. Pages 125–205 in R.L. Burgess and D.M. Sharpe, eds. "Forest Island Dynamics in Man-dominated Landscapes." Springer-Verlag, New York.

White, P.S. 1979. Pattern, process, and natural disturbance in vegetation. *Botanical Review* 45: 229–299.

White, P.S., and S.P. Bratton. 1980. After preservation: Philosophical and practical problems of change. *Biological Conservation* 18: 241–255.

Whitman, W. 1900 (1855, 1856, etc.). "Leaves of Grass, with Autobiography." McKay, Philadelphia, PA.

Whitney, G.G. 1987. Some reflections on the value of old-growth forests, scientific and otherwise. *Natural Areas Journal* 7: 92–99.

Whitney, G.G., and J.R. Runkle. 1981. Edge versus age effects in the development of a beech-maple forest. *Oikos* 37: 377–381.

Whittaker, R.H. 1956. Vegetation of the Great Smoky Mountains. *Ecological Monographs* 26: 1–80.

Whittaker, R.H. 1960. Vegetation of the Siskiyou Mountains, Oregon and California. *Ecological Monographs* 30: 279–338.

Whittaker, R.H. 1961. Vegetation history of the Pacific Coast states and the "central" significance of the Klamath Region. *Madrono* 16: 5–23.

Whittaker, R.H. 1972. Evolution and measurement of species diversity. *Taxon* 21: 213–251.

Wickman, B.E. 1991. Forest health in the Blue Mountains: The influence of insects and disease. U.S. Department of Agriculture, Forest Service, Pacific Northwest Research Station, LaGrande, OR.

Wiens, J.A. 1989. "The Ecology of Bird Communities." Vol. 2. "Processes and Variations." Cambridge University Press, New York.

Wilcove, D.S. 1985. Nest predation in forest tracts and the decline of migratory songbirds. *Ecology* 66: 1211–1214.

Wilcove, D.S., and D.D. Murphy. 1991. The spotted owl controversy and conservation biology. *Conservation Biology* 5: 261–262.

Wilcove, D., M. Bean, and P.C. Lee. 1992. Fisheries management and biological diversity: Problems and opportunities. *Transactions of the North American Wildlife and Natural Resources Conference* 57: 373–383.

Wilcove, D.S., C.H. McLellan, and A.P. Dobson. 1986. Habitat fragmentation in the temperate zone. Pages 237–256 in M.E. Soulé, ed. "Conservation Biology: The Science of Scarcity and Diversity." Sinauer, Sunderland, MA.

Wilcove, D.S., M. McMillan, and K.C. Winston. 1993. What exactly is an endangered species? An analysis of the U.S. Endangered Species list: 1985–1991. *Conservation Biology* 7: 87–93.

Wilcox, B.A. 1980. Insular ecology and conservation. Pages 95–117 in M.E. Soulé and B.A. Wilcox, eds. "Conservation Biology: An Ecological-Evolutionary Perspective." Sinauer, Sunderland, MA.

Wilcox, B.A., and D.D. Murphy. 1985. Conservation strategy: The effects of fragmentation on extinction. *American Naturalist* 125: 879–887.

Willers, B. 1992. Toward a science of letting things be. *Conservation Biology* 6: 605–607.

Williams, C.D., and J.E. Williams. 1992. Bring back the natives: A new strategy for restoring aquatic biodiversity on public lands. *Transactions of the North American Wildlife and Natural Resources Conference* 57: 416–423.

Williams, J.E. 1991. Preserves and refuges for native western fishes: History and management. Pages 171–189 in W.L. Minckley and J.E. Deacon, eds. "Battle Against Extinction: Native Fish Management in the American West." University of Arizona Press, Tucson, AZ.

Williams, J.E., and G.E. Davis. 1992. Strategies for conservation of fish communities. Paper presented at Symposium on Biodiversity in Managed Landscapes: Theory and Practice, Sacramento, CA, July 13–17, 1992.
Williams, J.E., and R.J. Neves. 1992. Introducing the elements of biological diversity in the aquatic environment. *Transactions of the North American Wildlife and Natural Resources Conference* 57: 343–354.
Williams, J.E., D.W. Sada, C.D. Williams and others. 1988. American Fisheries Society guidelines for introductions of threatened and endangered fishes. *Fisheries* 13(5): 5–11.
Williamson, G.B. 1975. Pattern and seral composition in an old-growth beech-maple forest. *Ecology* 56: 727–731.
Willson, Mary F. 1992. Biodiversity and ecological processes. Paper presented at Symposium on Biodversity in Managed Landscapes: Theory and Practice. Sacramento, CA, July 13–17, 1992.
Wilshire, H. 1992. The wheeled locusts. *Wild Earth* 2(1):27–31.
Wilshire, H.G., and Nakata, J.K. 1976. Off-road vehicle effects on California's Mojave Desert. *California Geology* 29(6):123–132.
Wilson, E.O. 1984. "Biophilia." Harvard University Press, Cambridge, MA.
Wilson, E.O. 1985. The biological diversity crisis. *BioScience* 35: 700–706.
Wilson, E.O. 1988. The current state of biological diversity. Pages 3–18 in E.O. Wilson, ed. "Biodiversity." National Academy Press, Washington, DC.
Wilson, E.O. 1992. "The Diversity of Life." Belknap Press of Harvard University Press, Cambridge, MA.
Wilson, E.O., and E.O. Willis. 1975. Applied biogeography. Pages 522–534 in M.L. Cody and J.M. Diamond. "Ecology and Evolution of Communities." Belknap Press of Harvard University Press, Cambridge, MA.
Wisdom, M.J., L.R. Bright, C.G. Carey, W.W. Hines, R.J. Pederson, D.A. Smithey, J.W. Thomas, and G.W. Witmer. 1986. "A Model to Evaluate Elk Habitat in Western Oregon." USDA Forest Service, Portland, OR.
Wolfe, S.H., J.A. Reidenauer, and D.B. Means. 1988. An Ecological Characterization of the Florida Panhandle. Biological Report 88(12). U.S. Fish and Wildlife Service, National Wetlands Research Center, Slidell, LA.
Wolke, H. 1991. "Wilderness on the Rocks." Ned Ludd Books, Tucson, AZ.
World Resources Institute. 1990. "World Resources 1990–91." Oxford University Press, New York.
World Resources Institute (WRI), The World Conservation Union (IUCN), United Nations Environment Programme (UNEP). 1992. "Global Biodiversity Strategy: Guidelines for Action to Save, Study, and Use Earth's Biotic Wealth Sustainably and Equitably." WRI, IUCN, UNEP, Washington, DC.
World Wildlife Fund–World Wide Fund for Nature. 1991. "The Importance of Biological Diversity." Yale Press, New Haven, CT.
Worster, D. 1977. "Nature's Economy: A History of Ecological Ideas." Cambridge University Press, Cambridge, UK.

Wright, H.E. 1974. Landscape development, forest fires, and wilderness management. *Science* 186: 487–495.

Wuerthner, G. 1992. Some ecological costs of livestock. *Wild Earth* 2(1): 10–14.

Yahner, R.H. 1988. Changes in wildlife communities near edges. *Conservation Biology* 2: 333–339.

Yahner, R.H., T.E. Morrell, and J.S. Rachael. 1989. Effects of edge contrast on depredation of artificial avian nests. *Journal of Wildlife Management* 53: 1135–1138.

Yoakum, J.D. 1978. Pronghorn. Pages 103–121 in "Big Game of North America—Ecology and Management." Stackpole Books, Harrisburg, PA.

Yoakum, J., and W.P. Dasmann. 1971. Habitat manipulation practices. Pages 173–231 in R.H. Giles, ed. "Wildlife Management Techniques." The Wildlife Society, Washington, DC.

Young, J.E. 1992. Mining the earth. Pages 100–118 in L. Starke, ed. "State of the World 1992." Worldwatch Institute. W.W. Norton and Co., New York.

Young, S.P. 1946. History, life habits, economic status, and control, Part 1. Pages 1–173 in S.P. Young and E.A. Goldman, eds. "The Puma, Mysterious American Cat." The American Wildlife Institute, Washington, DC.

Zahner, R. 1992. Benign neglect management: An old model for restoring health to the southern Appalachian national forests. *Wild Earth* 2(1): 43–46.

Zaslowsky, D. 1986. "These American Lands: Parks, Wilderness, and the Public Lands." The Wilderness Society and Henry Holt, New York.

Zedler, J.B. 1988. Restoring diversity in salt marshes—Can we do it? Pages 317–325 in E.O. Wilson, ed. "Biodiversity." National Academy Press, Washington, DC.

Zeiner, D.C., W.F. Laudenslayer, K.E. Mayer, and M. White. 1990. "California's Wildlife." Volume III. "Mammals." California Department of Fish and Game, Sacramento, CA.

Zeveloff, S.I. 1988. "Mammals of the Intermountain West." University of Utah Press, Salt Lake City, UT.

GLOSSARY

Abiotic Not biotic; often referring to the nonliving components of the ecosystem such as water, rocks, and mineral soil.

Adaptive management The process of implementing policy decisions as scientifically driven management experiments that test predictions and assumptions in management plans, and using the resulting information to improve the plans.

Allee effect The phenomenon of a population dropping below a threshold density or number of individuals from which it cannot recover; named after the animal ecologist W.C. Allee.

Alpha diversity Species diversity within a habitat.

Ancient forest Forest in late successional stages, or (at a landscape scale) a shifting mosaic of forest patches in various ages after natural disturbances.

Anthropocentric Placing humans above all other species in value and importance. *See also* **Humanism.**

Autecological Of or relating to the ecology or natural history of a single species.

Autecology Study of the ecology of a single species, its requirements, tolerances, and responses.

Baseline monitoring 1. Monitoring some element of biodiversity such as water quality that is not expected to change. 2. The initial set of measurements in an ongoing monitoring study, typically done before the system is perturbed by management.

Benchmark areas Areas of natural or minimally disturbed habitat that can serve as control or comparison areas to measure the effects of human activity and management on similar habitat in the same region.

Bergman's rule An ecogeographic rule stating that in homeotherms populations from cooler climates tend to have larger body size and hence have smaller surface-to-volume ratios than related populations living in warmer climates.

Beringia The geographic area of western Alaska, the Aleutians, and eastern Siberia that was connected in the Cenozoic by a land bridge when the Bering Sea region became emergent.

Beta diversity Species diversity between habitats or along an environmental gradient.

Biodiversity The variety of life and its processes; it includes the variety of living organisms, the genetic differences among them, the communities and ecosystems in which they occur, and the ecological and evolutionary processes that keep them functioning, yet ever changing and adapting.

Biogeography The scientific study of the geographic distribution of organisms.

Biosphere reserve A management model of the United Nation's Man and the Biosphere program where a core reserve is surrounded by buffer zones with human use increasing away from the core.

Biota All of the organisms, including animals, plants, fungi, and microorganisms, found in a given area.

Biotic Pertaining to any aspect of life, especially to characteristics of entire populations or communities.

Biotic impoverishment Loss of biota and biotic processes; virtually synonymous with loss of biodiversity.

Biotic succession *See* **Succession.**

Brood parasites Birds that lay eggs in nests of other species or in nests of other pairs of the same species so that the task of rearing young is done by other birds.

Cenozoic era An era of geological history that extends from the beginning of the Tertiary period (about 63–65 million years ago) to the present time and is marked by a rapid evolution of mammals and birds and of grasses, shrubs, and higher flowering plants and by little change in most invertebrate groups.

Centrarchid fish Fish of the sunfish family (Centrarchidae).

Chaining Removal of woody vegetation from rangelands by dragging a heavy anchor chain between two tractors.

Climax The theoretical culminating stage in plant succession for a given site, at which the vegetation is self-reproducing and thus has reached a stable condition through time.

Clone A population of individuals all derived asexually from the same single parent.

Coarse filter Biological inventory and land protection activities focused on communities, ecosystems, habitats, or landscapes.

Colonization The immigration of a species into a new habitat and the founding of a population.

Community All the organisms—plants, animals, and microbes—that live in a particular habitat and affect one another as part of the food web or through their various influences on the physical environment.

Connectivity The state of being functionally connected by movement of organisms, materials, or energy.

Conservation biology The field of biology that studies the dynamics of diversity, scarcity, and extinction.

Continental drift The gradual breaking up of the continents that has occurred steadily over the past 200 million years.

Core reserve The central, most protected area of a multiple-use module.

Corridor A route that allows movement of individuals or taxa from one region or place to another.

Cumulative effects The combined effects of all management and other human activities on a defined area of land, a body of water, or both.

Demography The study of birth rates, death rates, age distributions, sex ratios, and size of populations—a fundamental discipline within the larger field of ecology. Also the properties themselves, as in the demography (demographic traits) of a particular population.

Deoxyribonucleic acid (DNA) The fundamental hereditary material of all living organisms; the polymer composing the genes.

Desertification The impoverishment of arid, semiarid, and some subhumid ecosystems by the combined impact of human activity and drought.

Deterministic Not determined by chance.

Disturbance A significant change in structure or composition caused by natural events such as fire and wind or human-caused events such as logging.

Disturbance cycles Periodic recurrence of particular natural disturbances such as fire or flooding.

Diversity Ecological measure of the number of species and their relative abundance (evenness) in a community; a low diversity refers to relatively fewer species or more uneven abundance, whereas a high diversity refers to a higher number of species or more even abundance.

DNA *See* **Deoxyribonucleic acid.**

Ecological niche *See* **Niche.**

Ecosophy A set of ethics related to ecological and environmental matters.

Ecosystem A dynamic complex of plant, animal, fungal, and microorganism communities and their associated nonliving environment interacting as an ecological unit.

Ecosystem approach A strategy or plan to manage ecosystems to provide for all associated organisms, as opposed to a strategy or plan for managing individual species.

Ecosystem dysfunction Disruption of functioning of ecological processes in an ecosystem.

Ecosystem management Any land-management system that seeks to protect viable populations of all native species, perpetuate natural-disturbance regimes on the regional scale, adopt a planning timeline of centuries, and allow human use at levels that do not result in long-term ecological degradation. Also the public relations campaign of agencies that purport to serve these goals.

Edge effects The ecological changes that occur at the boundaries of ecosystems; these include changes in species composition, gradients of moisture, sunlight, soil and air temperature, wind speed, etc. Many edge effects have negative consequences. For example, forest-interior species have their populations reduced by edge effects.

Effectiveness monitoring Monitoring to determine if some human activity is having the desired effect.

Element *See* **Element of diversity.**

Element of diversity Heritage program terminology for natural community types, individual species, or other natural entities.

Element occurrence (EO) Heritage program terminology for the presence of an element of diversity at a specific location.

Endemic 1. (*n.*) A species or race native to a particular place and found only there. 2. (*adj.*) Restricted to a specified region or locality.

Endemism The relative number of endemic species found within a geographic area or region. High endemism indicates that there are many native species found only in that area or region. Low endemism indicates that most species found in that area or region are also found in other places.

Environmental ethic An ecosophy; *see* **Ecosophy.**

Environmental gradients The change in ecological or environmental features across space, such as changes in elevation, moisture, temperature, or soil chemistry.

Evenness The component of ecological diversity indices that refers to the variance in abundance of species.

Evolution In biology, any change in the genetic constitution of a population of organisms. Evolution can vary in degree from small shifts in the frequency of minor genes to the origins of complexes of new species.

Exotic *See* **Exotic species.**

Exotic species Species that occur in a given place, area, or region as the result of direct or indirect, deliberate or accidental introduction of the species by humans, and for which introduction has permitted the species to cross a natural barrier to dispersal.

Extant Now living; not destroyed or lost; not extinct.

Extinct No longer existing.

Extinction The human-caused or natural process whereby a species or population ceases to exist.

Extirpation Local extinction; a species or subspecies disappearing from a locality or region without becoming extinct throughout its range.

Filters 1. In biogeography, a geographic or ecological barrier that restricts some dispersal between regions and blocks passage of certain forms but not others. 2. In conservation, one of several methods of determining elements of biodiversity that require or have high priority for protection. *See also* **Coarse filter; Fine filter.**

Fine filter Biological inventory and land protection activities focused on individual rare species.

Fitness The relative survival value and reproductive capability of a given genotype in comparison with others of a population.

Flagship species Species that are popular and charismatic and which therefore attract popular support for their conservation.

Fragmentation *See* **Habitat fragmentation.**

Gamma diversity Diversity at the regional scale.

Gap analysis An assessment of the protection status of biodiversity in a specified region which looks for gaps in the representation of species or ecosystems in protected areas.

Gap dynamics The formation and replacement of patches or gaps in a landscape, as in the fall of trees and growth of new trees in that opening.

Gap formation The generation of patches in a landscape, such as the generation of openings in a forest as a result of trees falling down or dying and losing their canopy leaves.

Gap-phase replacement The regeneration of a forest or other vegetation type by succession in small patches or gaps.

Geographic information system A computer system capable of storing and manipulating spatial (e.g., map) data.

GIS *See* **Geographic information system.**

Global warming The projected increase in global temperature due to the release of the byproducts of fossil-fuel combustion into Earth's atmosphere.

Greater ecosystem A regional complex of ecosystems with common landscape-level characteristics such as the presence of wide-ranging mammal populations and periodic, large, natural disturbances such as wildfire. The concept of the greater ecosystem allows for integrated land management. The boundaries of greater ecosystems are somewhat arbitrary.

Greater Yellowstone Ecosystem (GYE) The greater ecosystem of 14 to 19 million acres which includes Yellowstone and Grand Teton National Parks and surrounding lands.

Group selection cut Removal of groups of trees ranging in size from a fraction of an acre up to about 2 acres. Area cut is smaller than the minimum feasible under even-aged management for a single stand.

GYE *See* **Greater Yellowstone Ecosystem.**

Habitat fragmentation Process by which habitats are increasingly subdivided into smaller units, resulting in their increased insularity as well as losses of total habitat area.

Halfway technology A technology that treats symptoms rather than eliminating root causes of a problem.

High-grading 1. In forestry, the practice of cutting the largest, healthiest, most vigorous, or most valuable trees in the forest. 2. In range ecology, the practice of grazing or browsing animals continuously selecting the most palatable or nutritious plants or plant parts from a range.

Holistic Relating to or concerned with wholes or with complete systems rather than with the analysis of, treatment of, or dissection into parts.

Hot spots In conservation, areas of high species richness or of high endemism, which are of high priority for protection.

Humanism A doctrine, attitude, or way of life centered on human interests or values; especially one which treats humans as superior to other forms of life. *See also* **Anthropocentric.**

Implementation monitoring Monitoring to determine if a planned action is being taken.

Infiltration The penetration of water through the ground surface into subsurface soil.

Interception The catching of precipitation before it reaches the surface of the ground.

Interspecific variation Variation among species.

Intraspecific variation Variation among individuals or populations of the same species.

Island biogeography The study of the distribution of living things on islands or fragments of habitat.

Keystone species A species that plays a pivotal role in an ecosystem and upon which a large part of the community depends.

LANDSAT/Thematic Mapper (TM) One of several satellites used to generate maps through remote sensing.

Landscape A heterogeneous land area composed of a cluster of interacting ecosystems that is repeated in similar form throughout.

Landscape diversity The pattern of habitats and species assemblages across a land area of thousands to millions of acres.

Landscape matrix *See* **Matrix.**

Laurasia The northern half of the supercontinent Pangaea, including North America, Europe, and parts of Asia.

Life forms Groups of species that use similar habitat for feeding, breeding, and other needs.

Little Ice Age The most distinctive climatic interval of the recent past (approximately A.D. 1350–1870), characterized by expansion of ice caps and glaciers and changes in distribution of trees and other species.

Macroinvertebrates Larger or more prominent invertebrates; abundance and species composition of macroinvertebrates in freshwater streams is often measured as an indication of impacts to the stream and watershed.

Managed forests Any forestland that is treated with silvicultural practices or harvested. The term is generally applied to land that is harvested on a scheduled basis.

Management Refers to managers' role which consists of the psychometric quantitative analyses of interdependent relationships of indigenous components as deflected by the theoretical analogy of perturbation and adumbration of sagacity.

Management indicator species A species that theoretically indicates the general health of an ecosystem. A decline in its population signals decline for other species living in the area.

Management thresholds Points along some measurable gradient where, by prior agreement, management will be changed.

Matrix The most extensive and most connected habitat type in a landscape, which often plays the dominant role in landscape processes.

Mesopredators Medium sized predators such as raccoons and foxes.

Mesozoic era An era of geological history including the interval between the Permian and the Tertiary periods (63 to 230 million years ago) and marked by the dinosaurs, marine and flying reptiles, ganoid fishes, cycads, and evergreen trees.

Metapopulation A set of partially isolated populations belonging to the same species. The populations are able to exchange individuals and recolonize sites in which the species has recently become extinct.

Metapopulation dynamics The processes of recolonization and extinction of subpopulations of a metapopulation.

Minimum dynamic area The smallest area with a natural disturbance regime, which maintains internal recolonization sources, and hence minimizes extinction.

Minimum viable population The low end of the viable population range; the smallest isolated population having an x percent chance of remaining extant for y years despite the foreseeable effects of demographic, environmental, and genetic stochasticity and natural catastrophes.

Multiple-use module A concentric design for nature reserves where the intensity of use increases outward from the core and intensity of protection increases inward.

MUM *See* **Multiple-use module.**

Mutagenic Causing mutations.

Mutation Broadly defined as any genetic change in an organism, either from an alteration of DNA composing individual genes or from a shift in the structure or number of chromosomes. Mutations create the new raw material for evolution.

Mutualism Symbiosis in which both of the partner species benefit.

Native 1. (n.) A species that has not been introduced from somewhere else by humans. 2. (adj.) Not introduced by humans.

Natural disturbance cycle The temporal pattern of natural disturbance events.

Natural disturbance events Recurring perturbations such as fires and floods that occur in ecosystems without human intervention.

Natural disturbance regime The temporal and spatial pattern of natural disturbance events.

Natural forests Forests consisting primarily of native species that are self-sustaining (requiring no cultural energy to be maintained) and whose dynamics would change little if humans were removed from the system.

Natural gap dynamics The manner in which patches (gaps) are formed and replaced in a landscape without human intervention.

Natural selection The differential contribution of offspring to the next generation by various genetic types belonging to the same population; the mechanism of evolution proposed by Darwin. Distinguished from artificial selection, the same process but carried out with human guidance. Selection may also act on groups, populations, species, higher taxa, and communities, but the mechanisms remain unclear.

Natural selection (forestry) A forest management system that emphasizes the removal of unhealthy, broken, and dying trees from a stand.

Neotropics The region from southern Mexico and the West Indies to South America.

New Forestry A type of forest management that seeks to incorporate ideas from conservation biology and landscape ecology into timber management; often distinguished by partial cutting to mimic natural disturbances.

New Perspectives An experimental program of the U.S. Forest Service that purports to incorporate ecosystem-management principles and increased public participation into forest planning.

Niche A vague but useful term in ecology, meaning the place occupied by a species in its ecosystem—where it lives, what it eats, its foraging route, the season of its activity, and so on. In a more abstract sense, a niche is a potential place or role within a given ecosystem into which species may or may not have evolved.

Node 1. An area of high species richness or diversity or of special value for biodiversity. 2. A core reserve area for a multiple-use module.

Old growth *See* **Old-growth forest.**

Old-growth dependent species Species that are dependent upon old-growth forest for survival.

Old-growth forest The older seral stages of natural forests.

Old-growth indicator species Old-growth associated species that indicate conditions similar to old growth are present in the area.

Old-growth species Plant and animal species that exhibit a strong association with old-growth forests.

Pangaea In plate tectonics, the supercontinent of the Permian period (230 to 280 million years ago) that was composed of essentially all continents and major continental islands.

Patch A nonlinear surface area differing in appearance from its surroundings, typically a small (less than 50 acres) portion of the landscape; small patches the size of an individual tree canopy are frequently called gaps.

Phenotype The observed traits of an organism, created by an interaction of the organism's genotype (hereditary material) and the environment in which it developed.

Phenotypically As a result of interaction of genetic material with the environment, rather than purely genetic.

Piling unmerchantable material (PUM) The practice of piling unmerchantable portions of trees (limbs, tops, etc.) rather than leaving them in place.

Plantation Stands of trees planted by humans.

Plate tectonics The study of the origin, movement, and destruction of plates and how these events have been involved in the evolution of the earth's crust.

Pleistocene The span of geological time preceding the Recent epoch, during which continental glaciers advanced and retreated and the human species evolved. The epoch began about 2.5 million years ago and closed with the end of the Ice Age 10,000 years ago (although some scientists argue that we are still in the Pleistocene).

PNV *See* **Potential natural vegetation.**

Population In biology, any group of organisms belonging to the same species at the same time and place.

Population genetics The study of genetics at the population level.

Population sinks Portions of a species geographic range where death rates exceed birth rates for local populations.

Population viability The probability that a population will persist for a specified period of time across its range despite normal fluctuations in demographic and environmental conditions.

Potential natural vegetation (PNV) The theoretical climax vegetation community that would occupy an area if left undisturbed by humans.

Preservationist A person concerned with natural things for their intrinsic value rather than for some utilitarian purpose; alternately, a person interested in stopping natural or human-induced change.

Primary natural *See* **Primary natural forest.**

Primary natural forest Virgin forests of any age arising through natural regeneration (primary or secondary) after natural disturbance.

Primary succession Succession beginning with bare rock or soil.

Primeval forest *See* **Virgin forest.**

Propagule The dispersing unit of a species, such as a seed, spore, or juvenile.

PUM *See* **Piling unmerchantable material.**

Recombination The formation by the processes of crossing-over and independent assortment of new combinations of genes in progeny that did not occur in the parents.

Reductionism The belief that a whole may be represented as a function of its constituent parts, the functions having to do with the spatial and temporal ordering of the parts and with the precise way in which they interact.

Regional reserve network A set of nature reserves or multiple-use modules (MUMs) functionally connected that provide greater protection than individual reserves or MUMs by themselves.

Regional reserve system *See* **Regional reserve network.**

Relaxation Loss of species in a given location after the area has been cut off from exchange with neighboring areas.

Representation Inclusion within a reserve network of the full spectrum of biological and environmental variation including all genotypes, species, ecosystems, habitats, and landscapes.

Representativeness Inclusion of all the archetypal communities in a reserve network.

Resourcism An anthropocentric view of the world that considers nonhuman beings to have value only as they can be used as goods and services for humans.

Retrogression On rangelands, movement away from the climax community to some lower seral stage, typically from overgrazing.

Salmonids Fish of the family Salmonidae, which includes the trout and salmon.

SCT *See* **Site conservation threshold.**

Second growth *See* **Secondary natural forest.**

Second-growth forest *See* **Secondary natural forest.**

Secondary natural forest Forests that have regenerated naturally on sites formerly logged, farmed, or otherwise cleared by humans.
Selection forestry One of several forest management systems based upon selective removal of trees with certain characteristics (e.g., *See* **High-grading; Natural selection forestry**).
Seral stage One of the stages of succession.
Serotinous Cones remaining closed on the tree with seed dissemination delayed or occurring gradually, often after fire.
Shannon index One of several mathematical indices of diversity that considers both species richness and evenness.
Simpson's index One of several mathematical indices of diversity that considers both species richness and evenness.
Single large or several small (SLOSS) A debate within the scientific community as to whether elements of biodiversity would be best protected by a single large reserve or several small ones of equivalent total area.
Site conservation threshold (SCT) The point at which erosion on rangelands accelerates due to human management.
SLOSS *See* **Single large or several small.**
Speciation The process of species formation: the full sequence of events leading to the splitting of one population of organisms into two or more populations reproductively isolated from one another.
Species–area effect *See* **Species–area relationship.**
Species–area relationship The relationship between the area of an island or some other geographic unit and the number of species living there; also, the pattern of increase in species richness with increasing area.
Species richness The number of species within a defined area.
State and transition model A model of plant dynamics for rangelands that postulates that vegetation communities can assume one of several possible stable states. According to this model, human-caused or natural events may push vegetation communities past thresholds beyond which they move rapidly to another stable state.
Stochastic Random within certain bounds of probability.
Stocks Fish populations that spawn in a particular river system (or portion of it) at a particular season and that do not interbreed to any substantial degree with any other fish population spawning in a different place or in the same place at a different season.
Structural diversity Diversity in a community that results from having many horizontal or vertical physical elements (e.g., layers or tiers of canopy). An increase in layering or tiering leads to an increase in structural diversity.
Structure The various horizontal and vertical physical elements of a community. Note: *Structure* is often used to describe only the vertical elements.
Succession The more or less predictable change in the composition of communities following a natural or human disturbance.
Tertiary period The first period of the Cenozoic era, beginning with the end of the Mesozoic era (Age of Reptiles) 63 to 65 million years ago and closing

with the start of the Pleistocene epoch about 2.5 million years ago; succeeded by the Quaternary period (Pleistocene plus Recent epochs).

Umbrella species Species that require large areas to maintain viable populations and by which protection of their habitat may protect the habitat and populations of many other more restricted or less wide ranging species.

Validation monitoring Scientific testing of the validity of the models and assumptions upon which a monitoring program is based.

Vernal pools Ephemeral pools that hold water only in the spring, after which they dry up; such pools often have unique flora and fauna.

Vertical structure The amount of vertical layering of vegetation.

Viability The ability of a population to maintain sufficient size so that it persists over time without significant human intervention in spite of normal fluctuations in numbers; usually expressed as a probability of maintaining a specific population for a specified period of time.

Viable population A population that contains an adequate number of individuals appropriately distributed to ensure a high probability of long-term survival without significant human intervention.

Virgin forest Forests of any age essentially uninfluenced by human activities; essentially equivalent to primary natural forests.

Virgin old growth Old-growth forests essentially uninfluenced by human activities.

Vulnerable species Species that are sensitive to human activity because of their life history, appearance, reputation, edibility, location, or other factors.

Weedy species A species that is growing or living where it is not wanted; in a management context, an exotic and/or invasive undesirable species that often requires concerted effort (labor and resources) to remove it from its current location, if it can be removed at all.

Xeric Characterized by, relating to, or requiring only a small amount of moisture.

Yarding unmerchantable material (YUM) Moving unmerchantable portions of trees from the stump to a central concentration area.

YUM *See* **Yarding unmerchantable material.**

SPECIES LIST

antelope—*See* pronghorn antelope
Apache trout—*Salmo apache*
Audubon bighorn—*Ovis canadensis auduboni*
auroch—*Bos* sp.
barbary sheep—*Ammotragus lervia*
beaver—*Castor canadensis*
big bluestem—*Andropogon gerardi*
big sagebrush—*Artemisia tridentata*
bighorn sheep—*Ovis canadensis*
bison—*Bison bison*
black bear—*Ursus americanus*
black cherry—*Prunus serotina*
black grama—*Bouteloua eriopoda*
black-capped vireo—*Virio atricapillus*
black-footed ferret—*Mustela nigripes*
black-tailed deer—One of several subspecies of mule deer (*Odocoileus hemionus*) found along the Pacific Coast from southern California to Alaska
blue grama—*Bouteloua gracilis*
bluebunch wheatgrass—*Agropyron spicatum*
bobcat—*Felis rufus*
Bohemian waxwing—*Bombycilla garrulus*
Brazilian pepper—*Schinus terebinthifolius*
brook trout—*Salvelinus fontinalis*
brown pelican—*Pelicanus occidentalis*
brown trout—*Salmo trutta*
brown-headed cowbird—*Molothrus ater*
buffalo—*See* bison
buffalo wolf—*Canis lupus nubilis*
bullfrog—*Rana catesbiana*
burro—*Equus asinus*
bush honeysuckle—*Lonicera tartarica* and *L. maackii*
California condor—*Gymnogyps californianus*
California white oak—*Quercus lobata*
Canada yew—*Taxus brevifolia*
carabid beetle—beetles of family Carabidae
Caribbean monk seal—*See* West Indian monk seal
Carolina parakeet—*Conuropsis carolinensis*
cedar waxwing—*Bombycilla cedrorum*
cheatgrass—*Bromus tectorum*
cladocerans—crustaceans of order Diplostraca, suborder Cladocera
climbing euonymus—*Euonymus fortunei*
Colorado squawfish—*Ptychocheilus lucius*
common buckthorn—*Rhamnus cathartica*
cougar—*See* mountain lion
cow—*Bos taurus*
coyote—*Canis latrans*
creosotebush—*Larrea tridentata*
crested wheatgrass—*Agropyron cristatum*
cutthroat trout—*Oncorhynchus clarki*
dall sheep—*Ovis dalli*
desert pupfish—*Cyprinodon macularius*
desert slender salamander—*Batrachoseps aridus*
desert tortoise—*Gopherus agassizii*
dog—*Canis familiarus* (or *C. lupus*)
domestic cat—*Felis domesticus*
domestic sheep—*Ovis aries*
Douglas-fir—*Pseudotsuga menziesii*
eastern cougar—*Felis concolor couguar*
eastern diamondback rattlesnake—*Crotalus adamanteus*
eastern hemlock—*Tsuga canadensis*
eastern white cedar—*Thuja occidentalis*

elk—*See* wapiti
Englemann spruce—*Picea engelmannii*
European starling—*Sturnus vulgaris*
Florida panther—*Felis concolor coryi*
fringed sagebrush—*Artemisia frigida*
garlic mustard—*Alliaria officinalis*
giant sequoia—*Sequoia gigantea*
Gila trout—*Oncorhynchus gilae*
glossy buckthorn—*Rhamnus frangula*
gopher tortoise—*Gopherus polyphemus*
grass carp—*Ctenopharyngodon idella*
gray catbird—*Dumetella carolinensis*
gray wolf—*Canis lupus*
great auk—*Pinguinus impennis*
green turtle—*Chelonia mydas*
Greenback cutthroat trout—*Oncorhynchus clarki stomias*
grizzly bear—*Ursus arctos*
gypsy moth—*Porthetria dispar*
heath hen—*Tympanuchus cupido cupido*
hoary elfin butterfly—*Incisalia polios obscurus*
hooded warbler—*Wilsonia citrina*
horse—*Equus caballus*
house mouse—*Mus musculus*
house sparrow—*Passer domesticus*
human—*Homo sapiens*
Hydrilla verticillata—an exotic water plant
Indian ricegrass—*Oryzopsis hymenoides*
indigo bunting—*Passerina cyanea*
ivory-billed woodpecker—*Campephilus principalis*
jack pine—*Pinus banksiana*
jaguar—*Felis onca*
Japanese honeysuckle—*Lonicera japonica*
Johnson grass—*Sorghum halpense*
Kirtland's warbler—*Dendroica kirtlandii*
kudzu—*Pueraria lobata*
Labrador duck—*Camptorhynchus labradorius*
leafy spurge—*Euphorbia esula*
least Bell's vireo—*Vireo bellii pusillus*
Little Kern golden trout—*Onchorhynchus aquabonita whitei*
loblolly pine—*Pinus taeda*
lodgepole pine—*Pinus contorta*
loggerhead shrike—*Lanius ludovicianus*
longleaf pine—*Pinus palustris*
marbled murrelet—*Brachyramphus marmoratus*
medusahead—*Taeniatherum* sp.
melaleuca—*Melaleuca quinquenervia*
Merriam's elk—*Cervus elaphus merriami*
mesquite—*Prosopis* sp.
Miller Lake lamprey—*Lampetra minima*
mink—*Mustela vison*
moose—*Alces alces*
mountain goat—*Oreamnos americanus*
mountain lion—*Felis concolor*
mountain sheep—*See* bighorn sheep
mule deer—*Odocoileus hemionus*
muskrat—*Ondatra zibethicus*
North Pacific plantain—*Plantago macrocarpa*
northern flicker—*Colaptes auratus*
northern flying squirrel—*Glaucomys sabrinus*
northern shrike—*Lanius excubitor*
northern spotted owl—*Strix occidentalis caurina*
opossum—*See* Virginia opossum
opossum shrimp—*Mysis relicta*
Pacific fisher—*Martes pennanti*
Pacific yew—*Taxus brevifolia*
Paiute cutthroat trout—*Oncorhynchus clarki seleniris*
panther—*See* mountain lion
passenger pigeon—*Ectopistes migratorius*
Pecos pupfish—*Cyprinodon pecosensis*
peregrine falcon—*Falco peregrinus*
pileated woodpecker—*Dryocopus pileatus*
ponderosa pine—*Pinus ponderosa*
Port Orford cedar—*Chamaecyparis lawsoniana*
pronghorn—*See* pronghorn antelope
pronghorn antelope—*Antilocapra americana*

puma—*See* mountain lion
raccoon—*Procyon lotor*
rainbow trout—*Oncorhynchus gairdneri*
red deer—*Cervus elaphus*
red herring—*Clupea harengus rufus*
red shiner—*Notropis lutrensis*
red wolf—*Canis rufus*
red-cockaded woodpecker—*Picoides borealis*
redwood—*Sequoia sempervirens*
rough fescue—*Festuca scabrella*
rough-leaved loosestrife—*Lythrum* sp.
rufous-sided towhee—*Pipilo erythrophthalmus*
saltcedar—*See* tamarisk
sand pine—*Pinus clausa*
sea mink—*Mustela vison macrodon*
sharp-tailed grouse—*Tympanuchus phasianellus*
sheepshead minnow—*Cyprinodon variagatus*
slash pine—*Pinus elliottii*
southern sea otter—*Enhydra lutris nereis*
spikemosses—*Selaginella* sp.
spotted knapweed—*Centaurea maculosa*
spotted owl—*Strix occidentalis*
starling—*See* European starling
steelhead—*See* steelhead trout
steelhead trout—sea-run stocks of rainbow trout
subalpine fir—*Abies lasiocarpa*
sugar maple—*Acer saccharum*
sundews—*Drosera* sp.
tamarisk—*Tamarix* sp.
tarbush—*Flourensia* sp.
tobosa—*Hilaria mutica*
tuliptree—*Liriodendron tulipifera*
turkey oak—*Quercus laevis*
valley oak—*See* California white oak
vermillion flycatcher—*Pyrocephalus rubinus*
Virginia opossum—*Didelphis virginiana*
wapiti—*Cervus elaphus*
West Indian monk seal—*Monachus tropicalis*
western diamondback rattlesnake—*Crotalus atrox*
western hemlock—*Tsuga heterophylla*
whitebark pine—*Pinus albicaulis*
white-footed mouse—*Peromyscus leucopus*
white-tailed deer—*Odocoileus virginianus*
whooping crane—*Grus americana*
wild turkey—*Meleagris gallopavo*
wiregrass—*Aristida stricta*
wolverine—*Gulo gulo*
wood stork—*Mycteria americana*
woundfin—*Plagopterus—argentissimus*
yellow birch—*Betula alleghaniensis*
yellow poplar—*See* tuliptree
yellow warbler—*Dendroica petechia*
zebra mussel—*Dreissena polymorpha*

INDEX

About the Authors

Reed F. Noss is editor of *Conservation Biology*, an international consultant in conservation, science director of The Wildlands Project, a research scientist at the University of Idaho, a research associate at Stanford University, and is on the Fisheries and Wildlife faculty at Oregon State University. He has an M.S. in ecology from the University of Tennessee, a Ph.D. in wildlife ecology from the University of Florida, and a fifth-degree black belt in Shito-Ryu karate. His twenty plus years in the environmental field include work with the Ohio Department of Natural Resources, Florida Natural Areas Inventory, and U.S. Environmental Protection Agency. He is a Pew Scholar in Conservation and the Environment (1993–96) and has published over one hundred papers. He lives with his wife, three children, two dogs, and a cat in the foothills of the Oregon Coast Range outside Corvallis.

Allen Y. Cooperrider has been a consultant in conservation biology with Big River Associates since 1991. He was educated in zoology and wildlife biology at the University of California, Berkeley, the University of Montana, and Syracuse University. He worked for many years as a wildlife biologist throughout the West, including seventeen years with the U.S. Bureau of Land Management. He and his wife live in rural Mendocino County in northern California on a river to which he hopes salmon will someday return.

Island Press Board of Directors